D1126159

Methods in Enzymology

Volume 219
RECONSTITUTION OF INTRACELLULAR TRANSPORT

METHODS IN ENZYMOLOGY

EDITORS-IN-CHIEF

John N. Abelson   Melvin I. Simon

DIVISION OF BIOLOGY
CALIFORNIA INSTITUTE OF TECHNOLOGY
PASADENA, CALIFORNIA

FOUNDING EDITORS

Sidney P. Colowick and Nathan O. Kaplan

Methods in Enzymology

Volume 219

Reconstitution of Intracellular Transport

EDITED BY

James E. Rothman

LABORATORY OF CELLULAR BIOCHEMISTRY
SLOAN-KETTERING INSTITUTE FOR CANCER RESEARCH
NEW YORK, NEW YORK

ACADEMIC PRESS, INC.

Harcourt Brace Jovanovich, Publishers

San Diego   New York   Boston
London   Sydney   Tokyo   Toronto

Academic Press, Inc.
1250 Sixth Avenue, San Diego, California 92101-4311

United Kingdom Edition published by
Academic Press Limited
24–28 Oval Road, London NW1 7DX

Library of Congress Catalog Number: 54-9110

International Standard Book Number: 0-12-182120-X

PRINTED IN THE UNITED STATES OF AMERICA
92 93 94 95 96 97   MM   9 8 7 6 5 4 3 2 1

Table of Contents

Section III. Identification of Transport Intermediates

Contributors to Volume 219

Article numbers are in parentheses following the names of contributors.
Affiliations listed are current.

STEPHEN A. ADAM (11), *School of Medicine, Northwestern University, Chicago, Illinois 60611*

W. E. BALCH (12, 25), *Departments of Cell and Molecular Biology, Scripps Research Institute, La Jolla, California 92037*

CON J. M. BECKERS (2, 31), *Section of Infectious Diseases, Department of Internal Medicine, Yale University School of Medicine, New Haven, Connecticut 06510*

MARY A. BITTNER (17), *Department of Pharmacology, University of Michigan Medical School, Ann Arbor, Michigan 48109*

MARC R. BLOCK (28), *Laboratoire d'Etude des Systemes, Adhesifs Cellulaires, A.T.I.P.E. du ANRS Associeé à l'URA 1178, Université J. Fourier, F53X-38041 Grenoble, France*

WILLIAM A. BRAELL (3), *Department of Biological Chemistry and Molecular Pharmacology, Harvard Medical School, Boston, Massachusetts 02115*

PATRICK BRENNWALD (33), *Department of Cell Biology, Yale University School of Medicine, New Haven, Connecticut 06510*

PHILIPPE CHAVRIER (37), *European Molecular Biology Laboratory, D-6900 Heidelberg, Germany*

DOUGLAS O. CLARY (30), *Howard Hughes Medical Institute, University of California, San Francisco, San Francisco, California 94143*

MARIA ISABEL COLOMBO (5), *Department of Cell Biology and Physiology, Washington University School of Medicine, St. Louis, Missouri 63110, and Instituto de Histologia y Embriologia, Universidad Nacional de Cuyo, 5500 Mendoza, Argentina*

IRÈNE CORTHÉSY-THEULAZ (16), *Department of Biochemistry, Stanford University School of Medicine, Stanford, California 94305*

H. W. DAVIDSON (12, 25), *Departments of Cell and Molecular Biology, Scripps Research Institute, La Jolla, California 92037*

SUSAN FERRO-NOVICK (14), *Department of Cell Biology, Yale University School of Medicine, New Haven, Connecticut 06510*

RAINER FRANK (37), *European Molecular Biology Laboratory, D-6900 Heidelberg, Germany*

DIETER GALLWITZ (35), *Department of Molecular Genetics, Max-Planck-Institute of Biophysical Chemistry, D-3400 Göttingen, Germany*

MICHELLE D. GARRETT (33), *Department of Cell Biology, Yale University School of Medicine, New Haven, Connecticut 06510*

LARRY GERACE (11), *Departments of Cell and Molecular Biology, Scripps Research Institute, La Jolla, California 92037*

YUKIKO GODA (15), *Molecular Neurobiology Laboratory, The Salk Institute, La Jolla, California 92037*

BASTIEN D. GOMPERTS (18), *Department of Physiology, University College London, London WC1E 6JJ, England*

MARY E. GROESCH (14), *Department of Cell Biology, Yale University School of Medicine, New Haven, Connecticut 06510*

LUDGER HENGST (35), *Department of Molecular Genetics, Max-Planck-Institute of Biophysical Chemistry, D-3400 Göttingen, Germany*

LINDA HICKE (32), *Department of Biochemistry, Biocenter, University of Basel, CH-4056 Basel, Switzerland*

RONALD W. HOLZ (17), *Department of Pharmacology, University of Michigan*

ix

Medical School, Ann Arbor, Michigan 48109

WIELAND B. HUTTNER (10), *Institute for Neurobiology, University of Heidelberg, D-6900 Heidelberg, Germany*

ALISA K. KABCENELL (33), *Department of Biochemistry, Boehringer Ingelheim Pharmaceuticals, Inc., Ridgefield, Connecticut 06877*

RICHARD A. KAHN (34), *Laboratory of Biological Chemistry, Division of Cancer Treatment, National Cancer Institute, National Institutes of Health, Bethesda, Maryland 20892*

JAMES M. LENHARD (5), *Department of Cell Biology and Physiology, Washington University School of Medicine, St. Louis, Missouri 63110*

J. PAUL LUZIO (7), *Department of Clinical Biochemistry, Addenbrooke's Hospital, University of Cambridge, Cambridge CB2 2QR, England*

LUIS S. MAYORGA (4, 5), *Instituto de Histologia y Embriologia, Universidad Nacional de Cuyo, 5500 Mendoza, Argentina*

JAMES M. McILVAIN, JR. (9), *Department of Cell Biology, Duke University Medical Center, Durham, North Carolina 27710*

IRA MELLMAN (20), *Department of Cell Biology, Yale University School of Medicine, New Haven, Connecticut 06510*

STEPHEN G. MILLER (23), *Department of Molecular and Cell Biology, University of California, Berkeley, Berkeley, California 94720*

HSIAO-PING H. MOORE (23), *Department of Molecular and Cell Biology, University of California, Berkeley, California 94720*

BARBARA M. MULLOCK (7), *Department of Clinical Biochemistry, Addenbrooke's Hospital, University of Cambridge, Cambridge CB2 2QR, England*

JOHN W. NEWPORT (8), *Department of Biology, University of California, San Diego, La Jolla, California 92093*

PETER NOVICK (33), *Department of Cell Biology, Yale University School of Medicine, New Haven, Connecticut 06510*

ROBERT PARTON (37), *European Molecular Biology Laboratory, D-6900 Heidelberg, Germany*

RUPERT PFALLER (8), *Department of Biology, University of California, San Diego, La Jolla, California 92093*

SUZANNE R. PFEFFER (15, 16), *Department of Biochemistry, Stanford University School of Medicine, Stanford, California 94305*

S. PIND (12), *Departments of Cell and Molecular Biology, Scripps Research Institute, La Jolla, California 92037*

ALAN PITT (4), *Department of Cell Biology and Physiology, Washington University School of Medicine, St. Louis, Missouri 63110*

H. PLUTNER (12), *Departments of Cell and Molecular Biology, Scripps Research Institute, La Jolla, California 92037*

BENJAMIN PODBILIEWICZ (20), *MRC Laboratory of Molecular Biology, Cambridge CB2 2QH, England*

MARC PYPAERT (21), *Cell Biology Laboratory, Imperial Cancer Research Fund, London WC2 3PX, England*

PAUL A. RANDAZZO (34), *Laboratory of Biological Chemistry, Division of Cancer Treatment, National Cancer Institute, National Institutes of Health, Bethesda, Maryland 20892*

THOMAS E. REDELMEIER (22), *Departments of Cell and Molecular Biology, Scripps Research Institute, La Jolla, California 92037*

MICHAEL REXACH (26), *Division of Biochemistry and Molecular Biology, Howard Hughes Medical Research Institute, University of California, Berkeley, Berkeley, California 94720*

GUENDALINA ROSSI (14), *Department of Cell Biology, Yale University School of Medicine, New Haven, Connecticut 06510*

JAMES E. ROTHMAN (1, 2, 27, 28, 29, 30, 31), *Laboratory of Cellular Biochemistry, Sloan-Kettering Institute for Cancer Research, New York, New York 10021*

RANDY W. SCHEKMAN (13, 26, 32), *Department of Molecular and Cell Biology, Howard Hughes Medical Research Institute, University of California, Berkeley, Berkeley, California 94720*

SANDRA L. SCHMID (22), *Departments of Cell and Molecular Biology, Scripps Research Institute, La Jolla, California 92037*

R. SCHWANINGER (12), *Departments of Cell and Molecular Biology, Scripps Research Institute, La Jolla, California 92037*

ALAN L. SCHWARTZ (4), *Department of Pharmacology and Pediatrics, Washington University School of Medicine, St. Louis, Missouri 63110*

RUTH A. SENTER (17), *Department of Pharmacology, University of Michigan Medical School, Ann Arbor, Michigan 48109*

TITO SERAFINI (27), *Department of Anatomy, University of California, San Francisco, San Francisco, California 94143*

MICHAEL P. SHEETZ (9), *Department of Cell Biology, Duke University Medical Center, Durham, North Carolina 27710*

ELIZABETH SMYTHE (22), *Departments of Cell and Molecular Biology, Scripps Research Institute, La Jolla, California 92037*

THIERRY SOLDATI (15), *Department of Biochemistry, Stanford University School of Medicine, Stanford, California 94305*

PHILIP D. STAHL (4, 5), *Department of Cell Biology and Physiology, Washington University School of Medicine, St. Louis, Missouri 63110*

RACHEL STERNE-MARR (11), *Jefferson Medical College, Thomas Jefferson University, Philadelphia, Pennsylvania 19107*

ELIZABETH SZTUL (6), *Department of Molecular Biology, Lewis Thomas Laboratory, Princeton University, Princeton, New Jersey 08544*

PETER E. R. TATHAM (18), *Department of Physiology, University College London, London WC1E 6JJ, England*

ARMAND TAVITIAN (36), *Faculté de Médecine Lariboisière–Saint Louis, INSERM U248, Paris 75010, France*

SHARON A. TOOZE (10), *Department of Cell Biology, European Molecular Biology Laboratory, D-6900 Heidelberg, Germany*

PETER WAGNER (35), *Department of Molecular Genetics, Max-Planck-Institute of Biophysical Chemistry, D-3400 Göttingen, Germany*

GRAHAM WARREN (21, 24), *Cell Biology Laboratory, Imperial Cancer Research Fund, London WC2 3PX, England*

M. GERARD WATERS (31), *Department of Molecular Biology, Princeton University, Princeton, New Jersey 08544*

OFRA WEISS (34), *Department of Endocrinology and Metabolism, Hadassah University Hospital, Jerusalem 91120, Israel*

FELIX T. WIELAND (19), *Institut für Biochemie I, Universität Heidelberg, D-6900 Heidelberg, Germany*

DUNCAN W. WILSON (29), *MCR Laboratory of Molecular Biology, University Postgraduate Medical School, Cambridge CB2 2QH, England*

PHILIP G. WOODMAN (24), *Department of Biochemistry and Molecular Biology, University of Manchester Medical School, Manchester M13 9PT, England*

LINDA J. WUESTEHUBE (13), *Division of Biochemistry and Molecular Biology, Howard Hughes Medical Institute, University of California, Berkeley, Berkeley, California 94720*

TOHRU YOSHIHISA (32), *Division of Biochemistry and Molecular Biology, Howard Hughes Medical Institute, University of California, Berkeley, Berkeley, California 94720*

AHMED ZAHRAOUI (36), *Faculté de Médecine Lariboisière–Saint Louis, INSERM U248, Paris 75010, France*

MARINO ZERIAL (37), *European Molecular Biology Laboratory, D-6900 Heidelberg, Germany*

Preface

This volume of *Methods in Enzymology* reflects both the breadth and intensity of current activities toward accomplishing and dissecting cell-free reconstitutions of numerous transport steps. Assembling this collection was made easy by the willingness and enthusiasm of the many investigators who so generously contributed their time and expertise, for which I am most grateful.

JAMES E. ROTHMAN

METHODS IN ENZYMOLOGY

VOLUME 73. Immunochemical Techniques (Part B)
Edited by JOHN J. LANGONE AND HELEN VAN VUNAKIS

VOLUME 74. Immunochemical Techniques (Part C)
Edited by JOHN J. LANGONE AND HELEN VAN VUNAKIS

VOLUME 75. Cumulative Subject Index Volumes XXXI, XXXII, XXXIV–LX
Edited by EDWARD A. DENNIS AND MARTHA G. DENNIS

VOLUME 76. Hemoglobins
Edited by ERALDO ANTONINI, LUIGI ROSSI-BERNARDI, AND EMILIA CHIANCONE

VOLUME 77. Detoxication and Drug Metabolism
Edited by WILLIAM B. JAKOBY

VOLUME 78. Interferons (Part A)
Edited by SIDNEY PESTKA

VOLUME 79. Interferons (Part B)
Edited by SIDNEY PESTKA

VOLUME 80. Proteolytic Enzymes (Part C)
Edited by LASZLO LORAND

VOLUME 81. Biomembranes (Part H: Visual Pigments and Purple Membranes, I)
Edited by LESTER PACKER

VOLUME 82. Structural and Contractile Proteins (Part A: Extracellular Matrix)
Edited by LEON W. CUNNINGHAM AND DIXIE W. FREDERIKSEN

VOLUME 83. Complex Carbohydrates (Part D)
Edited by VICTOR GINSBURG

VOLUME 84. Immunochemical Techniques (Part D: Selected Immunoassays)
Edited by JOHN J. LANGONE AND HELEN VAN VUNAKIS

VOLUME 85. Structural and Contractile Proteins (Part B: The Contractile Apparatus and the Cytoskeleton)
Edited by DIXIE W. FREDERIKSEN AND LEON W. CUNNINGHAM

VOLUME 86. Prostaglandins and Arachidonate Metabolites
Edited by WILLIAM E. M. LANDS AND WILLIAM L. SMITH

VOLUME 87. Enzyme Kinetics and Mechanism (Part C: Intermediates, Stereochemistry, and Rate Studies)
Edited by DANIEL L. PURICH

VOLUME 88. Biomembranes (Part I: Visual Pigments and Purple Membranes, II)
Edited by LESTER PACKER

VOLUME 89. Carbohydrate Metabolism (Part D)
Edited by WILLIS A. WOOD

VOLUME 90. Carbohydrate Metabolism (Part E)
Edited by WILLIS A. WOOD

VOLUME 91. Enzyme Structure (Part I)
Edited by C. H. W. HIRS AND SERGE N. TIMASHEFF

VOLUME 92. Immunochemical Techniques (Part E: Monoclonal Antibodies and General Immunoassay Methods)
Edited by JOHN J. LANGONE AND HELEN VAN VUNAKIS

VOLUME 93. Immunochemical Techniques (Part F: Conventional Antibodies, Fc Receptors, and Cytotoxicity)
Edited by JOHN J. LANGONE AND HELEN VAN VUNAKIS

VOLUME 94. Polyamines
Edited by HERBERT TABOR AND CELIA WHITE TABOR

VOLUME 95. Cumulative Subject Index Volumes 61–74, 76–80
Edited by EDWARD A. DENNIS AND MARTHA G. DENNIS

VOLUME 96. Biomembranes [Part J: Membrane Biogenesis: Assembly and Targeting (General Methods; Eukaryotes)]
Edited by SIDNEY FLEISCHER AND BECCA FLEISCHER

VOLUME 97. Biomembranes [Part K: Membrane Biogenesis: Assembly and Targeting (Prokaryotes, Mitochondria, and Chloroplasts)]
Edited by SIDNEY FLEISCHER AND BECCA FLEISCHER

VOLUME 98. Biomembranes (Part L: Membrane Biogenesis: Processing and Recycling)
Edited by SIDNEY FLEISCHER AND BECCA FLEISCHER

VOLUME 99. Hormone Action (Part F: Protein Kinases)
Edited by JACKIE D. CORBIN AND JOEL G. HARDMAN

VOLUME 100. Recombinant DNA (Part B)
Edited by RAY WU, LAWRENCE GROSSMAN, AND KIVIE MOLDAVE

VOLUME 101. Recombinant DNA (Part C)
Edited by RAY WU, LAWRENCE GROSSMAN, AND KIVIE MOLDAVE

VOLUME 102. Hormone Action (Part G: Calmodulin and Calcium-Binding Proteins)
Edited by ANTHONY R. MEANS AND BERT W. O'MALLEY

VOLUME 103. Hormone Action (Part H: Neuroendocrine Peptides)
Edited by P. MICHAEL CONN

VOLUME 104. Enzyme Purification and Related Techniques (Part C)
Edited by WILLIAM B. JAKOBY

VOLUME 105. Oxygen Radicals in Biological Systems
Edited by LESTER PACKER

VOLUME 106. Posttranslational Modifications (Part A)
Edited by FINN WOLD AND KIVIE MOLDAVE

VOLUME 107. Posttranslational Modifications (Part B)
Edited by FINN WOLD AND KIVIE MOLDAVE

[1] Introduction

By James E. Rothman

Insight into the complex maze of intracellular transport pathways has come in several waves, beginning with the pioneering work of the 1960s,[1] in which the fundamental relationships among the endoplasmic reticulum, Golgi, and secretory storage vesicles in regulated secretion were deduced, and the need for vesicular carriers to ferry cargo between topologically separate compartments was first recognized. These now central paradigms of cell biology were extended to the formation of plasma membranes, lysosomes, and other organelles in the 1970s, and continue to be refined as ever more sophisticated techniques of immunocytochemistry improve and better define subcompartments. In the 1980s elucidation of the molecular machinery of vesicular transport began with the reconstitution of intracellular transport in cell-free systems from animal cells[2] and the isolation of secretory mutants in yeast.[3]

Although important cell biological questions concerning the compartmental organization of the secretory pathway still remain, enough is now known to permit a meaningful dissection of the molecular machinery of individual segments. It is both fortunate and remarkable that these kind of pathways, whose essential purpose is to propagate the three-dimensional organization of the cytoplasm, can nonetheless be faithfully reproduced in dispersed cell-free systems without the benefit (or constraint) of preexisting spatial arrangements. This has opened the door to biochemistry, and superficial outlines of the steps involved in vesicle budding and fusion have already emerged. Crucial points remain. Among them, how, in step-by-step fashion, do coats assemble on membranes to yield a vesicle? What is the targeting signal that triggers uncoating and attachment of a vesicle? How can a protein machine fuse two lipid layers?

Answers to questions of molecular mechanism at this level have and will necessarily continue to emerge from cell-free systems, and thus need to be confirmed *in vivo*. Because so little is known, or can be learned, at this level from studies in whole cells, how can this be done? The answer is that the molecules discovered with cell-free systems, putatively performing the same roles in living cells, will provide the very tools to make the assessment of authenticity. As genes encoding these purified transport components are

---

[1] G. E. Palade, *Science* **189**, 347 (1975).
[2] E. Fries and J. E. Rothman, *Proc. Natl. Acad. Sci. USA* **77**, 3879 (1980).
[3] P. Novick, S. Fero, and R. Schekman, *Cell* **25**, 461 (1981).

METHODS IN ENZYMOLOGY, VOL. 219

manipulated, and antibodies microinjected, the predicted effects on cellular physiology can be scrutinized. This has already begun by synthesizing the results from animal cell-free biochemistry with those from yeast genetics, and the results are encouraging.[4-6]

The fruits of genetics and biochemistry are teaching us that many of the components of the secretory pathway are universal: the same machinery operates in yeast as in animals. The same enzyme system that fuses a vesicle with the Golgi also fuses endocytic vesicles. The coats that pinch off endocytic and Golgi vesicles have underlying similarities not evident from their morphology. And we learn that members of gene families (like the small GTP-binding proteins) are modular units that perform the same tasks in different places, combining a general mechanism with compartmental specificity. These and other themes, along with many important details, continue to emerge.

We can now look forward to an increased understanding of the secretory pathway so that, in the not too distant future, we will be able to discuss these complex events of macromolecular targeting with the same kind of sophistication with which we now describe the biosynthesis of macromolecules such as proteins and nucleic acids, and will be able to do so in a common language, that of protein biochemistry.

[4] D. W. Wilson *et al., Nature (London)* **339,** 355 (1989).
[5] D. O. Clary, I. C. Griff, and J. E. Rothman, *Cell* **61,** 709 (1990).
[6] C. A. Kaiser and R. Schekman, *Cell* **61,** 723 (1990).

Section I

Reconstitution in Cell-Free Extracts

[2] Transport between Golgi Cisternae

By CON J. M. BECKERS and JAMES E. ROTHMAN

Introduction

The transport of proteins between the different organelles or compartments that make up the secretory and endocytic pathways is thought to be mediated by small vesicles that bud off from one compartment and subsequently fuse with their target compartment. The actual biochemical reactions that occur during the budding of transport vesicles and their transfer to and fusion with their target compartment can now be analyzed using a number of cell-free systems. Three steps in the secretory pathway have been reconstituted to date: transport from the endoplasmic reticulum (ER) to the Golgi compartment,[1-3] transport from the cis- to the medial-Golgi compartment,[4] and transport from the medial- to the trans-Golgi compartment.[5] In the endocytic pathway the internalization of proteins from the plasma membrane in coated vesicles has been reconstituted,[6] as well as the fusion between endosomes.[7,8] In addition, Goda and Pfeffer[9] have established a cell-free system to study recycling of proteins between endosomes and the trans-Golgi network.

The cell-free system used to reconstitute cis- to medial-Golgi transport has provided us with most of the information to date regarding the biochemical pathway of vesicular transport and the specific cofactors involved. In this system, transport of the vesicular stomatitis virus glycoprotein (VSVG protein) is detected between the cis compartment of a mutant Golgi preparation (donor membranes) and the medial compartment of a wild-type Golgi preparation (acceptor membranes). Vesicular stomatitis virus G protein is used as a model protein because it is synthesized in large quantities in VSV-infected cells and because it is known to be transported along the secretory pathway to the plasma membrane in a fashion indistinguishable from normal cellular plasma membrane glycoproteins. The donor Golgi membranes are prepared from VSV-infected Chinese hamster

[1] C. J. M. Beckers, D. S. Keller, and W. E. Balch, *Cell* **50,** 523 (1987).
[2] D. Baker, L. Hicke, M. Rexach, M. Schleyer, and R. Schekman, *Cell* **54,** 335 (1988).
[3] H. Ruohola, A. K. Kabcenell, and S. Ferro-Novick, *J. Cell Biol.* **107,** 1465 (1988).
[4] W. E. Balch, W. G. Dunphy, W. A. Braell, and J. E. Rothman, *Cell* **39,** 405 (1984).
[5] J. E. Rothman, *J. Biol. Chem.* **262,** 12505 (1987).
[6] E. Smythe, M. Pypaert, J. Lucocq, and G. Warren, *J. Cell Biol.* **108,** 843 (1989).
[7] R. Diaz, L. S. Mayorga, and P. Stahl, *J. Biol. Chem.* **263,** 6093 (1988).
[8] W. A. Braell, *Proc. Natl. Acad. Sci. USA* **84,** 1137 (1987).
[9] Y. Goda and S. R. Pfeffer, *Cell* **55,** 309 (1988).

METHODS IN ENZYMOLOGY, VOL. 219

ovary (CHO) 15B cells.[10] These lack the medial-Golgi enzyme *N*-acetylglucosamine (GlcNAc) transferase I, which transfers GlcNAc from UDPGlcNAc onto the N-linked oligosaccharides found in VSV G protein. The acceptor Golgi membranes are prepared from uninfected wild-type CHO cells that do contain GlcNAc transferase I in their medial-Golgi compartment. Vesicular transport between the two Golgi populations can now be detected because GlcNAc will be incorporated into the N-linked oligosaccharides of VSV G protein as soon as it arrives in the acceptor compartment. If a transport reaction is therefore performed in the presence of UDP[^{3}H]GlcNAc, transport of VSV G protein between the donor and acceptor Golgi membranes can be measured by simply determining the incorporation of [^{3}H]GlcNAc in VSV G protein.

The incorporation of [^{3}H]GlcNAc, and therefore the transport of VSV G protein between the two Golgi membrane populations, was found to require intact membranes, ATP, and soluble and membrane-associated proteins.[4]

This system has contributed in a significant way to our understanding of vesicular transport. Not only has it made possible the identification and purification of a number of proteins that are required for transport,[11-16] it has also made possible the identification, for the first time, of a number of possible reaction intermediates in vesicular transport.[17-19]

Methods

Materials

All chemicals are obtained, unless indicated otherwise, from Sigma Chemical Company (St. Louis, MO). UDP[^{3}H]GlcNAc (5–25 Ci/mmol) is purchased from New England Nuclear (Boston, MA). The rabbit anti-mouse immunoglobulin G (IgG) (whole molecule; Cat. No. 0111-0082) is obtained from Cappel (Organon Teknika, West Chester, PA) and reconsti-

[10] C. Gottlieb, J. Baenziger, and S. Kornfeld, *J. Biol. Chem.* **250,** 3303 (1975).
[11] M. R. Block, B. S. Glick, C. A. Wilcox, F. T. Wieland, and J. E. Rothman, *Proc. Natl. Acad. Sci. USA* **85,** 7852 (1988).
[12] D. O. Clary and J. E. Rothman, *J. Biol. Chem.* **265,** 10109 (1990).
[13] B. S. Glick and J. E. Rothman, *Nature (London)* **326,** 309 (1987).
[14] M. G. Waters, T. Serafini, and J. E. Rothman, *Nature (London)* **349,** 248 (1991).
[15] M. G. Waters and J. E. Rothman, *J. Cell Biol.* in press (1992).
[16] P. J. Weidman, P. Melancon, M. R. Block, and J. E. Rothman, *J. Cell Biol.* **108,** 1589 (1989).
[17] V. Malhotra, L. Orci, B. S. Glick, M. R. Block, and J. E. Rothman, *Cell* **54,** 211 (1988).
[18] L. Orci, B. S. Glick, and J. E. Rothman, *Cell* **46,** 171 (1986).
[19] J. E. Rothman, *Nature (London)* **355,** 409 (1992).

tuted by the addition of 2 ml water. The hybridoma secreting the 8G5 monoclonal anti-VSV G protein antibody[20] is obtained from Dr. Binks Wattenberg (Upjohn, Kalamazoo, MI). The monoclonal antibody is purified from ascites with protein A-Sepharose (Pharmacia, Piscataway, NJ) using standard procedures.[21]

General Procedures

Protein concentrations are determined using the BCA protein assay reagent (Pierce, Rockford, IL) with bovine serum albumin (BSA) as a standard. All pH values mentioned are determined at room temperature and all sucrose concentrations are expressed as the ratio of grams of sucrose per total mass of solution (w/w).

Cells

Chinese hamster ovary wild-type (wt) and CHO 15B cells[10] are grown in α-modified minimum essential medium (αMEM) with 10% (v/v) fetal bovine serum (FCS; Gemini Bioproducts, Calabasas, CA). L cells and BHK21 cells are grown in Dulbecco's modified Eagle's medium (DMEM) with 10% (v/v) fetal bovine serum and 1 mM glutamine.

Virus

To obtain and maintain stocks of VSV (Indiana serotype, San Juan isolate) with sufficiently high titers, it is in our experience necessary to limit the number of times the virus is passed. To this end we usually start a virus preparation by three rounds of plaque purification of VSV on mouse L cells exactly as described by Bergmann.[22] A single plaque (3- to 5-mm diameter) is used to infect a single 10-cm plate of BHK21 cells (80–90% confluent) in 10 ml medium. After more than 90% of the cells have died and become detached from the plate (12–24 hr), the medium is harvested, cell debris removed by centrifugation (10 min at 800 g, 4°), and frozen in liquid nitrogen in 0.5-ml aliquots. A suitable aliquot of this "primary stock" [10 ml at $0.1–1 \times 10^8$ plaque-forming units (pfu)/ml] is used to infect four 15-cm plates (15 ml medium/plate) of BHK21 cells at 0.1 pfu/cell. After 90% of the cells have become detached from the plate, the medium is collected, clarified, and frozen as before. An aliquot of this "secondary stock" (60 ml at $0.5–1 \times 10^9$ pfu/ml) is used to infect twenty

[20] L. Lefrancois and D. S. Lyles, *Virology* **121,** 157 (1982).
[21] E. Harlow and D. Lane, "Antibodies: A Laboratory Manual." Cold Spring Harbor Lab., Cold Spring Harbor, New York, 1988.
[22] J. E. Bergmann, *Methods Cell Biol.* **32B,** 85 (1989).

15-cm plates of BHK21 cells at 0.1 pfu/cell. After 90% of the cells have become detached the medium is collected, clarified, and frozen. This virus stock (300 ml at $0.5-1 \times 10^9$ pfu/ml) is used to infect CHO 15B cells for the preparation of donor membranes.

Preparation of Donor Homogenate

Chinese hamster ovary 15B cells, grown on 15-cm plates until they are about 90% confluent ($2.5-3 \times 10^7$ cells/plate), are infected with VSV at 10 pfu/cell in serum-free αMEM (5 ml/plate), containing actinomycin D (10 μg/ml) and 25 mM N-2-hydroxyethylpiperazine-N'-2-ethanesulfonic acid (HEPES)–KOH (pH 7.2). After 1 hr at 37°, during which the plates are rocked every 10 min, complete growth medium is added (10 ml/plate) and the incubation continued at 37° for an additional 2.5 hr. Cells can be harvested either by mild trypsinization or by scraping with a rubber policeman. In either case the medium is poured off and the plates are washed twice with 10 ml Tris-buffered saline (TBS). Each plate is briefly rinsed with 5 ml of TBS containing 0.05% (w/v) trypsin and 2 mM ethylenediaminetetraacetic acid (EDTA) and subsequently incubated for 5–10 min at room temperature. The cells are washed off the plates in 5 ml ice-cold αMEM with 10% FCS (to inhibit the trypsin) by vigorous pipetting and collected in 50-ml conical centrifuge tubes on ice. Cells are pelleted (10 min at 800 g, 4°), washed once in TBS, and once in homogenization buffer (HB) (250 mM sucrose in 10 mM Tris-HCl, pH 7.4). The final pellet (2.5–3 ml/forty 15-cm plates) is resuspended after addition of 4 vol homogenization buffer and homogenized by passing the cell suspension 6–10 times through a ball-bearing homogenizer.[4] The extent of cell breakage can easily be monitored by examining a small aliquot of the homogenate by light microscopy. The resulting homogenate can be stored frozen in liquid nitrogen up to 6 months without a loss in activity.

Preparation of Acceptor Homogenate

Chinese hamster ovary wt cells used for acceptor preparations are usually grown in suspension in spinner flasks (Bellco, Vineland, NJ). In a typical preparation 3 liters of cell suspension (at $4-6 \times 10^5$ cells/ml) are harvested by centrifugation (10 min at 800 g and 4°), washed twice with TBS, and once with HB. The cell pellet (3–4 ml/3-liter culture) is resuspended after addition of 4 vol HB and homogenized as described for the preparation of the donor homogenate. The acceptor homogenate can also be stored frozen in liquid nitrogen.

Preparation of Donor and Acceptor Golgi Membrane Fractions by Sucrose Density Centrifugation

To 12 ml of either donor or acceptor homogenate, add 11 ml of 62% sucrose (w/w) in 10 mM Tris-HCl, pH 7.4 and 230°μl 100 mM EDTA, pH 7.4. After vigorous mixing, the final sucrose concentration is checked by refractometry and should be 37%. If necessary the sucrose concentration can be adjusted by addition of either 10 mM Tris-HCl, pH 7.4, or 62% sucrose in 10 mM Tris-HCl, pH 7.4. The mixture is placed in the bottom of a polycarbonate SW28 tube (Beckman, Palo Alto, CA) and overlaid with 15 ml 35% (w/w) sucrose in 10 mM Tris-HCl, pH 7.4, and 9 ml 29% (w/w) sucrose in 10 mM Tris-HCl, pH 7.4.

After centrifugation of the gradients for 2.5 hr at 25,000 rpm and 4° in an SW28 rotor, the membranes at the 35–29% interface are recovered by puncturing the side wall of the tube with a syringe. From 12 ml of homogenate one usually obtains 1.5–2 ml of the Golgi-enriched membrane fraction at a protein concentration of about 0.5–1 mg/ml. After the membranes are frozen in liquid nitrogen in suitable aliquots, they can be stored at −80° up to 1 year without a detectable loss of activity.

Preparation of Chinese Hamster Ovary Cytosol

The high-speed supernatant or cytosol can be prepared by centrifuging homogenates from either CHO wt or 15B cell for 60 min at 55,000 rpm at 4° in an SW55 rotor. The supernatant is recovered without disturbing either the pellets or the whitish lipid floating on top, and is subsequently desalted on a P-6DG (Bio-Rad, Richmond, CA) column (2.5 × 50 cm) equilibrated in 50 mM KCl, 25 mM Tris-HCl, pH 7.4, 0.5 mM dithiothreitol (DTT). The void volume fractions containing the bulk of the protein are pooled, frozen in aliquots in liquid nitrogen, and stored at −80°.

Assay Conditions

A standard transport reaction contains the following components:

HEPES–KOH, 25 mM, pH 7.4
KCl, 25 mM
Magnesium acetate, 2.5 mM
ATP, 50 μM
UTP, 250 μM
Creatine phosphate (CP), 2 mM
Rabbit muscle creatine phosphokinase (CPK), 7.3 IU/ml
UDP[^{3}H]GlcNAc, 0.4 μM (0.5 μCi)

Donor membranes
Acceptor membranes
Cytosol

This is prepared by mixing in a disposable glass tube [10×75 mm, e.g., Fisher Scientific (Pittsburgh, PA) 14-962-10A] on ice (per 50-μl reaction): 5 μl of a $10\times$ concentrated reaction buffer (250 mM HEPES–KOH, pH 7.4, 250 mM KCl, 25 mM magnesium acetate), 5 μl of an "ATP mix" [prepared fresh daily by mixing 5 μl CPK (1600 IU/ml, stored at $-80°$), 25 μl 200 mM CP, 5 μl 10 mM ATP (neutralized with NaOH), 10 μl 100 mM UTP (neutralized with NaOH), and 65 μl water], 5 μl (0.1 μCi/μl) UDP[^{3}H]GlcNAc (prepared fresh by gently drying down the appropriate amount of the ethanolic stock under a stream of nitrogen gas), and finally the appropriate amount of water. To this mixture were added gel-filtered cytosol and freshly thawed donor and acceptor membranes. Although the optimal amounts of the latter three components can vary between preparations, good results are usually obtained using 5 μl of gel-filtered cytosol and 5 μl of both donor and acceptor membranes. Transport is initiated by transfer of the reaction to 37° and stopped by returning it to ice.

Immunoprecipitation of Labeled Vesicular Stomatitis Virus G Protein

The labeled VSV G protein is immunoprecipitated with a combination of a monoclonal anti-VSV G protein antibody 8G5[20] and a polyclonal rabbit anti-mouse-IgG antiserum (to form the actual immune complexes). Again, the exact amounts and ratio of the monoclonal and polyclonal antibodies needed to obtain an optimal immunoprecipitation of the VSV G protein will depend on the specific preparations used, although in general we find that 1 μl of the monoclonal antibody (at 1.3–1.7 mg/ml protein) and 2 μl of the polyclonal antiserum per transport reaction are sufficient. Immune complexes are preformed by preincubating (per reaction) 1 μl monoclonal anti-VSV G protein antibody and 2 μl of the rabbit anti-mouse IgG antiserum for 5–10 min at room temperature, followed by the addition of 50 μl stop buffer [1% (v/v) Triton X-100, 1% (w/v) sodium cholate, 5 mM EDTA, 250 mM NaCl in 50 mM Tris-HCl, pH 7.5]. This is added to a transport reaction, mixed, and allowed to incubate for at least 1 hr at room temperature.

The immune complexes are collected on glass microfiber filters (Cat. No. 934-AH; Whatman, Clifton, NJ) by vacuum filtration. To prevent nonspecific adsorption of proteins the filters are preblocked for 5 min in a 2.5% (w/v) solution of nonfat dry milk (Carnation) in detergent buffer (DB) [1% (v/v) Triton X-100, 5 mM EDTA, 250 mM NaCl in 50 mM Tris-HCl, pH 7.5]. Each filter is first rinsed with 3 ml DB. The immunoprecipitation reaction is then diluted with 3 ml DB and rapidly filtered. The tube is rinsed once with 3 ml DB, which is again filtered. Finally each filter is washed rapidly with two 3-ml aliquots of DB and subsequently dried under a heat lamp.

The filters are transferred to scintillation vials, scintillation fluid (Scintiverse BD, Cat. No. SX18-4; Fisher Scientific) is added, and the samples are counted in a liquid scintillation counter.

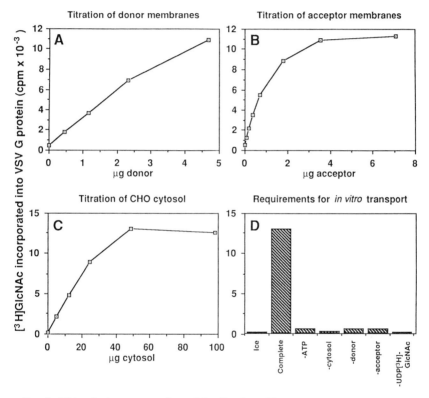

FIG. 1. All incubations were performed for 60 min at 37° as described in text. (A) Donor Golgi membranes were added at the amounts indicated to otherwise complete transport reactions, containing saturating amounts of cytosol (100 μg protein) and acceptor membranes (5 μg protein). (B) Increasing amounts of acceptor Golgi membranes were added to reactions containing saturating amounts of donor membranes (5 μg protein) and cytosol. (C) Increasing amounts of cytosol were added to transport incubations containing saturating amounts of donor and acceptor Golgi membranes. (D) Standard transport reactions were carried out but, where indicated, the reaction was left on ice, or was performed in the absence of the ATP mix, cytosol, donor, acceptor, or UDP[^{3}H]GlcNAc.

Demonstration of Requirements for *in Vitro* Transport

The cell-free transport system described above has been extensively studied over the years and has allowed the characterization of multiple factors, some better defined than others, that are required for transport.[19] The general requirements for vesicular transport that one would have predicted from *in vivo* observations in fact turn out to be essential for transport *in vitro:* transport occurs only at physiological temperatures, and requires both ATP and cytosolic factors in addition to Golgi membranes. The donor and acceptor Golgi membranes and UDP[^{3}H]GlcNAc are specific to the cell-free system used and are essential only to actually detect transport. Experiments demonstrating the different requirements for transport in the cell-free system are shown in Fig. 1.

[3] Detection of Endocytic Vesicle Fusion *in Vitro,* Using Assay Based on Avidin–Biotin Association Reaction

By WILLIAM A. BRAELL

Principle

The detection of the fusion of endocytic vesicles in a cell-free system generally involves an assay in which two probes are incorporated into two different preparations of endocytic vesicles, isolated from separate populations of cells. These probes are configured to generate a detectable signal if they contact each other when the two populations of vesicles are permitted to fuse together *in vitro.* The use of a detection scheme relying on the avidin–biotin association reaction for such a measurement possesses unique advantages. First, the association of avidin with biotin is rapid and extremely strong (K_d 10^{-15}), which makes the association effectively irreversible.[1] Thus, the signal will develop immediately on vesicle fusion, and will remain even if the product vesicles are subsequently lysed and the reacted probes are isolated. Moreover, any nondetectable biotinylated proteins added extravesicularly will act as scavengers for any avidin outside the vesicles, because the irreversibility of the association reaction precludes the scavenged avidin from subsequently detaching and reacting with the detectable biotinylated protein. This ensures that the only detectable signal will result from avidin–biotin associations occurring within the interior of the fused vesicles. Second, the association reaction is insensitive to varia-

[1] N. M. Green, *Biochem. J.* **89**, 585 (1963).

tions in pH, ionic strength, and temperature, over a range wider than would be reasonably expected to occur within endocytic vesicles.[2] This means that the amount of signal developed in the fusion detection scheme will depend only on the number of fusion events occurring, without regard to the quality of the internal environment of the vesicles themselves. Consequently, the level of fusion will be directly proportional to the amount of avidin – biotin complexes that are formed. The amount of these complexes are measured using an enzyme-linked immunosorbent assay (ELISA) protocol: one component (avidin) is conjugated to the reporter enzyme (here β-galactosidase), while the other component (biotin) is conjugated to the antigen (for example, transferrin) recognized by the antibodies immobilized on the ELISA substratum. For sensitivity, we use a fluorescent substrate for β-galactosidase, which speeds the detection process.

Preparation of Endocytic Probes

Avidin – β-Galactosidase

Avidin – β-Galactosidase (AvβGal) conjugates may be purchased commercially (Sigma Chemical Co., St. Louis, MO; Vector Laboratories, Burlingame, CA); however, such preparations can be significantly variable in their usefulness for endocytic assay work. Among the problems encountered are high nonspecific backgrounds due to "stickiness" of conjugates to cells [particularly Chinese hamster ovary (CHO) cells], or reduced ability to react with the biotinylated probes used in the assay. Therefore, it is highly recommended, when using these commercial sources, to test a sample from a given lot and, if satisfactory, to then purchase in bulk from that lot of material. Alternatively, these conjugates may be prepared as described below, using a method adapted from O'Sullivan and co-workers.[3]

Hen egg white avidin is dissolved at 3 mg/ml in buffer 1 (0.2 M NaCl, 50 mM sodium phosphate, pH 7), and dialyzed into the same buffer. From the literature,[4] a 1% (w/v) solution of avidin has an absorbance of 15.7 at 280 nm. A stock solution of m-maleimidobenzoyl-N-hydroxysuccinimide ester (MBS) is prepared at 0.1 M in dimethyl sulfoxide (DMSO). The MBS is added from this stock solution to a final concentration of 1 mM in the avidin solution, prewarmed to 30°; the mixture is incubated at 30° for

[2] N. M. Green, *Adv. Protein Chem.* **29,** 85 (1975).
[3] M. J. O'Sullivan, E. Gnemmli, D. Morris, G. Chieregatti, A. D. Simmonds, M. Simmons, J. W. Bridges, and V. Marks, *Anal. Biochem.* **100,** 100 (1979).
[4] M. D. Melamed and N. M. Green, *Biochem. J.* **89,** 591 (1963).

60 min and then cooled to ice temperatures. At this point, the avidin can be used for coupling, but it is preferable to acetylate the avidin first, to reduce nonspecific sticking of conjugates to cells. To accomplish this, acetic anhydride is added to the ice-cold solution to a final concentration of 3 mM, and the mixture is incubated on ice for 10 min. Unreacted reagents are then rapidly removed by gel filtration at 4° on a column of Sephadex G-50 equilibrated in buffer 1.

β-Galactosidase used for conjugation must have free sulfhydryl groups available; we have had best results with *Escherichia coli* β-galactosidase from Boehringer Mannheim (Indianapolis, IN; Cat. No. 567779). Prior to coupling, this protein is dissolved in buffer 2 [buffer 1 plus 1 mM MgCl$_2$, 0.1 mM ethylene glycol-bis(β-aminoethylether)-N,N,N',N'-tetraacetic acid (EGTA)] that has been carefully degassed and sparged with nitrogen, and dialyzed for 1 hr at 4° into this same buffer. The concentration is adjusted to 1 mg/ml, as calculated using the literature value of 19.1 for the 1% concentration absorbance at 280 nm.[5] Prior to coupling, the β-galactosidase and avidin are centrifuged at 10,000 g for 5 min at 4° to remove any particulates. The two proteins are mixed 1 : 1 (w/w) for coupling: the large molar excess of avidin minimizes formation of polymeric complexes. Coupling is performed for 60 min at 30°, then 60 min at 4°, then halted by adding 2-mercaptoethanol to a 12 mM final concentration. Separation of conjugates from free avidin is accomplished by gel filtration in coupling buffer plus 2-mercaptoethanol on Sepharose 6B or Sephacryl 300. Conjugates are easily concentrated by ammonium sulfate precipitation (50% saturation), and may be stored either at 4° as a precipitate, or redissolved in lyophilization buffer (0.1 M NaCl, 25 mM Tris-HCl, pH 7.5, 1.5 mM MgCl$_2$, 0.1 mM EGTA, 12 mM 2-mercaptoethanol, 10 mg/ml sucrose), lyophilized, and stored dry at $-20°$ under a nitrogen atmosphere.

Biotinylated Transferrin

Human apotransferrin (Sigma Chemical Co.) is first saturated with iron by the following procedure. Apotransferrin is dissolved in phosphate-buffered saline (PBS; 0.15 M NaCl, 20 mM NaPO$_4$, pH 7.4) at 10 mg/ml protein; a 1% solution of apotransferrin has a literature absorbance at 280 nm of 11.4.[6] To this is added 10 mM NaHCO$_3$ and 2 mM ferric ammonium citrate. This mixture is incubated at 30° for 30 min, during which time the color changes from pale yellow to orange. Residual iron is

[5] F. G. Loontiens, K. Wallenfels, and R. Weil, *Eur. J. Biochem.* **14**, 138 (1970).
[6] A. J. Liebman and P. Aisen, *Arch. Biochem. Biophys.* **121**, 717 (1967).

removed by dialysis at 4° against PBS. From the literature,[6] the absorbance (1%) of diferric transferrin at 280 nm is 14.1, and the ratio of absorbance at 470/280 nm is 0.043.

Biotinylation is accomplished by addition of N-hydroxysuccinimide (NHS)-activated biotin to the diferric transferrin, which reacts covalently with the protein. Best results for the assay are obtained with biotin that has a spacer arm incorporated in the linkage; for our purposes, suitable commercial preparations include biotin amidocaproate-NHS ester (Sigma), and NHS-LC-biotin or the cleavable disulfide reagent NHS-SS-biotin (Pierce, Rockford, IL). The biotin reagent is dissolved in DMSO at 62.5 mM, and added to the transferrin in PBS at 50-fold molar excess, thus leaving the DMSO level below 20%. This is incubated at 30° for 60 min, then placed on ice for 60 min. Excess unreacted reagent is removed by gel filtration on Sephadex G-25, followed by dialysis into PBS containing 3 nM NaN_3; any particulates are removed before storage by centrifugation in a microcentrifuge for 5 min at 10,000 g at 4°. This material may be stored at 4° for over 1 month, until used. When a radioactive probe is required, diferric transferrin (or biotinylated transferrin) may be radioiodinated by the chloramine-T method.[7]

Biotinylation of IgG is accomplished by the same procedure as for transferrin; however, a 25-fold molar excess of the biotinylation reagent is used, because IgG is larger than transferrin. For fusion assay purposes, we have usually used affinity-purified rabbit anti-goat IgG antibody, from Cooper Biomedical (Malvern, PA) or Sigma. Sigma also provides a biotinylated version of this IgG, which has proved adequate for fusion assay purposes. Biotin–insulin, used for scavenging extravesicular AvβGal, is generally (for convenience) purchased from Sigma, rather than being synthesized.

Biotin content of the protein is measured by a colorimetric assay at 500 nm, using the dye 2-(4′-hydroxyazobenzene)benzoic acid (HABA).[8] The HABA dye (neutralized with NaOH) is mixed with avidin in PBS at a 1:4 (w/w) ratio for the assay. A solution of 0.1 mM biotin is used to prepare the standard curve. As biotin displaces the HABA dye from the avidin, the absorbance at 500 nm decreases and the color changes from orange to yellow, until the avidin becomes saturated with biotin. For transferrin and other anionic proteins, the assay of biotin content should be performed in PBS supplemented to 0.5 M NaCl, to minimize electrostatic coprecipitation of the protein with the cationic avidin. Typically, each intact biotinylated transferrin (B-Tfn) protein molecule possesses four to five biotin moieties that are accessible to avidin.

[7] W. M. Hunter and F. C. Greenwood, *Nature (London)* **194,** 495 (1962).
[8] N. M. Green, *Biochem. J.* **94,** 23C (1965).

Biotinylated Mannose 6-phosphate – Bovine Serum Albumin

As transferrin populates only early endocytic compartments, other probes must be used to study subsequent endocytic processes. Proteins bearing mannose 6-phosphate (Man-6-P) moieties are endocytosed by the Man-6-P receptor, which enters later endocytic compartments, and can be used as probes for these compartments. We have used biotinylated Man-6-P – BSA for these purposes. Originally, we prepared Man-6-P – BSA by the procedure of Leichtner and Krieger,[9] using the pentamannosyl phosphate moiety prepared from the cell wall of the yeast *Hansenula holstii*. This procedure is, however, time consuming, so we have used an alternative method of preparing Man-6-P-labeled BSA, adapted from Sando and Karson,[10] which serves adequately. α-D-Mannopyranosylphenyl isothiocyanate is available commercially from Sigma. This reagent may be phosphorylated by the following protocol, after which it can be reacted with BSA to provide the Man-6-P groups. Fifty micromoles of α-D-mannopyranosylphenyl isothiocyanate is suspended in 0.95 mmol dry acetonitrile and 0.24 mmol dry pyridine. H_2O (110 μmol) is added, and the mixture is vortexed well. This suspension is cooled on ice in a fume hood, after which 220 μmol of phosphorus oxychloride is added and immediately vortexed. A vigorous reaction ensues, during which the sample warms and the mannose compound dissolves, so it is important to take appropriate safety precautions, and to keep the sample on ice. After 1 hr on ice, the reaction is stopped by adding 83.5 mmol H_2O at 4°, whereupon the compound precipitates. The pH of this mix is very low, and must be neutralized by addition of 10 N NaOH (do not exceed pH 7 here), by which point the compound has redissolved and assumed a pale yellow color. At this point, it is possible to add the reagent directly to the protein, as the by-product chemicals are compatible with protein-coupling procedures. Unused reagent may be stored frozen until used (adjusting the pH to 5 prior to freezing is recommended), although redissolving the thawed reagent may require warming. Reaction of BSA with excess reagent (40- to 100-fold molar excess provides maximal labeling) in 50 mM sodium carbonate buffer at pH 9 overnight will provide approximately 16 to 20 Man-6-P moieties per protein. The excess Man-6-P may be removed thereafter by dialysis or gel filtration; dialysis of the unbound Man-6-P is retarded somewhat relative to small ions, and overnight dialysis is required. If the BSA is also to be biotinylated, we recommend first labeling with biotin reagent at a 10-fold molar ratio to BSA, dialyzing to remove unused reagent, and then labeling with a 40-fold molar ratio of Man-6-P-phenyl

[9] A. M. Leichtner and M. Krieger, *J. Cell Sci.* **68**, 183 (1984).
[10] G. N. Sando and E. M. Karson, *Biochemistry* **19**, 3856 (1980).

isothiocyanate, because fewer sites on BSA accept biotin, compared to Man-6-P. These biotin–Man-6-P–BSA probes typically possess 5 biotin groups and 15 Man-6-P groups per molecule of BSA, and are bound by human Man-6-P receptors with an apparent K_s of 3 μM.

Preparation of Endocytic Vesicles

Cell Lines

Both CHO and K-562 cell lines may be propagated in α-modified MEM containing 10% (v/v) fetal bovine serum. K-562 is a suspension line; CHO may be propagated either in suspension or as monolayers. We generally prefer monolayer culture for our work with CHO, as the monolayer cells are more active for endocytosis. The K-562 line is preferred for work with biotin–transferrin, as these cells are rich in transferrin receptors (500,000/cell).[11]

For work with Man-6-P probes, the HL-60 suspension cell line is useful, because it is rich in Man-6-P receptors, and is easily cultured in RPMI 1640 medium supplemented with 20% (v/v) fetal bovine serum. This line is somewhat difficult to recover from frozen vials containing DMSO, because the cells prefer to grow at moderately high density (due to cross-feeding), and because DMSO tends to trigger terminal differentiation of the cells. We recommend recovery be performed in preconditioned medium, if possible, and that DMSO be removed once cells have recovered from freezing. Centrifugation to remove DMSO immediately after thawing is not recommended; the cells will not survive this rough treatment just after thawing.

Preparation of Vesicles Containing Endocytosed Probes

To minimize use of probes, the uptake is performed in suspension: either suspension-grown cells are used, or monolayer cells are detached from the substratum with trypsin, washed in serum-containing medium to inhibit and remove trypsin, and resuspended in ice-cold buffer. Ordinarily, cells are then washed by centrifugation three times with PBS at 4°. For uptake involving receptor-mediated endocytosis of transferrin, it is first necessary to preincubate cells in serum-free, HEPES-buffered medium containing 1 mg/ml BSA at 37° for 30 min, to facilitate removal of endogenous transferrin. Cells are then washed once in uptake buffer (0.15 M

[11] R. D. Klausner, J. Van Renswoude, G. Ashwell, C. Kempf, A. N. Schechter, A. Dean, and K. R. Bridges, *J. Biol. Chem.* **258**, 4715 (1983).

NaCl, 20 mM HEPES, pH 7.4, 5.5 mM glucose, 1 mg/ml BSA), and re-suspended at $4-8 \times 10^7$ cells/ml in uptake buffer containing the protein to be endocytosed. Fluid phase uptake of avidin–β-galactosidase conjugates or biotinylated IgG is done using 0.5 mg/ml of the probe in uptake buffer, either for 10–20 min at 37° or for 40–60 min at 20°. Receptor-mediated endocytosis of biotinylatd transferrin is performed at 125 nM (10 μg/ml) biotinylated transferrin in uptake buffer; probe is first bound to cells for 30 min at 4°, then allowed to endocytose for either 15 min at 37° or 60 min at 20°. Receptor-mediated uptake of the biotinylated Man-6-P–BSA is performed at 0.2–0.5 mg/ml probe; to avoid competition in the subsequent ELISA assay, the nonspecific protein used for preincubation and uptake incubations with the cells is changed from 1 mg/ml BSA to 1 mg/ml rabbit serum albumin for this probe.

Following uptake, cells are washed three times by centrifugation at 0° to remove excess probe, and resuspended in homogenization buffer (75 mM KCl, 25 mM NaCl, 20 mM HEPES, pH 7.4, 85 mM sucrose, 20 μM EGTA) at 8×10^7 cells/ml. This buffer has the advantage that the ionic conditions match those used for the cell-free fusion assay, thereby minimizing any adjustments of the buffer. Cells are homogenized either in a stainless steel Dounce (Wheaton, Millville, NJ) or using a ball-bearing homogenizer,[12] to approximately 80% breakage, as assessed by trypan blue dye exclusion. Postnuclear supernatant fraction is prepared by a centrifugation at 800 g for 5 min at 4°, and can be stored at $-80°$, following quick freezing in liquid nitrogen, for several months. Alternatively, small aliquots of postnuclear supernatant fraction may be prepared by freezing cells with liquid nitrogen directly in the homogenization buffer, and storing these at $-80°$; under these conditions, cell breakage is performed by quick-thawing cells, and immediately vortexing vigorously for 10 sec to effect cell breakage prior to preparing the postnuclear supernatant fraction. This latter technique is particularly useful for K-562 cells: early endosomes are readily released from the cells, but endoplasmic reticulum membrane is released in smaller amounts than is the case for homogenization with the ball-bearing homogenizer. For purposes of vesicle isolation, homogenates can also be prepared in isotonic sucrose buffer (0.25 M sucrose, 1 mM EDTA, 10 mM triethanolamine hydrochloride, pH 7.4), although these buffer conditions are not optimal for vesicle fusion unless adjusted.

Postnuclear supernatant fractions may be used directly, or after chemical treatment, to assay vesicle fusion; fusion is maximal when total protein exceeds 2 mg/ml in assays. Alternatively, vesicles maybe isolated and then assayed for fusion with newly added cytosolic protein. Vesicles active for

12 W. E. Balch, W. G. Dunphy, W. A. Braell, and J. E. Rothman, *Cell* **39**, 409 (1984).

fusion are isolated away from cytosol on isotonic Nycodenz (Nycomed AS, Oslo) step gradients. Postnuclear supernatant fractions, prepared from K-562 cells by homogenization in isotonic sucrose buffer, are overlaid on a shelf of 27.6% (w/v) Nycodenz in 10 mM triethanolamine hydrochloride, pH 7.4, 1 mM ethylenediaminetetraacetic acid (EDTA), and centrifuged in a Beckman (Palo Alto, CA) airfuge A-110 rotor at 4° for 5 min at 29 psi. Over 97% of the cytosolic protein is excluded from the Nycodenz layer containing the endocytic vesicles. Cytosolic protein is prepared by centrifugation of postnuclear supernatant fractions at 100,000 g for 1 hr at 4° in a Beckman 50-Ti rotor; this cytosolic fraction can be used, at 2 mg/ml protein, to reconstitute vesicle fusion with isolated endocytic vesicles.

Cell-Free Endocytic Vesicle Fusion Assay

Fusion Assay Conditions

Typical vesicle fusion reactions contain 2 mg/ml or more cell protein, 8 mM phosphocreatine, 1 mM ATP (pH 7), 1 mM MgCl$_2$, 50 μg/ml creatinine phosphokinase, 10 μg/ml biotin–insulin, 1 mM dithiothreitol, prepared in homogenization buffer, in a volume of 50 μl. ATP depletion experiments substitute 5 mM 2-deoxyglucose and 10 units/ml hexokinase for the ATP, phosphocreatine, and creatine phosphokinase. Vesicle fractions pretreated with N-ethylmaleimide are pretreated at 0° for 15–20 min with 1.5 mM N-ethylmaleimide (prior to adding the fusion reaction components); N-ethylmaleimide is then quenched with 2 mM dithiotreitol or 2-mercaptoethanol. Reactions are performed at 37° usually for 30 min, during which time the reaction rate is approximately linear; maximal fusion limits are attained after an ~120-min incubation. Reactions are halted by placing on ice, adding 1/10 vol of lysis buffer [10% (v/v) Triton X-100, 1% (w/v) sodium dodecyl sulfate (SDS), 50 μg/ml biotin–insulin], and mixing with 200 μl dilution buffer [0.05% (v/v) Triton X-100, 50 mM NaCl, 10 mM Tris-HCl, pH 7.5, 1 mg/ml heparin). Particulates are removed by centrifugation at 10,000 g for 2 min is a microcentrifuge, and the supernatant fraction is applied to pretreated microtiter wells for the fluorescent ELISA assay.

ELISA Procedure

Microtiter wells (Titertek high-binding wells; Flow Laboratories, McLean, VA) are precoated 3 hr or more with antibody prior to use, diluting antibody into 50 mM sodium carbonate buffer at pH 9.0. Other protein-binding microtiter wells have also proved satisfactory, although well-to-well reproducibility must be ascertained with individual brands, because the degree of binding may vary along a rack of wells. Binding of antibody to wells using phosphate-buffered saline for the dilution buffer has also proved successful. For fusion assays using biotin–transferrin, wells are coated with 1:500 diluted rabbit anti-human transferrin IgG (Boehringer Mannheim). For assays using as probes either biotinylated IgG or Man-6-P–BSA, coating is with 1:100 diluted goat anti-rabbit IgG or rabbit anti-BSA IgG (Sigma), respectively. Coated wells are rinsed with PBS, and preblocked by a 30- to 60-min incubation with wash buffer [1% (v/v) Triton X-100, 0.1% (w/v) SDS, 50 mM NaCl, 10 mM Tris-HCl, pH 7.5, 1 mM EDTA], then rinsed with PBS. Clarified fusion reaction lysates (200 μl/well) are applied to these wells in a humidified atmosphere for three or more hours to permit binding of reactants. Wells are washed three times with wash buffer, incubated briefly (5–10 min) with wash buffer, and rinsed three times with PBS. Fluorescence is developed on the wells by incubation in a humidified atmosphere for 60 min at 37° with 0.3 ml/well of ELISA reagent buffer (0.1 M NaCl, 25 mM Tris-HCl, pH 7.5, 1.5 mM MgCl$_2$, 12 mM 2-mercaptoethanol, 0.3 mM 4-methylumbelliferyl-β-galactoside) prewarmed to 37°. Aliquots of 200 μl reacted reagent buffer are diluted into 1 ml of glycine–carbonate buffer (133 mM glycine, 83 mM Na$_2$CO$_3$, pH 10.7), and fluorescence is read at 450 nm, with excitation at 366 nm. For quantitation, the fluorescence of a standard curve of 0 to 500 nM 4-methylumbelliferone in glycine–carbonate buffer is read; the molar extinction coefficient of 4-methylumbelliferone in ethanol is 15,850 at 325 nm.[13]

Density Gradient Analysis of Sorting

Following the fusion of early endocytic vesicles during the cell-free incubation, a separation of the fluid-phase AvβGal probe from the B-Tfn/transferrin receptors can be observed on Percoll density gradients. This separation can be observed even if AvβGal is endocytosed already bound to biotin-SS-Tfn, and then separated from the Tfn within the endocytic vesicles by *in situ* reduction of the disulfide linkage, prior to the cell-free incubation. As the two probes initially populate the same vesicles, this is evidence for cell-free sorting of the two probes in an endosome-like fashion. As incubation proceeds at 37° in the presence of ATP, the vesicles containing those AvβGal molecules not affiliated with receptor-bound B-Tfn are found to band at increasingly denser points on Percoll, while the receptor-bound B-Tf (and AvβGal attached to this B-Tf) remains in vesi-

[13] R. H. Goodwin and B. M. Pollack, *Arch. Biochem. Biophys.* **49**, 1 (1954).

cles of constant density. These separations are performed in 27% (v/v) Percoll in isotonic sucrose buffer (0.25 M sucrose, 10 mM triethanolamine–acetic acid, pH 7.4, 1 mM EDTA, 1 mg/ml polyvinylpyrrolidone) and centrifuged at 4° in a 50-Ti rotor for 2 hr at 17,000 rpm. For the purposes of isolation of the banded vesicles for subsequent analysis (*e.g.,* testing for fusion competence in the cell-free fusion assay) the inclusion of polyvinylpyrrolidone in the gradients, and in subsequent washings of vesicles with this buffer, facilitates the removal of the colloidal silica particles from the vesicles.

[4] Assays for Phagosome–Endosome Fusion and Phagosome Protein Recycling

By Alan Pitt, Luis S. Mayorga, Alan L. Schwartz, and Philip D. Stahl

Introduction

Endocytosis and phagocytosis are the two primary mechanisms employed by eukaryotic cells to internalize macromolecules from the extracellular milieu.[1,2] The most comprehensively studied endocytic pathway is receptor-mediated endocytosis via clathrin-coated pits and subsequent transport to lysosomes. This constitutive process intitiates at the cell surface, where specialized receptors and their associated ligands cluster in plasma membrane domains termed clathrin-coated pits. The clathrin-coats mediate pit invagination and vesiculation into the cytoplasm. Following enzymatic removal of the clathrin coat, the intracellular vesicle undergoes extensive fusion with other newly formed endosomes. Both the uncoated vesicles and the larger organelles formed by their multiple fusions are termed early endosomes. In the early organelles formed by their multiple fusions are termed early endosomes. In the early endosome, ligand–receptor dissociation and sorting occur. Soon afterward, receptors recycle to the plasma membrane while free ligand progresses to another population of vesicles termed late endosomes.[3–5] Late endosomes are both more

[1] A. L. Schwartz, *Annu. Rev. Immunol.* **8**, 195 (1990).

[2] S. C. Silverstein, S. Greenberg, F. Di Virgilio, and T. H. Steinberg, *in* "Fundamental Immunology" (W. E. Paul, ed.), p. 703. Raven, New York, 1989.

[3] P. D. Stahl, P. H. Schlesinger, E. Sigardson, J. S. Rodman, and Y. C. Lee, *Cell* **19**, 207 (1980).

[4] A. L. Schwartz, A. Bolognesi, and S. E. Fridovich, *J. Cell Biol.* **98**, 732 (1984).

[5] H. J. Geuze, J. W. Slot, G. J. A. M. Strous, and A. L. Schwartz, *Eur. J. Cell Biol.* **32**, 38 (1983).

dense and acidic and display a slight increase in hydrolytic activity compared to early endosomes. Also, late endosomes do not fuse *in vitro* to early endosomes.[6,7] From late endosomes, endocytosed material then migrates to very dense and hydrolase-rich lysosomes where complete hydrolysis occurs.[1]

Contrary to endocytosis, phagocytosis is a function of specialized cells. In vertebrates, mononuclear macrophages and polymorphonuclear granulocytes are the primary phagocytes.[2] While endocytosis internalizes individual molecules or small particles (< 100 nm), phagocytosis internalizes relatively large particles (> 500 nm) (e.g., bacteria). Phagocytic particle engulfment initiates on binding of phagocytic particles to Fc, complement (CR3), or mannose receptors.[2] The role of clathrin in phagocytic engulfment is not fully understood; however, much evidence indicates that a combination of regional actin polymerization and receptor–ligand interactions provides the mechanical force for engulfment.[8] Once engulfment is complete, the phagosome transforms into a degradative compartment termed a phagolysosome. In macrophages, phagolysosome formation requires extensive interactions among phagosomes and other vesicle populations. Most notably, phagosomes fuse with endosomes and then with lysosomes.[9] Thus, while endocytosis and phagocytosis employ different internalization mechanisms, the fate of internalized material is similar.

The biochemical mechanisms that regulate phagolysosome biogenesis are very complex. As evidenced by a number of investigations, both membrane fusion and membrane recycling play important roles in this process.[10,11] For example, many parasites that reside in phagosomes prevent their hydrolytic destruction by inhibiting hydrolase-containing compartments from fusing to phagosomes.[12] Also, following degradation and processing of phagocytosed antigen, vesicle-mediated protein trafficking is required for subsequent antigen presentation on the cell surface.[13] Therefore, to better understand the mechanisms employed by phagosomes to

---

[6] J. Gruenberg and K. E. Howell, *Proc. Natl. Acad. Sci. USA* **84**, 5758 (1987).

[7] R. Diaz, L. Mayorga, and P. Stahl, *J. Biol. Chem.* **263**, 6093 (1988).

[8] S. Greenberg, J. E. Khoury, F. Di Virgilio, E. M. Kaplan, and S. C. Silverstein, *J. Cell Biol.* **113**, 757 (1991).

[9] L. M. Mayorga, F. Bertini, and P. D. Stahl, *J. Biol. Chem.* **266**, 6511 (1991).

[10] W. A. Muller, R. M. Steinman, and Z. A. Cohn, *J. Cell Biol.* **86**, 304 (1980).

[11] T. Lang, C. de Chastellier, A. Ryter, and L. Thilo, *Eur. J. Cell Biol.* **46**, 39 (1988).

[12] K. A. Joiner, S. A. Fuhrman, H. M. Miettinen, I. Kasper, and I. Mellman, *Science* **249**, 641 (1990).

[13] C. V. Harding, D. S. Collins, J. W. Slot, H. J. Geuze, and E. R. Unanue, *Cell* **64**, 393 (1991).

process and traffic internalized material, we developed the following probes and cell-free assays.

Phagocytic Probe

The phagocytic particle in our assays is mouse ^{125}I-labeled anti-2,4-dinitrophenol (DNP) IgG – rabbit anti-mouse IgG – *staphylococcus aureus*.[9] This probe provides many advantages to the study of phagosome processing. The exposed Fc regions of anti-DNP IgG allow for Fc receptor-mediated phagocytosis. Also, assaying for the radiolabel of ^{125}I-labeled anti-DNP IgG throughout an experiment is both simple and sensitive. Mouse anti-DNP IgG is resistant to degradation within the hydrolase-containing compartments and the phagocytic probe is stable to pH 4. Finally, binding of anti-DNP IgG to DNP is not significantly affected by its radioiodination, thus allowing for its subsequent use in a vesicle fusion assay as described below. Endosomes in the phagosome – endosome fusion assay are labeled with purified β-glucuronidase that had previously been derivatized with dinitrofluorobenzene.[7] Thus, the endosomal probe is termed DNP-β-glucuronidase.

Synthesis of ^{125}I-labeled Anti-2,4-Dinitrophenol IgG-Rabbit Anti-Mouse IgG-Staphylococcus aureus

1. Isolate and iodinate anti-DNP IgG as previously described[14,15] to a specific radioactivity of 2000 cpm/ng anti-DNP IgG.

2. Wash formaldehyde-fixed *S. aureus* (IgGsorb; The Enzyme Center, Malden, MA) with HBSA {Hanks' balanced salt solution buffered with 10 mM 4-(2-hydroxyethyl)-1-piperazineethanesulfonic acid (HEPES) and 10 mM 2-[2-hydroxy-1,1-bis(hydroxymethyl)ethyl]aminoethanesulfonic acid (TES), pH 7.4} and supplemented with 1% (v/v) bovine serum albumin (BSA).

3. Incubate 200 μl of *S. aureus* of a 10% (v/v) suspension (approximately 4×10^7 particles/μl, 2 mg of IgG/ml-binding capacity) with 200 μg of rabbit anti-mouse IgG polyclonal antibody (IgG fraction; Organon Teknika Corporation, West Chester, PA) for 1 hr at 20°.

4. Wash the particles three times in HBSA and incubate with 25 μg of ^{125}I-labeled anti-DNP IgG for 1 hr at 20°.

5. Wash antibody-coated *S. aureus* three times and resuspend in HBSA to its original volume.

[14] R. K. Keller and O. Touster, *J. Biol. Chem.* **250**, 4765 (1975).
[15] F. L. Otsuka, M. J. Welch, K. D. McElvany, R. A. Nicolotti, and J. B. Fleischman, *J. Nucl. Med.* **25**, 1343 (1984).

As shown in Fig. 1, on internalization ^{125}I-labeled anti-DNP IgG remains associated with the *S. aureus* for approximately 5–10 min. Thereafter, the majority of ^{125}I-labeled anti-DNP IgG quickly dissociates from the particle. The low concurrent rise in trichloroacetic acid (TCA)-precipitable radioactivity indicates that the majority of the ^{125}I-labeled anti-DNP IgG that leaves the *S. aureus* is not degraded. Thus, ^{125}I-labled anti-DNP IgG marks two aspects of the phagosome. At early times after internalization (up to 10 min), ^{125}I-labeled anti-DNP IgG remains attached to the *S. aureus* and labels the phagocytic particle; later, it dissociates from the *S. aureus* and serves as a lumenal marker of the phagosome.

In Vitro Phagosome – Endosome Fusion

The probe discussed above has been used to investigate phagosome–endosome fusion.[9] The design of the fusion assay is similar to an endosome–endosome fusion assay previously developed in our laboratory.[7] Briefly, early endosomes in one population of cells are labeled with DNP–β-glucuronidase while the phagocytic probe described above is phagocytosed by a second population of cells. Postnuclear supernatants containing the two probes are obtained and mixed under fusogenic conditions (see below). On fusion and mixing of the vesicle contents, immune com-

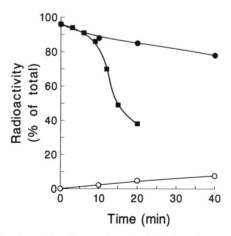

FIG. 1. Characterization of the phagocytic probe following phagocytosis. The phagocytic probe prepared as described above was internalized for 5 min followed by chase for up to 30 min. ●, Cell associated; ■, *S. aureus* associated; ○, TCA soluble.

plexes can form between the DNP–β-glucuronidase and anti-DNP IgG. Determination of the amount of β-glucuronidase activity associated with the antibody-coated *S. aureus* provides a measurement of complex formation.

Phagosome Labeling

1. Preincubate 100 μl of the phagocytic probe prepared as described above with 1×10^8 J774-E clone macrophages at 4° for 1 hr in a final volume of 0.5 ml.
2. Initiate phagocytosis by the addition of 2 ml of HBSA at 37° and incubate at 37° for 5 min.
3. Stop phagocytosis by the addition of 10 ml of HBSA at 4°.

With this protocol, approximately 90% of the antibody-coated *S. aureus* is cell associated after the 37° incubation and 70% is completely intracellular. At this early time of internalization, approximately 80% of the ^{125}I-labeled anti-DNP IgG is attached to the *S. aureus*.

Endosome Labeling

1. Preincubate 1×10^8 J774-E clone macrophages with DNP-β-glucuronidase (20 μg/ml) in 0.5 ml HBSA for 1 hr at 4°.
2. Initiate endocytosis by the addition of 2 ml HBSA at 37° for 5 min.
3. Stop internalization by the addition of excess HBSA at 4°.

Preparation of Postnuclear Supernatant

1. Wash the cells (centrifuge at 200 g for 3 min at 4°) once with HBSA, once with PBS with 5 mM EGTA, and once with homogenization buffer [250 mM sucrose, 0.5 mM ethylene glycol-bis (β-aminoethylether)-N,N,N',N'-tetraacetic acid (EGTA), 20 mM HEPES–KOH, pH 7.2]. Resuspend the cells in homogenization buffer to 5×10^7 cells/ml.
2. Homogenize the cells by either the syringe method as previously described[7] or by use of a ball-bearing homogenizer. Phagosomes are more fragile than endosomes and more prone to breakage during homogenization. Thus, to prevent overhomogenization, stop homogenizing when 80% of the cells are broken (15–30 passes through either device).
3. Remove nuclei by centrifuging homogenate at 200 g for 3 min at 4°.
4. Postnuclear supernatants (200-μl aliquots) may be frozen in liquid nitrogen for storage.

Endosome – Phagosome Fusion Assay

A. Phagosome Preparation

1. Quick-thaw a postnuclear supernatant aliquot (200 μl) containing phagocytic probe.
2. Dilute five-fold in homogenization buffer and centrifuge 1.5 min at 12,000 g at 4°.
3. Resuspend the pellet (phagosome-enriched fraction) in 100 μl homogenization buffer.
4. Add DNP–BSA (final concentration of 0.25 mg/ml) to quench mouse anti-DNP IgG present outside the vesicles.

B. Endosome Preparation

1. Quick-thaw postnuclear supernatants containing labeled endosomes.
2. Dilute 10-fold in homogenization buffer and centrifuge at 35,000 g for 1 min (lysosome-enriched fraction).
3. Centrifuge the resulting supernatant at 50,000 g for 5 min (endosome-enriched fraction).
4. Resuspend the endosome-enriched fraction in 100 μl homogenization buffer.

C. Fusion Assay

1. Combine 4 μl each of endosome-enriched fraction and phagosome-enriched fraction.
2. Add 2 μl cytosol (10 mg/ml) to make the reaction 2 mg/ml cytosol.
3. Add 1 μl of a 10X solution to make the reaction sample 250 mM sucrose, 0.5 mM EGTA, 20 mM HEPES–KOH, pH 7.2, 1 mM dithiothreitol, 1.5 mM MgCl$_2$, 100 mM KCl, 1 mM ATP, 8 mM creatinine phosphate, 31 units/ml creatinine phosphokinase, and 0.25 mg/ml DNP–BSA. Fusions performed in the absence of ATP include an ATP-depleting system (5 mM 2-mannose, 25 U/ml hexokinase).
4. Incubate the fusion reaction at 37° for 45 min.
5. Stop the reaction by returning the tube to 4°.
6. Quantitate the amount of DNP-β-glucuronidase associated with anti-DNP IgG-coated *S. aureus* by lysing samples in solubilization buffer [1% (v/v) Triton X-100, 0.2% (w/v) methylbenzethonium chloride, 1 mM ethylenediaminetetraacetic acid (EDTA), 0.1% (w/v) BSA, 0.15 M NaCl, 10 mM Tris-HCl, pH 7.4, 0.25 mg/ml DNP–BSA] followed by centrifugation at 2800 g for 5 min. Resuspend the pelleted phagocytic particles in 100 μl solubilization buffer plus 100 μl 4-methylumbelliferyl β-D-

glucuronide (1.64 mg/ml in 0.1 M acetate buffer, pH 4.5). Incubate at 37° for 2 hr.

7. Add 1.0 ml glycine buffer (130 mM glycine, 60 mM NaCl, 80 mM NaCO$_3$, pH 9.5) to stop the reaction and measure fluorescence as described.[7]

Phagosome–endosome fusion is cytosol, ATP, K$^+$, and NSF (the N-ethylmaleimide-sensitive factor active in vesicular transport in the secretory pathway) dependent.[7,16-18] At low concentrations of cytosol (0.1 mg/ml), GTPγS stimulates fusion while at high cytosol concentrations (1.5 mg/ml) GTPγS inhibits fusion markedly (Fig. 2A). Futhermore, salt concentrations also differentially affect fusion. At low salt concentrations (50 mM KCl or NaCl), maximal fusion is observed with low cytosol and GTPγS. At high salt concentrations (125 mM KCl or NaCl), maximal fusion is observed with high cytosol concentration without GTPγS (Fig. 2B). Electron microscopic examination of a phagosome–endosome fusion reaction reveals that multiple fusions among endosomes and phagosomes occur simultaneously (Fig. 3).

Analysis of Intracellular Pathway of Phagocytosed Mouse Anti-DNP

In Fig. 1, the ^{125}I-labeled anti-DNP IgG that dissociates from the *S. aureus* is not fully accounted for in the 10% (v/v) TCA-soluble fraction of radiolabeled protein. Assaying the cell media for radioactivity during prolonged periods of phagocytic uptake reveals that some ^{125}I-labeled anti-DNP IgG recycles intact out of the cell. Thus, an intracellular pathway exists to transport phagocytosed protein from the phagosome to the cell surface. To investigate whether nonphagosomal vesicle populations contribute to the recycling pathway, we fractionated cells following phagocytic uptake of our probe. J774-E clone macrophages internalized prebound antibody-coated *S. aureus* for 3 min as described above followed by chase at 37° in HBSA for increasing times. Cells were then fractionated as described below to determine the subcellular location of ^{125}I-labeled anti-DNP IgG.

Subcellular Fractionation following Phagocytic Particle Uptake and Chase

1. Following the procedure outlined above, allow J774-E clone macrophages to phagocytose antibody-coated *S. aureus* for 3 min at 37°.

[16] R. Diaz, L. S. Mayorga, P. J. Weidman, J. E. Rothman, and P. D. Stahl, *Nature (London)* **339**, 398 (1989).
[17] L. Mayorga, R. Diaz, and P. Stahl, *Science* **244**, 1475 (1989).
[18] L. S. Mayorga, R. Diaz, M. I. Columbo, and P. D. Stahl, *Cell Regul.* **1**, 113 (1989).

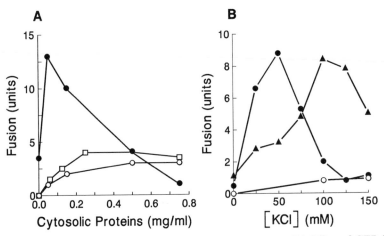

FIG. 2. *In vitro* fusion of phagosomes with early endosomes. (A) Effect of GTPγS at different cytosol concentrations: ●, GTPγS; ■, GTP; ○, control. (B) Effect of KCl: ▲, concentrated cytosol; ●, dilute cytosol + GTPγS; ○, no cytosol.

2. Wash the cells two times in HBSA and incubate again at 37° in HBSA for increasing lengths of time (0–30 min).

3. Obtain postnuclear supernatants as described above from cells representing each of the chase times.

4. Dilute the postnuclear supernatant five-fold in homogenization buffer and centrifuge 1.5 min at 12,000 *g* at 4°.

5. Save the pellet (Phagosome-enriched fraction) and centrifuge the supernatant at 35,000 *g* for 1 min at 4°.

6. Save the pellet (lysosome-enriched fraction) and centrifuge supernatant at 50,000 *g* for 5 min (endosome-enriched fraction).

As shown in Fig. 4A, ^{125}I-labeled anti-DNP IgG in the phagosome-enriched fraction decreases with chase times of up to 30 min. Assaying the remaining vesicles in the cell for ^{125}I-labeled anti-DNP IgG reveals that the endosome-enriched fraction displays a time-dependent increase in radioactivity (Fig. 4B). The majority of radioactivity in this fraction is detergent soluble, trypsin insensitive, and precipitable by 10% (v/v) TCA. The amount of ^{125}I-labeled anti-DNP IgG in the lysosome-enriched membrane pellet does not change with increasing chase times. Characterization of the ^{125}I-labeled anti-DNP IgG-containing vesicles (transport vesicles) in the endosome-enriched membrane pellet reveals that they share features characteristic of early endosomes. As determined by Percoll density gradient centrifugation, the transport vesicles display a buoyant density distribution

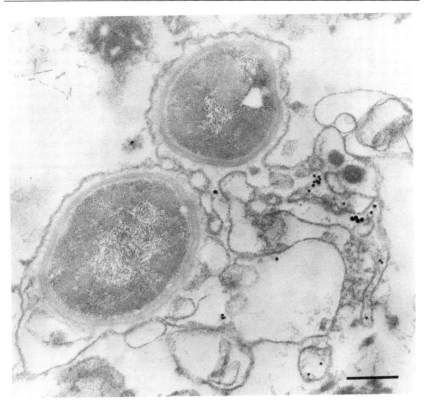

FIG. 3. Electron micrograph demonstrating multiple fusions among endosomes and phagosomes. Early endosomes (5 min at 37°) from three separate populations of J774-E clone macrophages were loaded with 5-, 10-, and 20-nm gold particles coated with mannose–BSA. Endosome fractions of postnuclear supernatants from these cells were incubated under fusogenic conditions (0.1 mg/ml cytosol, 20 μM GTPγS) with the phagosome fraction isolated as described in text. Bar: 0.2 μm.

similar to that observed for material endocytosed in the fluid phase similar to those of early endosomes (data not shown). Employing the DNP-binding property of ^{125}I-labeled anti-DNP IgG, we determined that the fusogenicity of transport vesicles is similar to that described previously for early endosome fusion. Early endosomes labeled with DNP-β-glucuronidase as described above were mixed under fusogenic conditions (i.e., ATP, cytosol, and KCl) with transport vesicles. Following incubation at 37° for 1 hr, fusion samples were lysed in the presence of the scavenger protein, DNP–BSA. Immune complexes were immunoprecipitated with immobilized rabbit anti-mouse IgG and immunoprecipitated β-

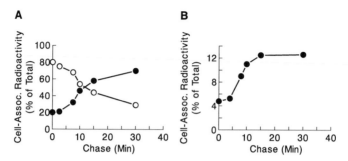

FIG. 4. Intracellular distribution of phagocytosed ^{125}I-labeled anti-DNP. (A) Loss of ^{125}I-labeled anti-DNP from phagosomal fraction (open circles) and concomitant appearance in the nonphagosomal fraction (closed circles). (B) Appearance of ^{125}I-labeled anti-DNP in endosome-enriched fraction obtained by differential centrifugation.

glucuronidase activity was assayed as described above. Figure 5 shows that with increasing times of chase, the signal for transport vesicle early endosome fusion increases. This result suggests that ^{125}I-labeled anti-DNP IgG released from phagosomes accumulates in a vesicular compartment competent to fuse with early endosomes. In light of the well-characterized communication that exists between early endosomes and the plasma membrane, this result suggests that recycling ^{125}I-labeled anti-DNP IgG

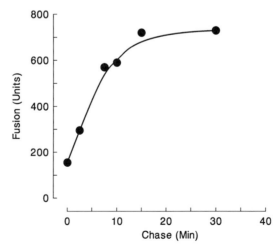

FIG. 5. *In vitro* fusion between early endosomes and transport vesicles. Transport vesicles were from cells that had internalized ^{125}I-labeled anti-DNP IgG-coated *S. aureus* for different lengths of time.

reaches the cell surface via transport through a vesicular compartment with endosomal characteristics.

Phagosome Purification

Due to both the large size and extremely dense nature of formaldehyde-fixed *S. aureus,* phagosomes can easily be purified to homogeneity by loading the phagosome-enriched pellet obtained as described above onto a sucrose cushion [60% (w/v) sucrose, 20 mM HEPES, pH 7.2, 0.5 mM EGTA, made in deuterium oxide] and centrifuging at 50,000 g for 5 min at 4°. As shown in Fig. 6, this method yields a very pure population of phagosomes. With purified phagosomes and the assays described above, additional interesting questions concerning phagolysosome biogenesis may be asked.

Thus, the phagocytic probe described in this chapter has proved useful in examining two important events in macrophage phagocytosis: phagosome – endosome fusion and protein recycling from the phagosome. With the phagocytic probe and the assays described herein, additional process associated with phagolysosome biogenesis will be addressed in the future.

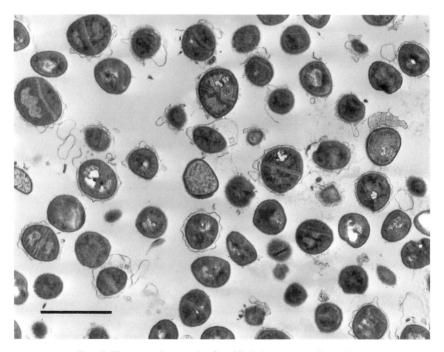

FIG. 6. Electron micrograph of purified phagosomes. Bar: 1.0 μm.

[5] Reconstitution of Endosome Fusion: Identification of Factors Necessary for Fusion Competency

By Maria Isabel Colombo, James M. Lenhard, Luis S. Mayorga, and Philip D. Stahl

Protein transport through the endocytic and exocytic pathways requires fusion of carrier vesicles with the appropriate target membrane.[1-3] The requirements for vesicle fusion have been partially delineated by the development of cell-free systems. Predictably, fusion events appear to have similar requirements. For example, ATP, temperature, and cytosolic and membrane-associated factors have all been found to be required for the various transport events. Nevertheless, the mechanism(s) and regulatory elements that support these processes are still not well understood. Elucidation of the molecular mechanisms involved in vesicle fusion require the identification and purification of individual components of the fusion machinery.

In Vitro Reconstitution of Endosome–Endosome Fusion

Preparation of Fusion Probes

The ligands used to study vesicle fusion in macrophages are designed to have high affinity for the macrophage mannose receptor.[4] These ligands are anti-dinitrophenol (DNP) IgG and DNP-β-glucuronidase. The antibody (HDP-1), a mouse IgG$_1$ monoclonal specific for DNP, is isolated and mannosylated as previously described by Diaz *et al.*[4] The conjugation reaction is performed in the presence of a low-affinity hapten, NBD-aminohexanoic acid (Sigma, St. Louis, MO), to protect the antibody-binding site from being modified. This hapten is subsequently removed by dialysis. β-Glucuronidase, a mannosylated glycoprotein, is isolated from rat preputial glands and derivatized with DNP using dinitrofluorobenzene.[4] Mannosylated anti-DNP IgG and DNP–β-glucuronidase form an immune complex when mixed together. The immune complex is stable over a broad pH and is fairly resistant to proteolytic degradation.

[1] W. Balch, *J. Biol. Chem.* **264**, 16965 (1989).
[2] J. Gruenberg and K. E. Howell, *Annu. Rev. Cell Biol.* **5**, 453 (1989).
[3] J. Rodman, R. W. Mercer, and P. Stahl, *Curr. Opinions Cell Biol.* **2**, 6 (1990).
[4] R. Diaz, L. Mayorga, and P. Stahl, *J. Biol. Chem.* **263**, 6093 (1988).

Uptake of Ligands and Vesicle Preparation

A macrophage cell line, J774 E-clone (mannose receptor positive), is grown to confluence in minimum essential medium containing Earle's salts supplemented with 10% (v/v) fetal calf serum. The cells are washed and resuspended in uptake medium (5×10^7 cells/ml) containing Hanks' balanced salt solution, 10 mg/ml bovine serum albumin (BSA), 10 mM N-2-hydroxyethylpiperazine-N'-2-ethanesulfonic acid (HEPES), and 10 mM N-tris(hydroxymethyl)methyl-2-aminoethanesulfonic acid (TES), pH 7.2. The cells are incubated for 5 min at 37° in uptake medium containing either mannosylated anti-DNP IgG (10 μg/ml) or DNP-β-glucuronidase (20 μg/ml). To stop the internalization step the cells are diluted with cold uptake medium and then washed with phosphate-buffered saline (PBS) containing 5 mM ethylenediaminetetraacetic acid (EDTA). The EDTA removes ligands that remain associated with the cell surface mannose receptor, because binding to this receptor is calcium dependent. After washing in homogenization buffer [250 mM sucrose, 0.5 mM ethylene glycol-bis(β-aminoethyl ether)-N,N,N',N'-tetraacetic acid (EGTA), 20 mM HEPES–KOH, pH 7], the cells are resuspended and homogenized (5×10^7 cells/ml) using a ball-bearing homogenizer.[5] Homogenates are then centrifuged at 800 g for 5 min at 4° to eliminate nuclei and intact cells. The supernatants are quickly frozen in liquid nitrogen and stored at $-80°$. For the experiments, two aliquots (200 μl) of postnuclear supernatants, each containing a different probe, are quickly thawed at 37°, diluted separately to a final volume of 3 ml with homogenization buffer, and pelleted for 1 min at 37,000 g at 4° in a Beckman (Palo Alto, CA) TL-100 centrifuge. The supernatants are centrifuged for an additional 5 min at 50,000 g at 4°. The pellets of the second centrifugation are enriched with endosomes.[4]

Preparation of Cytosol

J774 cytosol is prepared from postnuclear supernatants by centrifugation at 200,000 g for 15 min at 4° using a Beckman TL-100 centrifuge. Cytosol can either be immediately used or frozen in liquid nitrogen and stored at $-80°$. The cytosol (200 μl) is gel filtered just before use, through 1 ml Sephadex G-25 spin columns equilibrated with homogenization buffer. This step removes low molecular weight components (e.g., ATP, GTP, calcium) without significantly altering the protein content of the cytosol. The protein concentration after filtration is around 3 mg/ml. Cytosol prepared from other sources (e.g., L-929 mouse fibroblasts, rabbit

[5] W. Balch and J. Rothman, *Arch. Biochem. Biophys.* **240**, 413 (1985).

alveolar macrophages) also supports fusion of endosomes derived from J774.

Fusion Assay

In the standard assay for reconstitution of endosome–endosome fusion two populations of vesicles containing DNP–β-glucuronidase and mannosylated anti-DNP IgG are mixed at 4° to a final volume of 10–15 μl (1 mg/ml vesicle protein). The reaction mixture is in fusion buffer (250 mM sucrose, 0.5 mM EGTA, 20 mM HEPES–KOH, pH 7.0, 1 mM dithiothreitol (DTT), 1.5 mM MgCl$_2$, 75 mM KCl) containing 0.25 mg/ml DNP–BSA. The DNP–BSA is added as a scavenger to block the binding of mannosylated anti-DNP IgG and DNP–β-glucuronidase in the extravesicular compartment. To maintain a constant level of ATP during the fusion reaction, an ATP-regenerating system (1 mM ATP, 8 mM creatine phosphate, 31 U/ml creatine phosphokinase) is included. The medium is supplemented with 1–2 mg/ml of gel-filtered cytosolic proteins and incubated at 37° for 45 min. The fusion reaction is stopped on ice and the fusion-dependent immune complexes formed are immunoprecipitated as follows: vesicles are solubilized by addition of 100 μl of solubilization buffer [1% (v/v) Triton X-100, 0.2% (v/v) methylbenzethonium chloride, 1 mM EDTA, 0.1% (v/v) BSA, 150 mM NaCl, 10 mM Tris-HCl, pH 7.4] containing 50 μg/ml of DNP–BSA and 2 μl of a 10% suspension of *Staphylococcus aureus* (Staph A) coated with rabbit anti-mouse IgG.[4] The samples are incubated at 4° for 40 min, diluted with 1 ml of solubilization buffer, and pelleted by centrifugation at 1500 g for 5 min at 4°. The Staph A-bound immunocomplexes are then washed twice with 1 ml of solubilization buffer. The pellets are resuspended in 100 μl of solubilization buffer and 100 μl of β-glucuronidase substrate (4 mM 4-methylumbelliferyl-β-D-glucuronide in 0.1 M acetate buffer, pH 4.5). After incubation for 1 to 2 hr at 37°, the reaction is stopped with 1 ml of glycine buffer (133 mM glycine, 67 mM NaCl, 83 mM Na$_2$CO$_3$ adjusted to pH 9.6 with NaOH) and the fluorescence intensity is measured at 366-nm excitation and 450-nm emission.

Requirements for Endosome Fusion

Fusion, as assessed by immune complex formation, is time and temperature dependent, reaching a plateau after a 30-min incubation at 37°. Fusion is observed only at temperatures above 18° and reaches a maximal rate at 37°. The rate of fusion also increases with the total amount of vesicles present in the system, and is inhibited by dilution. Usually 5 to 10 μg of vesicle protein per 10 μl of mixture is required for optimal fusion.

Fusion depends on the presence of ATP, which is supplied by an ATP-regenerating system. The effect of several nucleotides on the fusion reaction is presented in Table I. When ATP is added in the absence of the regenerating system, only 60% of the fusion activity is observed. Inhibition of fusion is achieved by using an ATP-depleting system (25 units/ml hexokinase and 5 mM mannose). ADP does not support fusion, whereas GTP partially support fusion. Adenylimidodiphosphate, a nonhydrolyzable analog of ATP, does not support fusion, suggesting that ATP hydroly-

TABLE I
REQUIREMENTS FOR *in Vitro* FUSION OF ENDOSOMES

Experimental condition[a]	Relative fusion[b]
Regenerating system[c]	1.00
Depleting system[d]	0.02
ATP (1 mM)	0.62
ADP (mM)	0.00
GTP (1 mM)	0.34
PNP-AMP[e] (1 mM)	0.00
KCl (50 mM)	1.00
NaCl (50 mM)	0.98
Potassium gluconate (50 mM)	0.97
Sucrose	0.00
Cytosol (1.5 mg/ml)	1.00
Homogenization buffer	0.01
Bovine serum albumin (1.5 mg/ml)	0.04

[a] Fusion reactions were performed in the presence of complete fusion conditions for 30 min at 37° except when one of the components was omitted or substituted for another component. An ATP-regenerating system was always present, unless when the effect of nucleotides was tested. To assess the ionic requirement of fusion, other salts or sucrose was substituted for KCl in the fusion buffer.

[b] Fusion activity for each condition in a given experiment was compared to the fusion observed in the presence of standard fusion buffer and an ATP-regenerating system, which was assigned a value of 1. Values presented are average for a minimum of three experiments.

[c] The regenerating system consisted of 1 mM ATP, 8 mM creatine phosphate, and 31 units/ml creatine phosphokinase.

[d] The depleting system consisted of 25 units/ml hexokinase and 5 mM mannose.

[e] PNP-AMP, Adenylimidodiphosphate.

sis is required. Endosome fusion requires the presence of $50-75$ mM salts. Higher concentrations affect the integrity of the endocytic vesicles. KCl, NaCl, and potassium gluconate are equally effective (Table I). No fusion is observed in the absence of salt. The requirement for cytosol is saturable at about 1.5 mg/ml. When BSA or homogenization buffer is substituted for cytosol no fusion is observed (Table I).

GTP-Binding Proteins Involved in Endosomal Fusion Events

Guanosine 5'-triphosphate-binding proteins (GTP-binding proteins) have been implicated in transport along the endocytic and exocytic pathways.[6] Nonhydrolyzable analogs of GTP, such as GTPγS [guanosine 5'-(3-O-thio)triphosphate], have been shown to block transport from the endoplasmic reticulum (ER) to the Golgi[7] and among Golgi stacks[8] in cell-free systems. In addition, GTPγS affects the *in vitro* recycling of phosphomannosyl receptors to the Golgi apparatus.[9] GTPγS also modulates endosome–endosome fusion.[10-13] Gruenberg and co-workers[14] have presented evidence that the small GTP-binding protein Rab5 (26 kDa) is involved in the fusion of early endosomes. Recently, a role for GTP-binding proteins belonging to the heterotrimeric[15] and ADP-ribosylation factor[16] (ARF) gene families has also been implicated in endosome fusion.

Effect of GTPγS on Fusion of Early Endosomes

We have used the *in vitro* fusion assay to study the role of GTP-binding proteins in endosome–endosome fusion. Endosomal fractions containing the fusion probes are mixed with different concentrations of gel-filtered cytosol in the presence or absence of 20 μM GTPγS. Figure 1 shows that the effect of GTPγS on fusion is cytosol dependent. GTPγS has a dual

[6] W. E. Balch, *TIBS* **15**, 473 (1990).

[7] C. J. Beckers and W. Balch, *J. Cell Biol.* **108**, 1245 (1989).

[8] P. Melancon, B. Glick, V. Malhotra, P. Weidman, T. Serafini, M. Gleason, L. Orci, and J. Rothman, *Cell* **51**, 1053 (1987).

[9] Y. Goda and S. Pfeffer, *J. Cell Biol.* **112**, 823 (1991).

[10] L. Mayorga, R. Diaz, and P. Stahl, *Science* **244**, 1475 (1989).

[11] L. Mayorga, R. Diaz, M. I. Colombo, and P. Stahl, *Cell Regul.* **1**, 113 (1989).

[12] M. Wessling-Resnick and W. Braell, *J. Biol. Chem.* **265**, 16751 (1990).

[13] J. M. Lenhard, L. Mayorga, and P. D. Stahl, *J. Biol. Chem.* **267**, 1896 (1992).

[14] J.-P. Gorvel, P. Chavrier, M. Zerial, and J. Gruenberg, *Cell* **64**, 915 (1991).

[15] M. I. Colombo, L. S. Mayorga, P. J. Casey, and P. D. Stahl, *Science* **255**, 1695 (1992).

[16] J. M. Lenhard, R. A. Kahn, and P. D. Stahl, *J. Biol. Chem.* **267**, (1992).

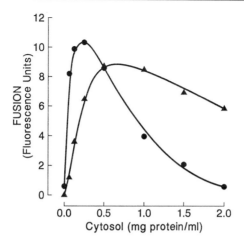

Fig. 1. Effect of GTPγS on *in vitro* fusion of endosomes at different cytosol concentrations. Endosomal vesicles containing the fusion probes were mixed in fusion buffer supplemented with different concentrations of gel-filtered cytosol. The samples were incubated in the presence (●) or the absence (▲) of 20 μM GTPγS. After a 45-min incubation at 37° the samples were solubilized and the amount of immunecomplex formed was quantitated.

effect on fusion among endosomes: it stimulates fusion at cytosol concentrations that are limiting for fusion, and inhibits fusion at saturating concentrations.[11] The concentration of GTPγS required for a half-maximal effect is around 2 μM. The effect is specific for guanosine triphosphonucleotides; 20 μM ATPγS [adenosine 5'-(3-*O*-thio)triphosphate] or 20 μM GDPβS [guanosine 5'-(2-*O*-thio)diphosphate] are ineffective. GTP can reverse the effect of GTPγS. Thus, gel filtration of the cytosol to eliminate the presence of endogenous nucleotides is critical for the assay.

The exact role of GTP-binding proteins in intracellular fusion events is unknown. Because endosome fusion occurs in the presence of GTPγS, GTP hydrolysis is not required for the process. However, it is likely that one or more GTP-binding proteins may be required for fusion. Alternatively, they may play a regulatory role in fusion. Previous results indicate that GTP-binding proteins facilitate the binding of cytosolic factors to the membranes.[11] That would explain the stimulatory effect of GTPγS at concentrations of cytosol that are limiting for fusion. On the other hand, an excess of factors irreversibly bound to the membranes may inhibit fusion at high concentrations of cytosol. It is also possible that two GTP-binding proteins participate in regulating fusion, one that participates in the stimulatory effect and another that participates in the inhibitory effect.

Identification of N-Ethylmaleimide-Sensitive Factor Required for
Endosome–Endosome Fusion

Minimal progress has been made in identifying proteins that participate
in endosome–endosome fusion. *In vitro* fusion assays, like the one used in
our laboratory, can provide information about specific cytosol and
membrane-associated proteins required for vesicle aggregation and fusion.

N-Ethylmaleimide Effect on Endocytic Vesicle Fusion

It has been shown that the sulfhydryl alkylating reagent *N*-
ethylmaleimide (NEM) blocks endosomal fusion.[4,17,18] Endosomes are
treated with 1 mM NEM for 15 min at 4°, followed by a 30-min incubation at 4° with 2 mM DTT, to quench unreacted NEM. As a control, NEM
treatment is carried out in the presence of 2 mM DTT. Treatment of both
vesicles and cytosol with NEM results in complete inhibition of the fusion
reaction (Table II). When NEM-treated cytosol is mixed with untreated
vesicles, fusion is almost fully inhibited (Table II). However, when NEM-
treated vesicles are incubated with untreated cytosol most of the fusogenic
activity is recovered. These results indicate that the NEM-sensitive factor is
present both in a soluble and membrane-associated form, but primarily
present in cytosol. It is also possible that other NEM-sensitive proteins
present only in the cytosol are inactivated by this treatment.

N-Ethylmaleimide-Sensitive Factor: A Common Protein in Several Fusion Events

In addition to endocytosis, other intracellular transport processes requiring membrane fusion (e.g., Golgi transport and transport from ER to
Golgi) have been shown to require NEM-sensitive factors. An NEM-
sensitive factor (NSF) has been identified and purified based on its ability
to restore vesicle-mediated transport to NEM-treated Golgi membranes.[19]
It has been shown that NSF is also required for endosome fusion *in vitro*[20]
and for the reconstitution of transport from ER to Golgi.[21] Addition of
purified NSF (25 ng/12 μl of fusion mixture) partially restores fusion of

[17] P. Woodman and G. Warren, *Eur. J. Biochem.* **173,** 101 (1988).
[18] W. A. Braell, *Proc. Natl. Acad. Sci. USA* **84,** 1137 (1987).
[19] M. Block, B. Glick, C. Wicox, F. Wieland, and J. Rothman, *Proc. Natl. Acad. Sci. USA* **85,** 7852 (1988).
[20] R. Diaz, L. Mayorga, P. Weidman, J. Rothman, and P. Stahl, *Nature (London)* **339,** 398 (1989).
[21] C. Beckers, M. Block, B. Glick, J. Rothman, and W. Balch, *Nature (London)* **339,** 397 (1989).

TABLE II
N-ETHYLMALEIMIDE SENSITIVITY OF ENDOSOME–ENDOSOME FUSION

Cytosol treatment	Vesicle treatment	Addition	Relative fusion[a]
None	None	—	1.00
NEM[b]	NEM	—	0.08
NEM	None	—	0.23
None	NEM	—	0.84
NEM	NEM	NSF[c]	0.55
4A6 IgM[d]	NEM	—	0.12
Control IgM[e]	NEM	—	0.87

[a] See Table I.
[b] Treatment with 1 mM N-ethylmaleimide (NEM) for 15 min at 4°. Excess NEM is quenched with 2 mM dithiothreitol (DTT). As a control, NEM treatment is carried out in the presence of 2 mM DTT. Vesicle fractions are mixed with 1.5 mg/ml cytosol and incubated for 30 min at 37° in complete fusion buffer.
[c] Addition of 25 ng (per 12-μl assay volume) of pure NSF.
[d] Cytosol (5 μg) is preincubated for 60 min at 4° with 1 μg of 4A6 IgM (anti-NSF IgM).
[e] Cytosol (5 μg) is preincubated for 60 min at 4° with 1 μg of IgM control.

NEM-treated endosomes (Table II). We have also tested the effect of a monoclonal IgM antibody that inactivates NSF.[19] J774 cytosol (5 μg) is preincubated for 60 min at 4°C with 1 μg of either control IgM or 4A6 IgM (anti-NSF IgM). Addition of cytosol preincubated with anti-NSF antibody inhibits the recovery of fusion of NEM-treated endosomes (Table II). Control IgM antibodies are not inhibitory. Furthermore, the restorative activity of pure NSF is inhibited by the monoclonal antibody.[20]

Thus, the ability of NSF to catalyze vesicle fusion in both the secretory and endocytic pathways indicates that a common mechanism for vesicle fusion exists.

Identification of Trypsin-Sensitive Factor Required for Endosome–Endosome Fusion

Similar to NEM treatment, preincubation of endosomes with trypsin completely abolishes fusion, but unlike NEM treatment, activity is not restored by the subsequent addition of untreated cytosol.[4,17] Therefore, fusion between endosomes requires trypsin-sensitive factors that reside on the cytoplasmic face of membranes and that are apparently not present in the cytosol.

Membrane-Associated Factor Required for Reconstitution of Endosome–Endosome Fusion after Trypsinization

To study the role of these trypsin-sensitive factors in endosomal fusion, vesicles containing each probe are resuspended in homogenization buffer (1 mg/ml) and treated with 10 μg/ml of trypsin for 30 min at 20°. To quench excess trypsin, the samples are incubated for 30 min (4°) with 20 μg/ml soybean trypsin inhibitor, 1 mM phenylmethylsulfonyl fluoride (PMSF), and 1 μM leupeptin. Control vesicles are preincubated with trypsin in the presence of protease inhibitors. Vesicles are then incubated in complete fusion buffer in the presence of 1 mg/ml cytosolic proteins. Treatment of both sets of vesicles with trypsin (TrV) results in a loss of fusogenic activity that is not restored by the addition of cytosol (Table III). However, when only one set of vesicles is trypsinized, fusion is virtually normal. Furthermore, addition of endosomes devoid of either fusion probe (blank vesicles) completely restores the fusion activity of TrV (Table III). This suggests that some factor(s) required for fusion may be transferred from one set of vesicles to the other. Alternatively, these factors need be present only in one set of fusing membranes. Because we have established that endosomes can fuse multiple times[22] it is possible that the blank vesicles fuse first with one set of TrV and then with the other. Table III shows that the restorative effect of blank vesicles was not observed in the absence of cytosol. These results indicate that untreated vesicles can provide the peripheral components needed to reconstitute endosomal fusion but not all of the soluble factors required for this process.

Restoration of Fusion Activity of Trypsinized Endosomes by High-Salt Extract Containing Peripheral Proteins

Preparation of High-Salt Extract (KCl Extract). J774 E-clone macrophages (2×10^9 cells) are homogenized in 40 ml homogenization buffer and pelleted at 800 g for 5 min at 4° to eliminate nuclei and intact cells. The supernatant is centrifuged at 50,000 g for 5 min at 4°. The pellet (endosomal fraction) is washed with a hypotonic buffer (20 mM HEPES–KOH, pH 7.0, containing 1 mM DTT, 0.5 mM EGTA, 0.5 mM PMSF, and 1 μM leupeptin) and the membranes are pelleted by centrifugation for 15 min at 200,000 g at 4°. The membrane fraction is then incubated at 4°

---

[22] R. Diaz, L. Mayorga, L. E. Mayorga, and P. Stahl, *J. Biol. Chem.* **264,** 13171 (1989).

TABLE III
PERIPHERAL MEMBRANE-ASSOCIATED PROTEIN(S) TO RESTORE
FUSION ACTIVITY TO TRYPSINIZED ENDOSOMES

Vesicle	Fusion reaction	Relative fusion[a]
CV–CV[b]	Complete[c]	1.00
CV–TrV[d]	Complete	1.08
TrV–TrV[e]	Complete	0.12
TrV–TrV	Complete + BV[f]	0.98
TrV–TrV	−Cytosol + BV	0.15
TrV–TrV	Complete + KE[g]	0.97
TrV–TrV	−ATP + KE	0.08
TrV–TrV	−Salt + KE	0.07
TrV–TrV	−Cytosol + KE	1.15
CV–CV	−Cytosol + KE	0.87
TrV–TrV	Complete + NKE[h]	0.78
TrV–TrV	Complete + TKE[i]	0.21
TrV–TrV	Complete + BKE[j]	0.14
TrV–TrV	Complete + NSF[k]	0.11

[a] See Table I.

[b] Untreated vesicles.

[c] Fusion reactions are performed in the presence of complete fusion buffer supplemented with 1 mg/ml of cytosolic proteins. To assess the requirements for fusion some components are omitted. Incubations are carried out for 45 min at 37°.

[d] One set of vesicles (containing one of the fusion probes) is incubated 30 min at 20° with 10 μg/ml trypsin. To quench the excess of trypsin the samples are then incubated 30 min at 4° with 200 μg/ml soybean trypsin inhibitor, 1 mM phenylmethylsulfonyl fluoride (PMSF), and 1 μM leupeptin.

[e] Both set of vesicles are trypsinized as described above.

[f] Addition of 0.7 mg/ml (final protein concentration) of untreated vesicles devoid of fusion probes.

[g] Addition of 1 mg/ml of proteins from a high-salt extract (KE) prepared from untreated endosomes.

[h] Addition of 1 mg/ml of KE preincubated 15 min at 4° with 1mM N-ethylmaleimide. Excess of N-ethylmaleimide was quenched with 2 mM DTT.

[i] Addition of 1 mg/ml of KE preincubated 15 min at 37° with 100 μg/ml of trypsin. Excess trypsin was quenched with 20 μg/ml soybean trypsin inhibitor, 1 mM PMSF, and 1 μM leupeptin.

[j] Addition of 1 mg/ml of KE boiled for 15 min.

[k] Addition of 25 ng of pure NSF to 12 μl fusion assay.

for 30 min with 0.5 M KCl in the hypotonic buffer and pelleted again for 15 min at 200,000 g at 4°. The supernatant, KCl extract (KE), is concentrated using Centricon 10 (Fisher Scientific, Pittsburgh, PA), quickly frozen in liquid nitrogen, and stored at −80°. Aliquots of KE (100 μl) are desalted before use by gel filtration through 1 ml Sephadex G-25 spin columns equilibrated with homogenization buffer containing 1 mg/ml of BSA. The BSA is included to avoid nonspecific binding of proteins to the column. Alternatively, KE can be desalted by dialysis against PBS. In this case the final salt concentration in the fusion assay must be adjusted to 50–75 mM.

Membrane-Associated Proteins Contained by KCl Extract that Restore Fusion Using Trypsinized Endosomes. Table III shows that addition of KE restores the fusion activity of TrV.[23] This indicates that membrane-associated protein(s) that are removed or inactivated by trypsin treatment can be restored by the addition of an extract containing peripheral membrane proteins. The requirements for fusion of TrV in the presence of KE are studied to assess whether the mechanism of fusion is the same as using untreated endosomes. Fusion of TrV requires ATP and is affected by the ionic strength of the fusion buffer, as has been described for normal vesicles. Strikingly, this fusion does not require cytosol (Table III). The restorative effect of KE, with no addition of cytosol, appears to be saturable at 1 mg/ml protein. It is known that some factors can cycle between cytosolic and membrane-associated forms.[19] Therefore, it is likely that most of the cytosolic factors required for fusion are present in KE. The KE is able to support fusion of nontrypsinized vesicles in the absence of cytosol (Table III). These results suggest that KE contains both the membrane-associated factor(s) required for reconstitution of fusion using TrV and the cytosolic components necessary for this process.

Protein(s) Distinct from N-Ethylmaleimide-Sensitive Factor Accounting for Restorative Activity of KCl Extract

Treatment of KE with 1 mM NEM for 15 min at 4° before addition to the fusion reaction only slightly inhibits the restorative effect using TrV (Table III). In contrast, preincubation of KE with trypsin results in almost complete inactivation of the ability to restore fusion. Incubation of KE at 100° also abolishes its activity. These results suggest that KE contains protein(s) that are responsible for its restorative activity.

As mentioned in the preceding section, NSF is a protein required for reconstitution of endosome fusion *in vitro*. The NSF is present both in

[23] M. I. Colombo, S. Gonzalo, P. Weidman, and P. Stahl, *J. Biol. Chem.* **266,** 23438 (1991).

membranes and cytosol. The restorative activity present in KE appears instead to be entirely membrane associated and largely NEM resistant. It is known that NSF is quickly inactivated by incubation at 37° in the absence of ATP.[19] Preincubation of KE under the above conditions does not result in inactivation of the restorative activity. Moreover, addition of pure NSF does not restore fusion using TrV (Table III). These observations suggest that NSF is not responsible for the fusion-promoting activity of KE. However, an NEM-sensitive factor(s) is required for the reconstitution of fusion using TrV, because treatment of TrV, cytosol, and KE with NEM completely abrogated fusion.[23]

These results indicate that even though NSF is required for the reconstitution of fusion using TrV, it does not account for the restorative effect of KE. Therefore, the fusion-promoting activity present in the KE is referred to as TSF (trypsin-sensitive factor) due to its sensitivity to proteolysis and its ability to restore fusion of trypsinized endosomes.

Trypsin-Sensitive Factor: Not One of Known Peripheral Membrane Factors Required for Vesicle Fusion

Large-Scale Preparation of KCl Extract from Bovine Brain. Bovine brains are stored at −80° prior to use. All the steps are performed at 4°. The meninges are removed and the tissue (280 g) is minced, rinsed, and suspended in 300 ml of homogenization buffer containing 1 mM DTT, 0.5 mM PMSF, and 1 μM leupeptin. The tissue is homogenized in a blender (four times, 15 sec) and centrifuged for 10 min at 7500 rpm in a JA-14 rotor (Beckman). The pellet is resuspended in homogenization buffer and the above procedure is repeated twice. The supernatants are combined (750 ml) and pelleted by centrifugation in a Beckman Ti 50.2 rotor at 40,000 rpm for 45 min. The supernatant is discarded. The membrane fraction is resuspended with 400 ml of hypotonic buffer (20 mM HEPES–KOH, pH 7.0 containing 1 mM DTT, 0.5 mM EGTA, 0.5 mM PMSF, and 1 μM leupeptin), incubated 30 min at 4°, and centrifuged again. The supernatant is discarded. The pellet is resuspended with 250 ml of 0.5 M KCl in the hypotonic buffer, incubated 30 min at 4°, and centrifuged. Typically, the supernatant (B-KE), has a protein concentration of 0.7 – 1 mg/ml.

The B-KE provides large quantities of starting material that has restorative activity similar to KE prepared from J774 cells. Preliminary fractionation studies indicate that the restorative activity of KE consists of one or more high molecular weight proteins (200,000) that can be resolved from a large protein peak by gel-filtration chromatography.[23] Trypsin-sensitive factor is distinct from Fr1,[20] a high molecular weight cytosolic protein(s) that is required for transport through the Golgi stacks.[24] Because of its large size, TSF is unlikely to correspond to the low molecular weight proteins, soluble NSF-associated protein (SNAP)[24,25] (involved in the Golgi vesicular transport), and to the GTP-binding protein Rab5[14] (26,000).

The use of trypsinized endosomes and the restoration of fusion like KE have led to the identification of a novel protein(s) termed TSF that seems to be distinct from other previously described factors required for fusion. This protein(s) appears to play an essential role in fusion among early endosomes.

Acknowledgments

We would like to thank Susana Gonzalo for excellent technical assistance. This work was supported in part by Department of Health, Education, and Welfare Grants GM 42259 and AI 20015. Dr. Lenhard is a recipient of a research fellowship from the Cancer Research Institute. Dr. Mayorga is supported by a Rockefeller Foundation Biotechnology Career Fellowship and by an Antorch Reentry Grant.

[24] D. Clary, I. Griff, and J. Rothman, *Cell* **61**, 709 (1990).
[25] D. Clary and J. Rothman, *J. Biol. Chem.* **265**, 10109 (1990).

[6] Transcytotic Vesicle Fusion with Plasma Membrane

By ELIZABETH SZTUL

Introduction

Cell-free assays provide superior experimental systems in which to identify and characterize the components and the molecular mechanisms that underlie membrane transport events. General parameters and specific components required for protein transport have been defined in a number of reconstituted systems. Such *in vitro* assays have been developed to analyze traffic from the endoplasmic reticulum (ER) to the Golgi,[1-3] between Golgi cisternae,[4] from the trans-Golgi network (TGN) to the plasma

[1] C. J. M. Beckers, D. S. Keller, and W. E. Balch, *Cell* **50**, 23 (1987).
[2] M. F. Rexach and R. W. Schekman, *J. Cell Biol.* **114**, 219 (1991).
[3] H. Ruohola, A. K. Kabcenell, and S. Ferro-Novick, *J. Cell Biol.* **107**, 1465 (1988).
[4] W. E. Balch, W. G. Dunphy, W. A. Braell, and J. E. Rothman, *Cell* **39**, 405 (1984).

membrane (PM),[5,6] from the endosome to the TGN,[7] from the PM to an endosomal compartment,[8] and between endosomal compartments.[9]

A membrane traffic pathway that remains largely uncharacterized at the molecular level is the pathway that functionally connects distinct PM domains in polarized epithelial cells. In such cells, the lateral diffusion and intermixing of basolateral and apical PM components is prevented by junctional complexes and specific proteins are transported from one to the "opposite" PM domain by vesicular traffic. Proteins destined to be trans-cytosed are internalized via coated pits and vesicles from the cell surface (basolateral or apical) and delivered to an early endosomal compartment associated with the corresponding PM domain. Within the early endo-some, the proteins are sorted and packaged into specialized carriers, the transcytotic vesicles, which are then translocated[10] across the cell and specifically fuse with the "opposite" PM.

The molecular mechanisms that control the selectivity of transcytotic fusion as well as the actual interactions occurring during fusion events are currently not understood. To approach these enigmas experimentally, we have developed an *in vitro* system that reconstitutes the fusion of transcy-totic vesicles with the PM. In our system, we analyze the fusion of transcy-totic vesicles originating from the basolateral endosome (and containing proteins internalized from the basolateral PM) with the apical PM.

Choice of Experimental System

Two basic criteria must be met when establishing *in vitro* fusion assays: the availability of the compartments involved in the fusion event and the availability of a specific marker protein that undergoes a covalent modifi-cation following fusion of the distinct compartments. Both of these re-quirements are fulfilled by a cell-free system that is based on the transcyto-tic transport of polymeric immunoglobulin A (pIgA) by the rat liver. Clearance of circulating pIgA by hepatocytes represents the single most extensive transcytotic transport of molecules across epithelial cells and, consequently, the transcytotic compartment of the hepatocyte (although not directly measured by morphometric analysis) is expected to be highly

[5] I. De Curtis and K. Simons, *Proc. Natl. Acad. Sci. USA* **85**, 8052 (1988).

[6] S. A. Tooze, U. Weib, and W. B. Huttner, *Nature (London)* **347**, 207 (1990).

[7] Y. Goda and S. R. Pfeffer, *J. Cell Biol.* **112**, 823 (1991).

[8] B. Podbilewicz and I. Mellman, *EMBO J.* **9**, 3477 (1990).

[9] J. Gruenberg and K. E. Howell, *Annu. Rev. Cell Biol.* **5**, 453 (1989).

[10] Translocation in the basolateral to apical direction is microtubule dependent while transcytosis in the opposite direction does not seem to involve microtubules.

[10a] W. Hunziker, P. Male, and I. Mellman, *EMBO J.* **9**, 3515 (1990).

developed. Considering the fact that numerous protocols for subcellular fractionation of rat liver have been developed and allow the isolation of fractions significantly enriched in specific organelles, it is not surprising that a fraction containing the transcytotic compartment can be easily prepared.[11] The apical PM of hepatocytes, probably because of its increased resistance to shearing due to added stability provided by the junctional complexes, can also be isolated in a relatively undamaged, "sealed" state, retaining its "cytosolic face out" conformation.[12] It is clear, therefore, that the compartments involved in transcytotic fusions can be easily prepared from rat liver.

Choice of Marker Protein

The transcytosis of pIgA is mediated by the integral membrane protein, the polymeric IgA receptor (pIgA-R). In contrast to other receptors that undergo repeated cycles of ligand internalization, delivery, and release, the pIgA-R does not recycle. Following transcytosis, it is proteolytically cleaved and secreted from the cells. Consequently, because its half-life is relatively short,[13] it is synthesized at rates approaching those of secretory proteins (i.e., significantly higher than those of other integral membrane proteins) and can be radiolabeled following an *in vivo* biosynthetic pulse. Previous work[13,14] has defined the intracellular pathway and posttranslational processing of the receptor. As shown in Fig. 1, pIgA-R is synthesized in the ER as a core-glycosylated 105K precursor. Following transport through the Golgi, where the receptor is terminally glycosylated (which increases the apparent molecular weight to 116K), the protein is delivered to and inserted into the basolateral PM. The receptor is phosphorylated on serine residues (a modification that increases the apparent molecular weight of the receptor to 120K),[15] endocytosed from the basolateral PM, and internalized into an "early" endosomal compartment. The receptor is then sorted into transcytotic vesicles and these carriers translocate across the cell by a microtubule-dependent mechanism[16] and then specifically interact and fuse with the acceptor apical PM. The fusion results in the insertion of the pIgA-R into the apical PM where the receptor comes in contact with an apical PM protease.[17,18] The protease cleaves within the

[11] E. S. Sztul, A. Kaplin, L. Saucan, and G. Palade, *Cell* **64,** 81 (1991).
[12] A. L. Hubbard, D. A. Wall, and A. Ma, *J. Cell Biol.* **96,** 217 (1983).
[13] E. S. Sztul, K. E. Howell, and G. E. Palade, *J. Cell Biol.* **100,** 1248 (1985).
[14] E. S. Sztul, K. E. Howell, and G. E. Palade, *J. Cell Biol.* **100,** 1255 (1985).
[15] J. M. Larkin, E. S. Sztul, and G. E. Palade, *Proc. Natl. Acad. Sci. USA* **83,** 759 (1986).
[16] B. M. Mullock, R. S. Jones, J. Peppard, and R. H. Hinton, *FEBS Lett.* **120,** 278 (1980).
[17] L. S. Musil and J. U. Baenziger, *J. Cell Biol.* **104,** 1725 (1987).
[18] L. S. Musil and J. R. Baenziger, *J. Biol. Chem.* **263,** 15799 (1988).

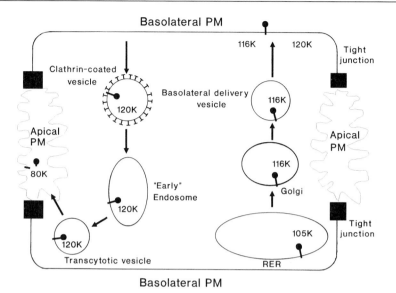

FIG. 1. Intracellular pathway of the pIgA receptor. Both the biosynthetic exocytic pathway forms of the receptor (105K, 116K) as well as the endocytic–transcytotic pathway forms (predominantly 120K with a small amount of 116K present) are shown.

ectodomain of the receptor, producing a fragment (80K) called secretory component (SC), which is secreted into the apical lumen, and a fragment containing the transmembrane and the cytosolic domains, which is probably degraded intracellularly. The cleavage of the 120K form of the pIgA-R to generate the 80K fragment can, therefore, be used as a measure of membrane fusion, because only fusion places both the pIgA-R and the protease within a single membrane compartment. This specific cleavage forms the basis for a cell-free assay to reconstitute the fusion event between transcytotic vesicles and the apical PM.

Description of Cell-Free Assay

Preparation of Donor Fraction

As the donor, we prepare a transcytotic vesicle fraction containing biosynthetically labeled pIgA receptor (120K form). Previous work has shown that 60 min after an *in vivo* radiolabeling pulse, labeled pIgA-R is present predominantly within transcytotic vesicles,[19] and that these vesicles

[19] E. S. Sztul, K. E. Howell, and G. E. Palade, *J. Cell Biol.* **97**, 1582 (1983).

are enriched in a fraction recovered within 1.15 M sucrose after differential sucrose density gradient centrifugation.[11]

To prepare the donor fraction, a 120- to 160-g male Sprague-Dawley rat is anesthetized with intraperitoneally injected Nembutal (Abbott Laboratories, North Chicago, IL) at 0.1 ml/100 g animal weight and kept under anesthesia throughout the entire radiolabeling period. An injection is made into the saphenous vein of the rat of 200 μl of solution containing 3 mCi of Trans[^{35}S] label (ICN, Irvine, CA) in phosphate buffered saline (PBS). After 60 min, the liver is removed, weighed, and placed in ice-cold 0.25 M sucrose. All the subsequent steps are performed in the cold room. The liver is minced with a pair of scissors and washed with ice-cold 0.25 M sucrose until free of blood. The minced tissue is made up to ~30% (w/v) with 0.25 M sucrose and homogenized.

The homogenate is centrifuged for 10 min at 7000 g_{av} (12,500 rpm, 4°) in a Ti 70 rotor and the supernatant collected. The pellet, containing predominantly cell debris, nuclei, and mitochondria, is discarded. The supernatant is centrifuged for 90 min at 105,000 g_{av} (40,000 rpm, 4°) in a Ti 70 rotor. The supernatant (cytosol) is removed, diluted to a protein concentration of 10 mg/ml, aliquoted, and stored at −80° while the pellet, containing all cellular membrane compartments (total microsomal fraction), is resuspended in 10 ml of 1.22 M sucrose. The resuspended fraction is then placed on the bottom of a centrifuge tube and overlaid with 10 ml of 1.15 M sucrose, followed by 10 ml of 0.86 M sucrose and 10 ml of 0.25 M sucrose. This gradient is then centrifuged for 3 hr at 82,500 g_{av} (27,000 rpm, 4°) in an SW 28 rotor. The gradient fractions are collected from the top and the fraction recovered in the 1.15 M sucrose is removed, aliquoted in 0.5-ml samples, and frozen in liquid nitrogen. The fraction is kept at −80° until use.

The protein concentration of the 1.15 M sucrose fraction is measured by the Bio-Rad microprotein assay (Bio-Rad, Bethesda, MD). Usually the protein concentration is between 4 and 6 mg/ml and fractions are used without dilution. When protein concentrations are higher than 6 mg/ml, the fraction is diluted (to 4 mg/ml) with 0.25 M sucrose.

Preparation of Acceptor Fraction

The procedure used to isolate the acceptor fraction is based on that described by Hubbard *et al.*[12]

Male Sprague-Dawley rats (120–170 g) are starved overnight and anesthetized with Nembutal (as above). Livers are excised intact and perfused through the portal vein with ice-cold 0.154 M NaCl until blanched. All subsequent procedures are carried out at 4°. Individual livers are weighed, minced with scissors, added to 4–4.5 vol of 0.25 M STM (0.25 M

sucrose, 5 mM Tris-HCl, pH 7.4, 1.0 mM MgCl$_2$), and homogenized by 10 up-and-down strokes in a 40-ml Dounce-type glass homogenizer with a loose-fitting pestle. Three livers are routinely used and weigh a total of 20–30 g. The homogenate is adjusted to 20% (liver wet weight to total volume) with 0.25 M STM and filtered through four layers of moistened gauze. The filtrate is centrifuged (25–30 ml/50-ml plastic tube) at 280 g_{av} for 5 min (1100 rpm) in an International Equipment Company (Needham Heights, MA) PR-6000 or Beckman (Palo Alto, CA) TJ-6R. The supernatant is saved and the pellet is resuspended by three strokes of the loose Dounce homogenizer in 0.25 M STM to one-half the initial homogenate volume. The suspension is again centrifuged as above. The first and second supernatants are combined and centrifuged (25–30 ml/50-ml tube) at 1500 g_{av} (2600 rpm) for 10 min. The resulting pellets are pooled and resuspended by three strokes of the loose Dounce in ~1–2 ml of 0.25M STM per gram of liver (initial wet weight). 2.0M STM is added to obtain a density of 1.18 g/cm^3 (1.42 M) and sufficient 1.42M STM is added to bring the volume to approximately twice that of the original homogenate (i.e. 10% wt/vol). 35-ml aliquots of the sample are added to cellulose nitrate tubes and overlaid with 2–4 ml 0.25M sucrose. After centrifugation for 60 min at 82,000 g_{av} in a Beckman L5-65 centrifuge (25,000 rpm, SW 28 rotor, no brake), the pellicle at the interface is collected with a blunt-tipped Pasteur pipette and resuspended in a loose Dounce in sufficient 0.25M sucrose to obtain a density of 1.05 g/cm^3. This suspension is centrifuged at 1500 g_{av} for 10 min and the final pellet [designated the plasma membrane (PM) fraction] resuspended in 0.25M sucrose in a loose Dounce homogenizer. The fraction is diluted to 2 mg/ml protein, aliquoted, and kept at 80° until use.

Fusion Assay

Standard fusion assay consists of 200 μl (800 μg protein) donor fraction, 150 μl (300 μg protein) acceptor fraction, 50 μl (500 μg protein) rat liver cytosol, 50 μl ATP-regenerating system, and 50 μl fusion buffer. The ATP-regenerating system is prepared fresh by mixing 100 μl of 40 mM ATP, 100 μl of 200 mM creatine phosphate, and 20 μl of 2000 U/ml creatine phosphokinase. The fusion buffer is diluted from a 10X stock (kept at −80°) containing 500 mM N-2-hydroxyethylpiperazine-N'-2-ethanesulfonic acid (HEPES)–KOH, pH 7.2, 900 mM KCl, and 25 mM MgCl$_2$. To initiate fusion, the components are added in the following order: fusion buffer, ATP-regenerating system, cytosol, donor fraction, and acceptor fraction, and incubated at 37° for 1 hr. In some experiments, the entire reaction mixture is supplemented with 1 mM N-ethylmaleimide (NEM) and incubated at 4° for 60 min. Excess NEM is quenched with

2 mM dithiothreitol (DTT) prior to warming to and incubation at 37°. After the incubation, the tubes are placed on ice and the entire sample is subjected to immunoprecipitation.

Analysis of Results

Immunoprecipitation

The samples (500 μl) are solubilized by the addition of 20 μl 10% (w/v) sodium dodecyl sulfate (SDS; final concentration of 0.4%, w/v), boiled for 2 min, and then supplemented with 50 μl of Triton X-100 solution containing 20% (v/v) Triton X-100, 1.5 M NaCl, 20 mM ethylenediaminetetraacetic acid (EDTA), and 300 mM Tris, pH 7.4. The samples are spun for 2 min in an Eppendorf microfuge and the supernatants are removed and incubated with 20 μl of anti-pIgA-R serum (raised in rabbits against the biliary 80K form of the rat receptor) for 2 hr at room temperature. The antigen–antibody complexes are recovered by adding 100 μl of a 50% (v/v) suspension of protein A–Sepharose beads in PBS, incubating for 1 hr at room temperature and centrifuging for 10 sec in a microfuge. The beads are washed three times (1 ml each) with NETS buffer [150 mM NaCl, 10 mM EDTA, 0.5% (v/v) Triton X-100, and 0.25% (w/v) SDS]. The proteins are eluted off the beads by adding 30 μl of "cracking buffer" [125 mM Tris, pH 6.8, 2% (w/v) SDS, 10% (v/v) glycerol, 5% (v/v) 2-mercaptoethanol, and 0.01% (v/v) bromphenol blue] and the eluted sample analyzed by SDS polyacrylamide gel electrophoresis (PAGE).

SDS-PAGE and Fluorography

The SDS-PAGE is carried out as previously described[19] except that 0.8-mm gels are used. Electrophoresis is carried out at room temperature at a constant voltage of 150 mV for 30 min and then at a constant power of 15 W for 1.5 hr. Gels are processed for fluorography with EN[3]Hance (New England Nuclear, Boston MA) according to the directions of the manufacturer. Gels are dried and used to expose Kodak (Rochester, NY) diagnostic X-OMAT (XAR5) film at −80° for approximately 3 days.

Fusion Requirements

We modeled our system on that previously developed to reconstitute fusion of early endosomal elements.[20] Previous workers have shown that membrane–membrane fusions within the endocytic pathway require ATP and cytosolic factors and are temperature and NEM sensitive. Consequently, we tested whether similar requirements apply in our cell-free

[20] J. Gruenberg and K. E. Howell, *Proc. Natl. Acad. Sci. USA* **84,** 5758 (1987).

system. Donor fraction, containing the radiolabeled 120K form of the
pIgA-R, was supplemented with the acceptor fraction and various combi-
nations of reagents and incubated at 37° for 1 hr. After incubation, the
samples were solubilized and pIgA-R forms immunoprecipitated and analysed by SDS-PAGE and fluorography. Fusion was scored by the conversion of the 120K form of the pIgA-R to the 80K fragment. As shown in Fig.
2 (lane 3), incubation of the donor fraction in the presence of the acceptor
fraction, cytosol, and an ATP-regenerating system, resulted in the conversion of approximately 50% of the 120K pIgA-R to the 80K form. A slower
migrating band of ~100K was sometimes observed (arrowhead in Fig. 2)
and might represent an intermediate in the conversion. Incubations of the
donor fraction under analogous conditions but without added acceptor
(Fig. 2, lane 1) did not result in cleavage of pIgA-R, thus showing that the
processing enzyme resides in the acceptor fraction and is not a component
of transcytotic vesicles. Depletion of the reaction mixture of ATP by
substituting the ATP-regenerating system with an ATP-depleting system
(composed of 5 mM glucose and 500 U/ml hexokinase) (Fig. 2, lane 2) or
treatment of the reaction mixture with 1 mM NEM (Fig. 2, lane 6) prevented the formation of the 80K pIgA-R. Similarly, omitting cytosol from
the reaction mixture resulted in the inhibition of conversion. (Fig. 2, lane
4). Incubation of the complete reaction mixture at 4° also prevented the
conversion of the receptor (Fig. 2, lane 5). Thus, conditions required for
fusion in the *in vitro* system are analogous to those described previously for
other membrane–membrane fusion events and this reconstituted system
can now be used to analyze components required exclusively during transcytotic fusion events.

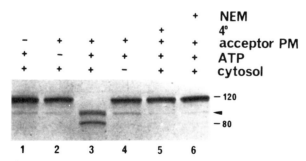

FIG. 2. Requirements for *in vitro* fusion. Donor fraction was incubated without (lane 1) or
with (lanes 2–6) acceptor fraction under different conditions. After incubation for 1 hr at the
appropriate temperature, the entire sample was solubilized and pIgA-R was immunoprecipitated. The immunoprecipitates were analyzed by SDS-PAGE and fluorography. Lane 2,
reaction mixture containing an ATP-depleting system; lane 3, reaction mixture containing
the complete fusion medium; lane 4, reaction mixture without cytosol; lane 5, reaction
mixture containing the complete fusion medium, incubated at 4°; and lane 6, reaction
mixture containing the complete fusion medium supplemented with 1 mM NEM.

[7] Reconstitution of Rat Liver Endosome–Lysosome Fusion in Vitro

By Barbara M. Mullock and J. Paul Luzio

The demonstration of fusion between two membrane-bound compartments in cell-free systems is usually achieved by measuring the mixing and interaction of the contents of the compartments. Marker pairs for content mixing assays have often been appropriate enzymes and substrates or antibodies and antigens, although other interacting compounds have also been used. In the case of endosome–lysosome fusion, assessment of content mixing is made difficult both by the destructive nature of the interior of lysosomes *in vivo* and by their tendency to lose their proton-pumping ability during purification. The former property makes it difficult to load lysosomes with reagents for measuring content fusion and the latter results in variable enzyme activity within purified lysosomes *in vitro*. Hence cell-free endosome–lysosome interaction in partially purified systems has so far been demonstrated by examining changes in the isopycnic density of endocytosed ligand.[1] Better purification of the components of the system has made it possible not only to examine the characteristics of the system with more precision but also to demonstrate membrane fusion by a fluorescence dequenching method. In the long term, demonstration of content mixing will probably have to be based on immunoprecipitation of a protein that is relatively resistant to lysosomal enzymes.

For the isopycnic density change assay, liver endosomes are loaded *in vivo* by intravenous injection of radioiodine-labeled protein ligand, usually asialofetuin (ASF), into the rat. The simplest possible system is that in which the interval between injection and killing the animal is so short that the ASF has not reached the lysosomes. A postmitochondrial supernatant from the liver of a rat thus loaded can be analyzed by centrifugation on Nycodenz (Nycomed, Birmingham, England) gradients either untreated or after incubation at 37° with an ATP-regenerating system. Nycodenz gradients separate endosomes from lysosomes; after the 37° incubation the radiolabel is found in the lysosomal position.[1] However, the postmitochondrial supernatant contains not only several different compartments on the endosomal pathway but also much irrelevant material. This chapter will therefore describe methods to separate endocytic compartments after *in vivo* loading with ^{125}I-labeled ASF. Such preparations are then incubated

---

[1] B. M. Mullock, W. J. Branch, M. van Schaik, L. K. Gilbert, and J. P. Luzio, *J. Cell Biol.* **108**, 2093 (1989).

with lysosomes, rapidly prepared from an unlabeled rat, and, after incubation under appropriate conditions, the fate of the ^{125}I-labeled ASF is assessd by centrifugation of Nycodenz gradients. Strictly, such an assay shows only close association between endosomes and lysosomes rather than true fusion; however with these purified preparations of endosomes and lysosomes, measurement of membrane fusion becomes possible and will be described.

Separation of Endocytic Compartments

Centrifugation of postmitochondrial supernatant on shallow, isoosmotic, isopycnic gradients of Ficoll in a vertical rotor gives a rapid separation of three different endocytic compartments from each other and from sinusoidal plasma membrane.[2,3] Immediately after a pulse dose of ^{125}I-labeled ASF, radioactivity appears in sinusoidal plasma membrane, but with increasing time between the pulse and fractionation the radioactive label moves sequentially to light endosomes, dense endosomes, and finally to the region on the cushion at the bottom of the gradient. This contains lysosomes, remaining small mitochondria, and rough endoplasmic reticulum as well as very dense endosomes. Light and dense endosomes are contaminated primarily by smooth endoplasmic reticulum, which can be largely removed by recentrifugation on metrizamide. Light (early) endosomes do not participate in endosome–lysosome interaction[1] so only the separation of dense endosomes is described. These endosomes probably correspond to the carrier vesicles[4] described in Chinese hamster ovary (CHO)[5] and (Madin–Darby canine kidney (MDCK)[6] cells. The separation of very dense endosomes, which can also participate in cell-free endosome–lysosome interaction, can also be achieved by using a mixed step gradient of Ficoll and Nycodenz, followed by recentrifugation on metrizamide to remove endoplasmic reticulum contaminants. However, unlike dense endosomes, which are stable under the conditions used, very dense endosomes are rendered partially leaky if incubated at 37° after prolonged centrifugation. The purified very dense endosomes contain approximately 15% of the total N-acetyl-β-glucosaminidase activity of the

[2] W. J. Branch, B. M. Mullock, and J. P. Luzio, *Biochem. J.* **244**, 311 (1987).

[3] J. H. Perez, W. J. Branch, L. Smith, B. M. Mullock and J. P. Luzio, *Biochem. J.* **251**, 763 (1988).

[4] B. M. Mullock, J. H. Perez, J. P. Luzio, and B. M. Mullock, *Proc. Eur. Workshop Endocytosis, 2nd* (P. J. Courtoy, ed.), p. 123. Springer-Verlag, Berlin and New York, 1992.

[3] J. Gruenberg, G. Griffiths, and K. E. Howell, *J. Cell Biol.* **108**, 1301 (1989).

[6] M. Bomsel, R. Parton, S. A. Kuznetsov, T. A. Schroer, and J. Gruenberg, *Cell* **62**, 719 (1990).

homogenate as well as possessing mannose 6-phosphate receptors, suggesting that the fraction contains the prelysosomal compartment.[7,8]

Preparation of Dense Endosomes

Preparation of Gradients. All gradient solutions contain 10 mM N-tris(hydroxymethyl)methyl-2-aminoethanesulfonic acid (TES) and 1 mM ethylenediaminetetraacetic acid (EDTA), pH 7.4. Stock solutions of 200 mM TES and 0.5 M EDTA are adjusted to pH 7.4 with NaOH. Ficoll 400 (Pharmacia, Piscataway, NJ) is dissolved by adding 1 ml of water/g at room temperature, dialyzed against a large volume of distilled water for 2 hr, then adjusted to give a 25% (w/v) stock solution, which can be stored in aliquots at $-20°$. From this stock, 22 and 1% Ficoll solutions are prepared, each containing 0.25 M sucrose in addition to TES and EDTA, because Ficoll exerts little osmotic pressure. Nycodenz is prepared as a stock 45% (w/v) solution containing TES and EDTA, and may also be stored at $-20°$. Four-milliliter cushions of this solution are placed in the bottom of each Beckman (Palo Alto, CA) 1 × 3.5 in. polyallomer Quick Seal centrifuge tube. Linear gradients made from 15 ml/tube of each of the 22 and 1% Ficoll solutions are then poured over the cushions at 4°. The simplest gradient maker consists of two open chambers connected by a tap at the base and with an exit from the bottom of one chamber that can be connected to a peristaltic pump. The chamber with the exit is stirred and becomes the mixing chamber. Ficoll (22%) is placed in the mixing chamber and 1% Ficoll in the other chamber. The tap between the chambers is opened and the pump started simultaneously, running at about 1 ml/min. With a multichanneled pump, several gradients may be made at once. The mixture is delivered to the centrifuge tubes through bent, large-bore hypodermic needles, which rest in the centrifuge tube openings and drip freely. It is difficult, and in practice has proved unnecessary, to arrange for the gradient to run down the walls of vertical rotor tubes with their narrow openings. The gradients may be left at 4° overnight; they will tend to smooth by diffusion.

Step gradients of metrizamide (centrifugation grade; Nycomed, Birmingham, England) are prepared immediately before use by underlaying 11 ml of 12% (w/v) metrizamide with 14 ml of 18% (w/v) metrizamide and the latter with 4 ml of 45% (w/v) Nycodenz solution in each Quick Seal tube. All solutions contain TES and EDTA and are at 4°. Underlaying is

[7] G. Griffiths, R. Matteoni, R. Back, and B. Hoflack, *J. Cell Sci.* **95**, 441 (1990).
[8] Y. Deng, G. Griffiths, and B. Storrie, *J. Cell Sci.* **99**, 571 (1991).

done with a syringe attached to a narrow steel tube long enough to reach to the base of the centrifuge tube.

Preparation of ^{125}*I-Labeled Asialofetuin Postmitochondrial Supernatant.* Asialofetuin in phosphate-buffered saline (PBS) is labeled with ^{125}I by the Iodogen (Pierce, Rockford, IL) method[9] and free iodide removed by passage through Sephadex G-25. Approximately 4–5 μg of such ASF (30–40 μCi) in 0.3 ml is injected into the jugular vein of an anesthetized rat. After 10 min the liver is perfused thoroughly with ice-cold 0.25 M sucrose containing 10 mM TES and 1 mM Mg^{2+}, pH 7.4 (STM) and removed to more ice-cold STM. The liver is rapidly blotted, weighed, and transferred to 3 ml/g of fresh, cold STM. After being roughly chopped with scissors, the liver is homogenized with three strokes of a Teflon pestle rotating at 2400 rpm in a Potter–Elvejhem homogenizer. The homogenizer vessel is kept in iced water. The homogenate is centrifuged at 1500 g for 10 min at 4° and the postmitochondrial supernatant poured off.

Gradient Centrifugation and Collection of Dense Endosomes. For fractionation, 5 ml of postmitochondrial supernatant is pumped onto each prepared linear Ficoll gradient. The tubes are heat sealed and centrifuged at 200,000 g for 1 hr in a vertical rotor at 4° with slow initial acceleration and no braking in the final stages of deceleration. Fractions are collected at 4° by clamping the centrifuge tube vertically, passing a 2-mm diameter stainless steel tube to the bottom, and sucking out the gradient using a peristaltic pump running at approximately 1 ml/min. This method of gradient collection is usually said to be poorer than upward displacement or tube piercing but with the gradients used in this work it gives reproducible results quickly and simply. Fractions of about 0.7 ml are collected and refractive indices of small samples of sufficient fractions measured to identify those with indices between 1.3661 and 1.3730, the range within which dense endosomes appear. Five or six fractions from the center of this region are pooled and diluted with 10 mM TES, 1 mM NaEDTA to a refractive index of less than 1.354. Up to 10 ml of this mixture is then loaded over a metrizamide step gradient (material from three Ficoll gradients is usually loaded to two metrizamide gradients). The tubes are centrifuged and unloaded as for Ficoll gradients. Fractions containing most ligand (around and above the interface between the two metrizamide solutions) are pooled, diluted 3-fold with STM, and centrifuged at 170,000 g at 4° to sediment endosomes, which are then resuspended by 20 strokes with a hand-held glass–glass homogenizer.

---

[9] P. Salacinski, J. Hope, C. MacClean, B. Clement-Jones, J. Sykes, J. Price, and P. J. Lowry, *J. Endocrinol.* **81**, 131P (1979).

Recovery and Purification of Dense Endosomes. This method yields approximately 1 mg of protein/g liver. About 25% of the total ^{125}I-labeled ASF in dense endosomes is recovered (assuming that all the radioactivity in the nuclear pellet is trapped postmitochondrial supernatant). The final preparation is purified approximately 50-fold relative to the homogenate with respect to protein and approximately 40-fold with respect to the lysosomal enzyme *N*-acetyl-*β*-glucosaminidase, but only 7-fold with reference to the smooth endoplasmic reticulum marker glucose-6-phosphatase. Transmission electron microscopy shows that the preparation consists of a reasonably homogeneous population of vesicles, mostly of diameter 0.1– 0.2 μm.

Rapid Preparation of Lysosomes

Intact lysosomes can be prepared in 2 hr using the scheme in Fig. 1.[10] All operations are at 4°. The rat may be starved overnight to reduce the glycogen content of the final preparation. The final wash may be omitted but gives better packed pellets and more reproducible results. Ninety percent of the *N*-acetyl-*β*-glucosaminidase activity of the final product is latent. Organelles (including endosomes) other than lysosomes are present at less than 2% of their homogenate concentrations.[1,10] This method yields approximately 2 mg protein/g liver from a starved rat.

Reaction between Endosomes and Lysosomes Analyzed by Centrifugation

For maximum activity ^{125}I-labeled ASF-loaded dense endosomes from approximately 0.3 g liver and lysosomes from 0.6 to 0.9 g of an unlabeled liver are resuspended in 0.5 ml of cytosol. Endosomes are usually prepared on the previous day and left at 4° overnight before the final centrifugation. Lysosomes are freshly made. Cytosol is prepared by centrifuging postmitochondrial supernatant at 288,000 g for 45 min at 4°. Small molecules are usually removed by filtration through BioGel P-6 (Bio-Rad, Richmond, CA) equilibrated with STM and aliquots are stored in liquid nitrogen. The mixture is incubated at 37° for 15–30 min with an ATP-regenerating system consisting of 10 mM phosphoenolpyruvate and 35 U of pyruvate kinase. If the cytosol has been filtered, 1.3 mM GTP is also required for maximum activity. After incubation the mixture is chilled and loaded over a linear gradient made from 2.4 ml of 35% (w/v) Nycodenz, 0.055 M sucrose and 2.4 ml 0.25 M sucrose, both containing 10 mM TES and

[10] G. A. Maguire and J. P. Luzio, *FEBS Lett.* **180**, 122 (1985).

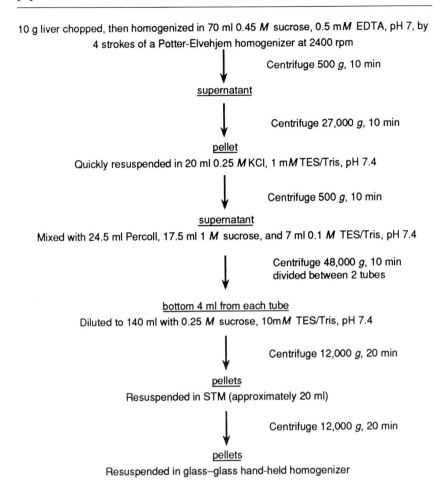

10 g liver chopped, then homogenized in 70 ml 0.45 M sucrose, 0.5 mM EDTA, pH 7, by
4 strokes of a Potter-Elvehjem homogenizer at 2400 rpm

Centrifuge 500 g, 10 min

supernatant

Centrifuge 27,000 g, 10 min

pellet
Quickly resuspended in 20 ml 0.25 M KCl, 1 mM TES/Tris, pH 7.4

Centrifuge 500 g, 10 min

supernatant
Mixed with 24.5 ml Percoll, 17.5 ml 1 M sucrose, and 7 ml 0.1 M TES/Tris, pH 7.4

Centrifuge 48,000 g, 10 min
divided between 2 tubes

bottom 4 ml from each tube
Diluted to 140 ml with 0.25 M sucrose, 10mM TES/Tris, pH 7.4

Centrifuge 12,000 g, 20 min

pellets
Resuspended in STM (approximately 20 ml)

Centrifuge 12,000 g, 20 min

pellets
Resuspended in glass–glass hand-held homogenizer

Fig. 1. Preparation of lysosomes. STM is 0.25 M sucrose, 10 mM TES, 1 mM Mg^{2+}. The
0.1 M TES/Tris buffer has 2.29 g TES and 0.77 g Tris/100 ml. All operations are at 4°.

1 mM EDTA in a 6-ml polyallomer Ultracrimp tube. The tube is sealed
and centrifuged in a Sorvall (Norwalk, CT) TV-86S vertical rotor for
20 min at 370,000 g at 4°. Fractions (5 drop) are collected using a 1-mm
diameter stainless steel tube and a peristaltic pump to suck the gradient
from the bottom of the centrifuge tube. After γ-counting the fractions, the
positions of lysosomal and endosomal peaks are established by measurement of the refractive indices of small samples of appropriate fractions.
The positions of these peaks with reference to refractive index are very

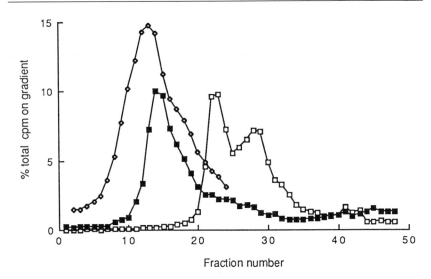

FIG. 2. Movement of ^{125}I-labeled ASF from the endosomal to the lysosomal position on Nycodenz gradients. ^{125}I-Labeled ASF-loaded dense endosomes were incubated with unlabeled lysosomes, gel-filtered cytosol, GTP, and an ATP-regenerating system at 37° for 30 min. □, Mixture kept in ice; ■, mixture incubated at 37°; ◇, N-Acetyl-β-glucosaminidase (measured on incubated sample; arbitrary units).

constant but N-acetyl-β-glucosaminidase measurements[11] can also be made if desired. A typical result is shown in Fig. 2.

It should be noted that, using purified dense endosome fractions, almost 50% of the optimal endosome–lysosome interaction can still be observed in the centrifugation assay in the absence of cytosol.[4] This is probably because some dense endosomes are already primed for fusion when prepared, whereas the remainder require priming by incubation with cytosol and ATP.

Reaction between Dense Endosome Membrane and Lysosomal Membrane Measured by Fluorescence Dequenching

Octadecylrhodamine B chloride (R18) (Molecular Probes, Inc., Eugene, OR) added to a suspension of membranes becomes incorporated in the membranes at self-quenching concentrations. The self-quenching is

[11] G. A. Maguire, K. Docherty, and C. N. Hales, *Biochem. J.* **212,** 211 (1983).

relieved by fusion of the membranes with unlabeled membranes and the increase in fluorescence can be measured.[12,13] Cytosol, which contains phospholipid transfer proteins[14] and fatty acid-binding proteins,[15] produces dequenching in the absence of unlabeled membranes, so that the dense endosome–lysosome interaction cannot be studied under conditions found to be optimal using the centrifugation assay.

For R18 loading, dense endosomes are suspended in 1 ml of STM at about 0.5 mg protein/ml and warmed to 30°. Three microliters of 20 mM R18 in ethanol, also at 30°, is added with rapid and thorough mixing. The mixture is incubated for 15 min at room temperature with gentle mixing and protected from light. The free R18 is then removed by applying the mixture to a 5-cm column of Sepharose CL-4B equilibrated with STM, in a Pasteur pipette. The column is washed with STM and the colored eluate containing the R18-loaded membranes collected in a volume of 1.2–1.3 ml. Incubation mixes containing unlabeled receptor membrane (e.g., lysosomes) and, for optimal activity, vanadium-free ATP are prepared in a final volume of 0.9 ml STM. These and the R18-loaded membranes are equilibrated at 37° for 2 min. The reaction is started by adding 0.1 ml of the R18-loaded membrane suspension to each incubation mixture and mixing thoroughly. Fluorescent emission is read at 590 nm using an excitation wavelength of 560 nm. The samples should be protected from direct light during incubation. At the end of the experiment 60 μl of 20% (w/v) Triton X-100 is added to each incubation mix and the emission again read to give the total fluorescence in that mix. Emissions are calculated as a percentage of the total emission, allowing for the volume increase on Triton addition. Any leakage of R18 is corrected for by including a mixture with no acceptor membranes; this has never exceeded 2%/hr. Figure 3 shows that dequenching is proportional to the quantity of lysosomes added as receptor. A crude mitochondrial preparation (made by washing and resuspending the 0.25 M KCl pellet from a lysosome preparation) was much less efficient than lysosomes in dequenching (Fig. 3); unlabeled dense endosomes and light endosomes were similarly ineffective as receptors.[4] No dequenching with any receptor membrane occurred at 4°. For these membrane fusion assays, aliquots of the subcellular fractions in STM may be stored in liquid nitrogen.

[12] D. Hoekstra, T. de Boer, K. Klappe, and J. Wischut, *Biochemistry* **23**, 5675 (1984).

[13] D. Hoekstra, *Hepatology* **12**, 615 (1990).

[14] J. E. Rothman, *Nature (London)* **347**, 519 (1990).

[15] J. Storch and N. M. Bass, *J. Biol. Chem.* **265**, 7827 (1990).

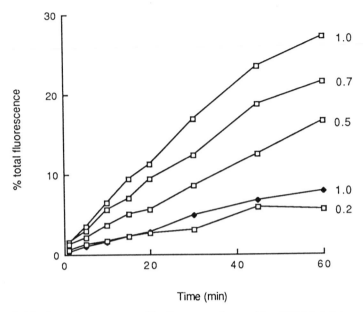

FIG. 3. Membrane fusion measured by fluorescent dequenching. R18-labeled dense endo-somes were incubated with 2 mM vanadium-free ATP and amounts of lysosomes (□) or mitochondria (◆), as shown (mg protein/incubation).

[8] Nuclear Envelope Assembly following Mitosis

By RUPERT PFALLER and JOHN W. NEWPORT

Entry into mitosis of higher eukaryotic cells is accompanied by a series of reversible morphological rearrangements that are indispensible to achieve cell duplication. The most prominent among them includes chromosome condensation, spindle formation, and breakdown of the cell nucleus. All the cellular components recruited to perform mitosis-specific functions are recycled at the end of mitosis, resulting in the reformation of interphase cells.

In recent years *in vitro* systems have been developed to investigate these cell functions on a molecular level. One of the most valuable systems to study the cell cycle *in vitro* is derived from unfertilized eggs from *Xenopus laevis* frogs. Depending on the way these extracts are prepared, they can be used for the *in vitro* study of nuclear assembly (interphase extracts),[1] nuclear disassembly (mitotic extracts),[2] or to perform several rounds of nuclear assembly and disassembly (cycling extracts).[3] Here we describe preparation and use of interphase extracts from *Xenopus* eggs that promote assembly of nuclei around a chromatin substrate and, therefore, constitute the experimental basis to study nuclear architecture, function, and dynamics on a molecular level. We will focus on membrane assembly of the nuclear envelope although it should be noted that the potential of *Xenopus* extracts has also been exploited in investigating structure, function, and dynamics of the nuclear pore complex and the nuclear lamina, the other components of the nuclear envelope.

Fractionation of *Xenopus laevis* Eggs

Progesterone-stimulated maturation induces oocytes to proceed from prophase of meiosis I to metaphase of meiosis II. At this stage the mature oocytes appear as unfertilized eggs after passing down the oviduct of the frog. Unfertilized *Xenopus* eggs are arrested in the second metaphase of meiosis by a calcium-sensitive activity called CSF (cytostatic factor). Cytostatic factor stabilizes the activity of the central cell cycle regulator MPF (M-phase promoting factor), a kinase specifically activated during mitosis. Activation of the eggs (either by using Ca^{2+}-ionophores or breaking eggs in the absence of protein synthesis) leads to inactivation of MPF activity and extracts can be prepared that promote reconstitution of nuclei around a DNA substrate (interphase extracts). Extracts from nonactivated eggs, prepared in the presence of protein synthesis, retain their MPF activity and promote nuclear breakdown (mitotic extracts). Using *Xenopus* interphase extracts, distinct steps in the reformation of the cell nucleus at the end of mitosis can be resolved and studied in detail. The advantage of this system is that all the components are naturally present in the mitotic state and their transition into the interphase state during nuclear reconstitution can be investigated.

Hormone Induction of Unfertilized Xenopus laevis Eggs

Hormone induction employing gonadotropin causes female *Xenopus laevis* frogs to lay eggs.[4] This is done in two steps. In a first step, frogs are primed with pregnant mare serum gonadotropin. Two hundred units (U)

---

[1] J. Newport, *Cell* **48**, 205 (1987).

[2] J. Newport and T. Spann, *Cell* **48**, 219 (1987).

[3] A. W. Murray and M. W. Kirschner, *Nature (London)* **339**, 275 (1989).

[4] J. Newport and M. Kirschner, *Cell* **30**, 675 (1982).

of the hormone (Calbiochem, La Jolla, CA), dissolved in 0.5 ml water, is injected. In a second step, 2–7 days after priming, the frogs are injected with 500 U human chorionic gonadotropin (hCG) (Sigma, St. Louis, MO) the night before harvesting the eggs.

Eggs are washed in fourfold-diluted MMR buffer [MMR contains 100 mM NaCl, 2 mM KCl, 1 mM MgCl$_2$, 2 mM CaCl$_2$, 0.1 mM ethylene glycol-bis(β-aminoethyl ether)-N,N,N',N'-tetraacetic acid (EGTA), 5 mM N-2-hydroxyethylpiperazine-N'-2-ethanesulfonic acid (HEPES), pH 7.8] and subsequently treated with a 2% (w/v) solution of cysteine, pH 7.8 (about 200 ml/amount of eggs laid by one frog) for about 5 min. During the cysteine treatment the jelly coat is removed and a tighter packing of the eggs occurs. The dejellied eggs are then washed quickly three times with MMR (200 ml each) to remove the cysteine and are transferrd onto a petri dish. Eggs that display irregular pigment distribution (the brown-colored pigment should be evenly concentrated on one-half of the surface of the eggs are removed under a light microscope using a Pasteur pipette. The eggs are then washed twice with 100 ml of lysis buffer (250 mM sucrose, 50 mM KCl, 2.5 mM MgCl$_2$, 10 mM HEPES, pH 7.4) containing 50 μg cycloheximide/ml. Cycloheximide is included to prevent any further protein synthesis, especially the cyclin subunit of MPF.[5] Finally, the eggs are transferred to a 15-ml Falcon tube (Becton Dickinson, Oxnard, CA) with dithiothreitol (DTT; 1 mM final concentration), the inhibitor of actin gelation cytochalasin B (5 μg/ml; Sigma), and a protease inhibitor mix containing aprotinin and leupeptin (10 μg/ml final concentration of each inhibitor; Sigma) are added. Eggs are packed at room temperature by gentle centrifugation for 15 sec at 100 g in a clinical centrifuge and excess buffer is removed. Nuclear assembly *in vitro* is dilution sensitive and, therefore, it is essential to attain a highly concentrated egg cytoplasm.

Preparation of Crude Egg Cytoplasm

The packed eggs are crushed by centrifugation for 12 min at 12,000 g in a Sorvall (Norwalk, CT) HB-4 rotor at 4°. After centrifugation, three layers can be distinguished: a bottom layer containing yolk, a layer containing the wanted crude cytoplasmic fraction (crude interphase extract) in the middle, and a yellow layer of low density lipid on top (Fig. 1). The crude interphase extract is removed with a syringe. If supplemented with an ATP-regenerating system, it will support nuclear assembly around protein-free DNA (for example, DNA from bacteriophage λ).[1] These *in vitro* assembled nuclei are morphologically and functionally indistinguishable from intact, isolated nuclei.

[5] A. W. Murray, M. J. Solomon, and M. W. Kirschner, *Nature (London)* **339,** 280 (1989).

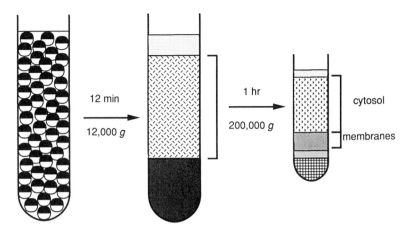

Fig. 1. Fractionation of eggs from *Xenopus laevis*. Dejellied eggs are crushed by centrifugation for 12 min at 12,000 g. The crude egg cytoplasm contained in the middle layer is removed and centrifuged for 1 hr at 200,000 g. From the cytoplasmic fractions separated in this step a cytosolic and a membrane fraction can be recovered that together promote *in vitro* assembly of nuclei around demembranated frog sperm chromatin.

Fractionation of Crude Interphase Extract

Further fractionation of the crude interphase extract can be achieved by ultracentrifugation.[6] The crude extract is supplemented with cytochalasin B (10 μg/ml) and aprotinin/leupeptin (10 μg/ml each) and spun for 75 min at 200,000 g at 4° in a Beckman (Palo Alto, CA) TLS 55 rotor. In case of large-scale preparations, centrifugation also can be carried out for 2 hr at 150,000 g at 4° in a Beckman SW 50.1 rotor. After ultracentrifugation the cytoplasmic fractions are separated from the cytosolic fraction according to their density (Fig. 1). The bottom layer consists predominantly of glycogen. On top of it there are two layers of membrane vesicles of different density (which can be distinguished by their different color) followed by the clear cytosolic fraction. Finally, there is a yellow layer of residual lipid. The lipid layer is carefully removed by suction. Then the clear cytosolic layer is removed and respun for 20 min at 200,000 g at 4° to remove residual particulate material. The resulting interphase cytosol is frozen in liquid nitrogen and kept at −80°, where it is stable for several months.

The upper of the two membrane fractions is removed, diluted with 5–10 vol of lysis buffer and supplemented with aprotinin/leupeptin (10 μg/ml) and 1 mM DTT. It should be noted that only the upper mem-

[6] K. L. Wilson and J. Newport, *J. Cell Biol.* **107**, 57 (1988).

brane layer is required for nuclear reconstitution and the lower membrane layer may actually have inhibitory effects. The washed membranes are reisolated by sedimentation through 0.2 ml of lysis buffer containing 0.5 M sucrose (Beckman TLS 55 rotor, 20 min at 30,000 g, 4°). Finally, the membrane pellet is resuspended in lysis buffer containing 0.5 M sucrose in a final volume corresponding to 10% of the volume of the crude cytoplasmic fraction. Aliquots of 5- to 10-μl volume are rapidly frozen in liquid nitrogen and stored at $-80°$. Freezing of small aliquots of membranes has proved to be important to retain their ability for nuclear assembly.

Preparation of Mitotic Extracts

Extracts from *Xenopus* eggs can be obtained under conditions that preserve MPF activity and keep it in a mitotic state.[2,7] Preparation of mitotic extracts occurs similar to interphase extracts, with the following changes. Eggs are washed in all steps with 0.1 M NaCl instead of MMR. Before breaking the dejellied eggs, they are washed in threefold concentrated EB buffer (240 mM β-glycerophosphate, 60 mM MgCl$_2$, 80 mM EGTA, pH 7.3), which does not contain cycloheximide. Omission of cycloheximide is important to allow continuous protein synthesis, in particular synthesis of the cyclin subunit of MPF. After breaking the eggs by centrifugation, the crude mitotic extract is further fractionated by ultracentrifugation as described above.

The MPF activity can be further enriched from mitotic extracts by ammonium sulfate precipitation.[7,8] Mitotic cytosol is diluted with an equal volume of EB buffer (80 mM β-glycerophosphate, 15 mM MgCl$_2$, 20 mM EGTA, pH 7.3) containing 1 mM DTT and 0.5 mM ATPγS and spun for 30 min at 200,000 g and 4° in a Beckman TLS 55 rotor. To the supernatant, 0.43 vol of 3.6 M ammonium sulfate solution in EB buffer is added dropwise to give a final ammonium sulfate concentration of 1.08 M. After incubation on ice for 30 min, precipitated protein is sedimented by centrifugation (Sorvall HB-4 rotor, 10 min at 10,000 rpm and 4°). Precipitated protein is dissolved in EB, 1 mM DTT, 0.5 mM ATPγS (in 20–30% of the volume of the undiluted cytosol) and dialyzed against 100 vol EB, 1 mM DTT, 0.1 mM ATPγS for 5 hr. The crude MPF preparation is frozen in liquid nitrogen and stored at $-80°$, at which it is stable for months. The MPF activity can be tested either by measuring phosphorylation of histone H1, a specific substrate of MPF kinase,[9,10] or measuring its ability to induce

[7] W. G. Dunphy and J. W. Newport, *J. Cell Biol.* **106,** 2047 (1988).
[8] M. Wu and J. G. Gerhart, *Dev. Biol.* **79,** 465 (1980).
[9] D. Arion, L. Meijer, L. Brizuela, and D. Beach, *Cell* **55,** 371 (1988).
[10] W. G. Dunphy and J. W. Newport, *Cell* **58,** 181 (1989).

nuclear envelope breakdown of *in vitro* assembled nuclei (see below).[7] Ammonium sulfate precipitation enriches MPF activity 5- to 10-fold and the yield is about 80%.

In Vitro Assembly of Nuclei

Employing cytosol and membrane fractions isolated from *Xenopus* eggs in combination with a chromatin substrate, nuclei can be formed that, by morphological and functional criteria, are indistinguishable from isolated interphase nuclei.[6] They contain a double membrane, a nuclear lamina, and nuclear pore complexes.[11] The DNA is enclosed in the nuclear envelope in a decondensed form. These nuclei can perform DNA replication and specific transport through the nuclear pore complex.[12]

As already mentioned, when crude cytoplasm is used assembly of nuclei *in vitro* can also be achieved around protein-free DNA. The crude egg extract obviously contains all basic components necessary to form chromatin. To study specific aspects of nuclear envelope assembly, however, chromatin from demembranated *X. laevis* frog sperm is the preferred DNA substrate because cytosol and membrane fraction derived from crude interphase extracts are already sufficient to drive nuclear reconstitution.

Preparation of Demembranated Frog Sperm Chromatin

A convenient and reliable chromatin substrate for nuclear envelope assembly is demembranated frog sperm chromatin. The following procedure is described for isolation of sperm chromatin from the testes of one frog.[6,13] Testes are removed from an adult male *X. laevis* frog and, after all surrounding tissue has been removed, they are put in 1 ml of buffer X [200 mM sucrose, 80 mM KCl, 15 mM NaCl, 5 mM ethylenediaminetetraacetic acid (EDTA), 7 mM MgCl$_2$, 15 mM piperazine-N,N'-bis(2-ethane sulfonic acid) (PIPES)–KOH, pH 7.2] on ice. To release the sperm, the testes are minced vigorously using tweezers. Isolation of the sperm is then achieved by repeated sedimentation and washing. First, big pieces of tissue are removed by centrifugation in a clinical centrifuge at 100 g for 10 sec. The supernatant containing released sperm is removed and kept on ice. To increase the yield, the pellet is reextracted with 0.5 ml buffer X as described above. The resulting supernatant is combined with the supernatant of the first extraction and centrifuged in a clinical centrifuge for 2 min

[11] J. W. Newport, K. L. Wilson, and W. G. Dunphy, *J Cell Biol.* **111**, 2247 (1990).
[12] D. R. Finlay and D. J. Forbes, *Cell* **66**, 17 (1990).
[13] M. J. Lohka and Y. Masui, *Science* **220**, 719 (1983).

at 350 g at 4° to isolate the sperm. The sperm pellet is usually contaminated with some somatic and red blood cells, which sediment preferentially at the bottom of the pellet. Further purification, therefore, can be achieved by washing the crude sperm pellet, avoiding the tightly packed pellet of red blood cells at the bottom. The sedimented, white sperm are resuspended in 1 ml of buffer X, avoiding the reddish bottom pellet, transferred to new tubes, and sedimented by centrifugation as described above. The sperm should be washed this way at least three times.

After the washing procedure is finished, the sperm are demembranated using lysolecithin. The sperm pellet is resuspended in 270 μl buffer X, warmed to room temperature, and 30 μl of a 0.5% (w/v) solution of lysolecithin (Sigma) in buffer X is added. The sperm then is suspended thoroughly and incubated at room temperature for 5 min. Then, 0.9 ml buffer X containing 3% (w/v) bovine serum albumin (BSA) is added to neutralize the lysolecithin. The demembranated sperm chromatin is reisolated by centrifugation for 2 min at 350 g at 4° and washed once in 1.2 ml buffer X containing 3% (w/v) BSA. The chromatin is then resuspended in 50 μl buffer X. The concentration of sperm chromatin is determined by counting the number of sperm contained in a 100-fold diluted aliquot using a hemacytometer. Aliquots of 5–10 μl (at a concentration of 50,000 sperm/μl) are frozen in liquid nitrogen and stored at −70°. Successful removal of the sperm membrane is checked by combined light and fluorescence microscopy employing the DNA-specific fluorescence dye Hoechst 33258 (Hoechst-Roussel Pharmaceuticals, Somerville, NJ). Only chromatin molecules that no longer have an intact membrane will bind the dye and can be visualized. They can, therefore, be distinguished from intact sperm molecules, which can be detected by light microscopy. Treatment with lysolecithin usually yields more than 95% demembranated sperm chromatin.

Assembly of Synthetic Nuclei

A reaction mixture for nuclear reconstitution contains interphase cytosol and interphase membranes, which can be used either from frozen stocks (as described above) or from freshly prepared extracts.[6] Demembranated frog sperm chromatin is used as DNA substrate from frozen stocks. Reconstitution of nuclei is strongly ATP dependent, therefore interphase cytosol is supplemented with an ATP-regenerating system consisting of 2 mM ATP, 10 mM creatine phosphate, and 50 μg creatine kinase/ml cytosol. ATP and creatine phosphate are added from 0.2 M stock solutions (in 10 mM potassium phosphate, adjusted to pH 7.0) and creatine kinase (Sigma) is added from 5-mg/ml stock solutions [dissolved in 50 mM NaCl, 50% (v/v) glycerol, 10 mM potassium phosphate, pH 7.0, and stored at $-20°$]. As already mentioned, nuclear assembly *in vitro* is sensitive to dilution of the cytosolic fraction, which will support formation of a nuclear envelope. Nuclear formation is completely blocked when the cytosol is diluted greater than twofold.

A typical assay is composed of 100 μl interphase cytosol supplemented with an ATP-regenerating system, 10 μl of interphase membranes, and demembranated frog sperm chromatin (typically 500–1000/μl final concentration). Formation of a nuclear envelope occurs at room temperature and is followed by combined phase-contrast and fluorescence microscopy. For observation on a microscope equipped for light and fluorescence microscopy, 2–3 μl of sample is added on a microscope slide to the same volume of the DNA-specific dye Hoechst 33258 (10 μg/ml), dissolved in 3.7% (v/v) formaldehyde (if fixation of the sample is required) or lysis buffer. The appropriate magnification for observing nuclei is about 250- to 400-fold.

Typically during 1–2 hr of incubation the sperm chromatin becomes enclosed in a phase-dense nuclear envelope (visualized by light microscopy) and the DNA decondenses (visualized by fluorescence microscopy using the DNA dye Hoechst 33258) (Fig. 2). The size of the nuclei formed depends on the ratio of membrane versus chromatin concentration in the assay. At the membrane concentration used here the maximal size, expressed by the surface area of the nuclei formed, is about 400 μm^2/nucleus.[6] It will decrease at concentrations of frog sperm chromatin higher than 2000/μl.

Steps of Nuclear Reconstitution

Using the isolated fractions described above some specific steps of nuclear envelope assembly can be studied. For the *Xenopus* system, so far, the following steps have been characterized. Membrane vesicles derived from interphase extracts can bind to demembranated frog sperm chromatin in the absence of cytosolic components.[14,15] To allow membrane binding, however, the chromatin must be partially decondensed. Partial decondensation is promoted by nucleoplasmin[14,16] (contained in a heat-stable cytosolic fraction) or functionally equivalent poly(L-glutamic acid).[15] With

[14] J. W. Newport and W. G. Dunphy, *J. Cell Biol.* **116**, 295 (1992).
[15] R. Pfaller, C. Smythe, and J. W. Newport, *Cell* **65**, 209 (1991).
[16] A. Philipott, G. H. Leno, and R. A. Laskey, *Cell* **65**, 569 (1991).

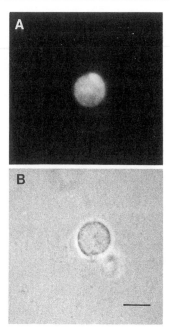

FIG. 2. Nuclear assembly *in vitro*. Assembly of nuclei *in vitro* was carried out as described in the text. The image of a reconstituted nucleus, visualized by DNA-specific fluorescence microscopy (A) and phase-contrast microscopy (B) is shown. Bar, 3 μM.

an experimental system for the initial interaction of nuclear membrane vesicles with chromatin at hand, two aspects of nuclear envelope assembly can be investigated, namely subsequent steps in nuclear envelope assembly and regulation of these interactions during the course of the cell cycle.

1. Chromatin-bound membrane vesicles can be fused to form basic double-membrane structures on the chromatin surface in the presence of ATP and GTP. Membrane fusion is inhibited by *N*-ethylmaleimide and GTPγS, indicating mechanisms similar to membrane fusion in vesicular protein transport are involved.[14] Completion of nuclear envelope assembly from chromatin-bound membranes requires additional factors from interphase cytosol to allow assembly of nuclear pore complexes and the nuclear lamina.

2. Binding of membrane vesicles to chromatin is regulated by a phosphorylation regulatory system.[15] A regulatory kinase and phosphatase appear to be contained in interphase cytosol. Inhibition of the okadaic acid-sensitive phosphatase of either type 1 or 2A leads to release of

chromatin-bound membrane vesicles. Membrane release is caused by the regulatory kinase that, most likely, phosphorylates a membrane-localized component. In interphase extracts the equilibrium of activities of the regulatory kinase and phosphatase favors membrane binding. In mitotic extracts, this equilibrium is shifted toward membrane release and no requirement for okadaic acid is observed.[14]

Controlled Decondensation of Frog Sperm Chromatin

In sperm chromatin the DNA is organized in a highly condensed form. To observe membrane binding, the DNA must be partially decondensed. This can be achieved in either of two ways. The first approach employs heat-stable components of the cytosolic fraction.[14] Cytosolic proteins are denatured by heating at 95° for 5–10 min and sedimented by centrifugation for 20 min at 50,000 g at 4° in a Beckman rotor TLS 55. The protein components of the supernatant (termed "heat extract") consist predominantly of the three acidic proteins nucleoplasmin, N1, and N2, which are involved in nucleosome assembly and chromosome decondensation.[17] In the heat extract, a partial decondensation of sperm chromatin can be achieved corresponding to a 25- to 30-fold volume increase. The extent of decondensation is dependent on the chromatin concentration and is maximal below a concentration of 3000 DNA molecules/μl heat extract.

In the second approach, poly(L-glutamic acid) is substituted for the heat extract.[15] Nucleoplasmin contains tracts of polyglutamic acid that are thought to be functionally involved in chromatin decondensation.[18] Using commercially available poly(L-glutamic acid) (Sigma; average M_r 13 K) at concentrations of 0.5–2.0 mg/ml results in efficient, partial decondensation of sperm chromatin. Sperm chromatin decondensed by polyglutamic acid provides a defined system to investigate the interaction of nuclear envelope membranes with chromatin *in vitro*.

Formation of Nuclear Envelope from Membrane Vesicles Bound to Frog Sperm Chromatin

When demembranated frog sperm chromatin (1000/μl) is incubated in 100 μl heat extract containing 10 μl added membrane suspension (about 150 μg protein), the membranes bind to the decondensed sperm. After incubation for 1 hr at room temperature the sample is layered in a 0.6-ml borosilicate culture tube (Kimble, Vineland, NJ) on top of 1 ml lysis buffer

[17] R. A. Laskey, B. M. Honda, A. D. Mills, and J. T. Finch, *Nature (London)* **275**, 416 (1978).
[18] C. Dingwall, S. M. Dilworth, S. J. Black, S. E. Kearsey, L. S. Cox, and R. A. Laskey, *EMBO J.* **6**, 69 (1987).

containing 1 M sucrose and centrifuged for 5 min in a clinical centrifuge at 500 g at 4°. Under these conditions, chromatin containing bound membrane vesicles is pelleted on the bottom and separated from unbound membranes. The resulting pellet is gently resuspended in 250–400 μl of interphase cytosol containing an ATP-regenerating system and incubated at room temperature for 1 hr. The formation of a phase-dense nuclear envelope is then assessed by combined light and fluorescence microscopy as described above.

Reversible Binding of Nuclear Membrane Vesicles to Frog Sperm Chromatin

Binding of Membrane Vesicles to Frog Sperm Chromatin. To a solution of 180 μl poly(L-glutamic acid) (0.5 mg/ml) in lysis buffer demembranated frog sperm chromatin (1500/μl) and membranes (0.3 mg/ml final protein concentration) are added. ATP and GTPγS (Boehringer Mannheim, Indianapolis, IN) are added to 2 and 0.1 mM final concentration, respectively. GTPγS, dissolved in 0.1 M HEPES, pH 7.2, containing 1 mM DTT, is added from a 20 mM stock solution. After incubation for 30 min at room temperature, binding of membrane vesicles to chromatin is assessed by fluorescence microscopy employing a combination of DNA-specific and membrane-specific fluorescence dyes. Hoechst 33258 is used to stain DNA (see above) and 6,6′-dihexylcarbocyanine (Kodak, Rochester, NY) (2 μg/ml final concentration) to stain membranes. Both dyes are dissolved together in 3.7% formaldehyde and samples for microscopy are prepared by adding 2–3 μl of sample to the same volume of dye solution on a microscope slide. Samples are best observed at about 300- to 600-fold magnification. Chromatin molecules, densely packed with bound membrane vesicles, can be easily distinguished from background fluorescence caused by unbound membrane vesicles (see Fig. 3).

Release of Chromatin-Bound Membrane Vesicles Employing Okadaic Acid and Interphase Cytosol. Aliquots (30 μl) are mixed with 5 μl interphase cytosol containing an ATP-regenerating system and 0.5 mM GTPγS. Okadaic acid (Moana Bioproducts, Honolulu, Hawaii), dissolved in H$_2$O, is added from a 0.1 mM stock solution to a final concentration of 1 μM. Relatively high concentrations of GTPγS are required to inhibit fusogenic activities in interphase cytosol. To demonstrate that membrane release occurs as a result of the inhibition of a cytosolic protein phosphatase by okadaic acid, in one control sample only interphase cytosol and no okadaic acid is added and in a second control sample, which contains okadaic acid, lysis buffer is added instead of the interphase cytosol. Binding of membranes to chromatin is assessed by fluorscence microscopy as described

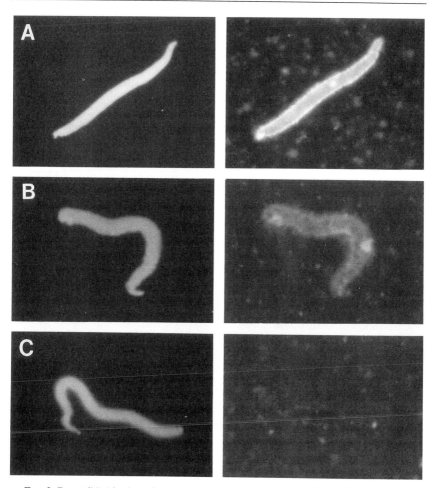

FIG. 3. Reversible binding of membrane vesicles to frog sperm chromatin. As outlined in the text, membrane vesicles derived from *Xenopus laevis* egg extracts are first bound to demembranated frog sperm chromatin that was partially decondensed by poly(L-glutamic acid). Then aliquots of this substrate of chromatin-bound membrane vesicles are combined with okadaic acid (A), interphase cytosol (B), and both interphase cytosol and okadaic acid (C), respectively. At different time points, aliquots are removed and binding of membranes to chromatin is assessed by fluorescence microscopy employing fluorescence dyes specific for DNA (left) and membranes (right), respectively. After incubation for 1 hr, in the sample containing the combination of interphase cytosol and okadaic acid, the characteristic staining of DNA-associated membranes has disappeared due to release to chromatin-bound vesicles.

above. Release of chromatin-bound membranes occurs within 30 to 60 min after addition in the sample containing interphase cytosol and okadaic acid while neither interphase cytosol nor okadaic acid alone affect membrane binding (Fig. 3). Depending on the quality, interphase cytosol can be diluted up to 20-fold and still promote membrane release within 1 hr.

Membrane Release Employing Mitotic Extracts. Binding of membrane vesicles is carried out as described above, except EB buffer is used instead of lysis buffer to preserve MPF activity. After incubation for 30 min at room temperature, 5 μl of mitotic extract, fractionated and enriched for MPF activity by ammonium sulfate precipitation (see above), is added to a 30-μl aliquot of the binding reaction. Release of membrane vesicles bound to chromatin is assessed by combined fluorescence microscopy for membranes and DNA, as described above. To achieve membrane release within 1 hr, the MPF-containing protein fraction can be diluted about 20-fold. Dependence of the observed membrane release on the mitotic state of the added protein fraction can be demonstrated by employing an ammonium sulfate-precipitated protein fraction isolated from interphase cytosol instead of mitotic cytosol. Although this protein fraction does not lead to release of chromatin-bound membrane vesicles by itself, it will in combination with okadaic acid, thus demonstrating that it contains both the regulatory kinase and phosphatase.

[9] Formation of Endoplasmic Reticulum Networks in Vitro

By James M. McIlvain, Jr., and Michael P. Sheetz

Introduction

The endoplasmic reticulum (ER) is an abundant tubulovesicular or planar membrane system that comprises approximately 50% of the total membrane content of a cell. Because of the density of ER membranes in the cytoplasm, it is difficult to visualize individual strands of ER except in the peripheral regions of highly spread cells. The peripheral ER strands in living cells are highly dynamic.[1,2] This supports the dynamic functions that have been proposed for the ER, i.e., the processing of exported molecules,

[1] C. Lee and L. B. Chen, *Cell* **54**, 37 (1988).
[2] C. Lee, M. Ferguson, and L. B. Chen, *J. Cell Biol.* **109**, 2045 (1989).

the movement of hydrophobic materials within the cytoplasm, and the uptake and release of Ca^{2+}.

In this chapter, we describe in detail a method for reconstituting ER membrane networks *in vitro*.[3] The networks are formed by microtubule-dependent transport and exhibit the natural morphology of ER observed in living cells. However, the density of membrane is much lower, allowing for the ease of addition or subtraction of components by medium exchange. To the extent that native membrane functions can be preserved *in vitro,* networks provide an excellent system for studying the molecular bases of three major ER functions: protein translation and translocation, lateral membrane transport, and Ca^{2+} dynamics.

Commercially available *in vitro* translation kits have made routine the *in vitro* translation and translocation of proteins into dog pancreas microsomes.[4] However, the microsomes do not morphologically resemble native ER. They are small spherical vesicles and questions relating to the lateral movement of the translation complex, focal concentration of the translational activity, or other factors dependent on normal ER morphology cannot be addressed. *In vitro* formed ER networks can be used to study these issues in detail because they can be fixed and permeabilized to determine the distribution of components during protein processing.

Because the ER runs in a continuous fashion from one side of the cell to the other, it has been proposed that the ER plays a major role in transporting lipophilic agents and extracytoplasmic components throughout the cytoplasm. To determine if there is free diffusion within the networks we have employed the technology of single particle tracking (SPT) with gold-tagged antibodies.[5-7] This has enabled us to look for the presence of barriers to diffusion as well as to measure the rate of diffusion within the network.

Intracellular Ca^{2+} levels are thought to be regulated by the ER, but the controlling mechanisms are poorly understood. It has been shown that ER networks in the cortex of sea urchin eggs can be loaded with the Ca^{2+} indicator dye, Fluo-3, to measure the level of Ca^{2+} within the ER directly.[8] Extension of those studies to the *in vitro* formed networks holds the promise of being able to measure directly many aspects of ER Ca^{2+} transport.

[3] S. L. Dabora and M. P. Sheetz, *Cell* **54**, 27 (1988).
[4] Promega Corp., Madison, Wisconsin.
[5] M. R. deBrander, R. Nuydens, G. Geuens, M. Moeremans, and J. de Mey, *Cell Motil. Cytoskeleton* **6**, 105 (1986).
[6] M. P. Sheetz, S. Turney, H. Qian, and E. L. Elson, *Nature (London)* **340**, 284 (1989).
[7] H. Qian, M. P. Sheetz, and E. L. Elson, *Biophys. J.* **60**, 910 (1991).
[8] M. Terasaki and C. Sardet, *J. Cell Biol.* **115**, 1031 (1991).

In addition to the major ER functions, the networks lend themselves to the investigation of numerous other questions, such as the requirements for fusion in network formation, osmotic properties of ER membranes,[3] and the role of lipid composition in ER membrane structure. Because large amounts of networks can be formed *in vitro,* the samples can be followed biochemically as well. Thus there are many advantages to the *in vitro* system with relatively few disadvantages.

Materials

Choice of Cells and Tissues

A variety of cell lines and tissues have been utilized but only chick embryo fibroblasts (CEFs), African green monkey kidney cells (CV-1s), Madin–Darby canine kidney cells (MDCK), and *Xenopus* egg extracts[9] have reliably given networks by this procedure. The only whole-tissue extract that gave networks was the dense microsome fraction from embryonic chick brain. A variety of other tissues and cell lines were tested, including Chinese hamster ovary (CHO) cells, several lymphocyte and macrophage lines, HeLa cells, rat and chicken liver, and dog pancreas microsomes.

Tissue Culture. Primary CEF cells are obtained from 12-day-old embryos,[10] allowed to grow to confluency (2–3 days) in minimum essential medium (MEM) containing Earle's salts (GIBCO, Grand Island, NY), 5% (v/v) fetal bovine serum, and 40 units/ml penicillin and 40 μg/ml streptomycin. The primaries are then either split to secondaries (1 : 1) in preparation for networks or frozen and stored in liquid N_2 for later use.

Stock Solutions

Buffer

PMEE': 35 mM piperazine-N,N'-bis(2-ethanesulfonic acid) (PIPES), 5 mM MgSO$_4$·7H$_2$O, 1 mM ethylene glycol-bis (β-aminoethyl ether)-N,N,N',N'-tetraacetic acid (EGTA), 0.5 mM ethylene diaminetetraacetic acid (EDTA), pH 7.4 with KOH. Dithiothreitol (DTT) is added prior to use at a final concentration of 1 mM

---

[9] V. J. Allan and R. D. Vale, *J. Cell Biol.* **113,** 347 (1991).
[10] P. M. Kelley and M. J. Schlesinger, *Cell* **15,** 1277 (1978).

Protease Inhibitors

Protease inhibitor cocktail (200×): Pepstatin (0.2 mg/ml), TAME (2.0 mg/ml), tolylsulfonyl phenylalanyl chloromethyl ketone (TPCK;2.0 mg/ml), leupeptin (0.2 mg/ml), soybean trypsin inhibitor (2.0 mg/ml) in 50% ethanol. *Note:* The cocktail is a slurry; resuspend each aliquot before addition to PMEE'. Keeps for at least 1 month at $-20°$
Phenylmethylsulfonyl fluoride (PMSF) (100×): 100 mM in 2-propanol

Other Stocks

ATP, 10 mM
GTP (200×), 200 mM
Taxol, 4 mM in 100% dimethyl sulfoxide (DMSO)

Methods

Cell Preparation

Harvesting. Two-day-old confluent secondary CEF cells are harvested from four 850-cm² roller bottles by trypsinizing with 12.5 ml of 0.05% (v/v) trypsin and 0.53 mM 4NaEDTA (GIBCO) per roller bottle for 10–15 min at 37°. The cells are collected, sieved through a syringe filter unit (no filter, support screen only), collected in 50-ml disposable conical tubes on ice (25 ml/tube). Each roller bottle is subsequently rinsed with 12.5 ml of culture medium containing serum, sieved as above, and added to each 50-ml conical tube. The cells are then pelleted by centrifugation at 1000 g for 10 min at 4°. The supernatant is discarded, cells resuspended in 30 ml of PMEE' containing 1 mM DTT, 1× protease inhibitor cocktail, and 1 mM PMSF in a single tube, and pelleted by centrifugation at 1000 g for 10 min at 4°. The supernatant is discarded, the packed cell volume (PCV) determined (generally, 700–800 μl), and cells resuspended in an equal volume of PMEE' buffer giving a total volume of approximately 1.5 ml.

Homogenization. The cells are then homogenized in an ice-cold cell buster using a 8.004-mm diameter ball.[11] It is critical for maximum network formation that no more than four full strokes (one stroke is equal to one down/up cycle of a syringe) be used for each aliquot of cells homogenized. The homogenized suspension is then centrifuged at 1000 g for 15 min at 4°.

Membrane and Motor Fractions. The supernatant (S1) is carefully collected from below a lipid layer that is often present; the nuclear pellet

[11] W. Balch and J. E. Rothman, *Arch. Biochem. Biophys.* **240**, 413 (1985).

(P1) is discarded. The S1 is then centrifuged at 100,000 *g* for 30 min at 4°. The supernatant (S2) from the high-speed spin is collected and set aside, and the pellet (P2) is resuspended in 30–40 μl of PMEE′ buffer by trituration; this becomes the stock solution of microsomal membranes. Taxol and GTP are added to the S2 for a final concentration of 20 and 1 m*M*, respectively, and incubated at 37° for 15 min to polymerize endogenous microtubules. The S2 is then centrifuged for 5 min at 100,000 *g* in an airfuge. The final supernatant (S3) is collected and used neat in the final assay. The pellet of endogenous microtubules (MTs) is discarded.

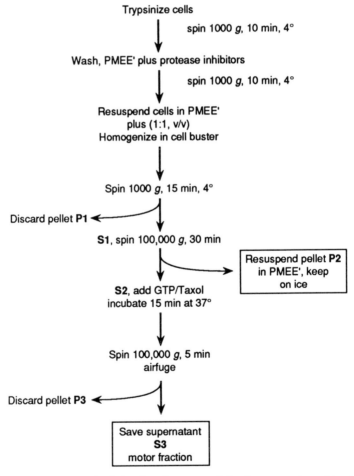

Fig. 1. Flow diagram of cell fractionation protocol for the preparation of *in vitro* ER networks.

See Fig. 1 for an overview of the initial isolation of network components from whole cells.

Network Preparation

Microtubule Density. Microtubule density is critical to network formation; too few MTs will result in little to no network formation. Phosphocellulose-purified bovine tubulin[12] (5 mg/ml) polymerized in 20 μM taxol/1 mM GTP at 37° for 15 min is diluted 1 in 5 with PMEE′ buffer. The 5× dilution provides a sufficient microtubule array to maximize network formation. Preparations in which too few microtubules are present are characterized by networks having a high degree of Brownian motion along the length of the membrane tubules.

Membrane Density. Membrane density is less critical to network formation. The choice of membrane density will depend on the method of analysis to be utilized. Low dilutions of membrane stock (less than 1 in 10) will provide a dense polygonal network with individual tubules spaced less than 5 μm apart. Tracking of individual 40-nm gold particles requires a low signal-to-noise ratio, therefore dilutions of 1 in 40 are normally used.

Incubation. The standard network assay mixture consists of (in order): 1 μl of MTs, 3 μl of S3, 1.2 μl of 10 mM ATP, and 1 μl of P2 diluted 1 in 40 in PMEE′. Networks are formed on cover slips (No. 0, 24 × 60 mm) within an area separated by two parallel lines of silicone grease and placed in a humidified chamber for 1–2 hr at room temperature or 30–60 min at 37°.

Methods of Analysis

Video Microscopy

Although the networks can be observed with standard differential interface contrast (DIC) optics, monitoring MT density and organelle motility requires video-enhanced contrast DIC microscopy (Fig. 2).[13,14]

Single-particle tracking (SPT) using antibodies conjugated to 40-nm gold particles is used to track the movement of membrane-bound proteins.[5] Computer hardware and software record the x, y position of individual gold particles with time. We then characterize individual particle movements as described in Gelles *et al.*[15] Several commercial tracking

[12] R. C. Williams and J. C. Lee, this series, Vol. 85, p. 376.
[13] R. D. Allen, N. S. Allen, and J. L. Travis, *Cell Motil.* 1, 291 (1981).
[14] B. J. Schnapp, this series, Vol. 134, p. 561.
[15] J. Gelles, B. J. Schnapp, and M. P. Sheetz, *Nature (London)* 331, 450 (1988).

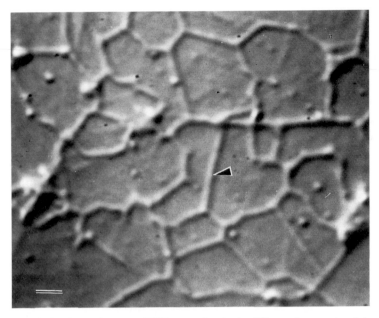

FIG. 2. Video-enhanced contrast DIC image of an *in vitro* ER membrane network (arrowhead) formed from CEF cells. Bar: 1 μm.

routines are now available that can be used to track individual particle movements.

Fluorescence

Fluorescence microscopy provides a powerful alternative method for visualizing the membrane networks. Vital dyes such as $DiOC_6(3)$ have been used to stain ER in living cells.[1,2] $DiOC_6(3)$-stained networks are strikingly similar in appearance to that observed in living cells.[3] Other fluorescent dyes, such as the Ca^{2+} indicator Fluo-3, can be loaded into the networks.

The ability to label specific proteins of interest within the membrane network makes immunofluorescent labeling the most useful aspect of fluorescence microscopy.

Successful immunostaining requires that close attention be paid to the effects of fixation on the system being labeled. The membranous nature of the networks makes them particularly sensitive to osmotic disruption. The following fixation protocol has been found to give satisfactory fixation with the minimum disruption of the networks.

Stock Solutions

Washing buffer (PMEE'): 35 mM PIPES, 5 mM MgSO$_4 \cdot$7H$_2$O, 1 mM
    EGTA, 0.5 mM EDTA, pH 7.4, with 0.5 M sucrose
Fixing solution: 1% (v/v) formaldehyde, 0.0125% (v/v) glutaraldehyde
    [electron microscopy (EM) grade] in washing buffer
Quenching solution: 10 mM glycine in washing buffer
Permeabilizing solution (used only for luminal staining): 0.1% (v/v)
    saponin, 1% (v/v) bovine serum albumin (BSA) in washing buffer
Blocking solution: 1% (v/v) BSA, 0.05% (w/v) saponin, in washing buffer

Fixation Protocol. The networks to be fixed are set at a slight angle to
facilitate the movement of the solutions across them and are gently washed
three times with 50–100 μl of washing buffer, followed by 15 min in fixing
solution. The fixative is removed by washing three times with washing
buffer and quenched for 1 hr in the quenching solution. If luminal proteins
are to be labeled, permeabilize the networks for 15 min with permeabilizing solution. Wash three times with washing buffer. Label for 1 hr with
primary antibody in blocking solution and wash three times with washing
buffer followed by 1 hr with the fluorescently conjugated secondary antibody in blocking solution or 10% (v/v) host serum. Remove the excess
secondary with three final washes in washing buffer, and mount on a glass
slide.

Discussion

In vitro ER networks provide an accessible ER preparation, which can
be utilized to study a variety of ER functions. Because they are stabilized
on a cover slip, they can be microscopically visualized even while the
buffer is being exchanged. Biochemical amounts of network can be prepared for study by preparing networks that cover areas as large as 1 cm^2 on
a single cover slip.

There is considerable interest in the regional specialization of the ER.
The dimensional similarity between the *in vitro* networks and *in vivo* ER
networks supports the notion of functional similarity. With the *in vitro* ER
networks, it is easy to observe lateral specialization. However, the limited
number of cell extracts that support network formation restricts the applicability of this technique, although we have been able to fuse some exogenous microsome fractions to existing networks. This opens the possibility
of studying ERs from many other systems, including the sarcoplasmic
reticulum from muscle.

Relatively few probes exist for the cytoplasmic surface of the ER, which
has limited the number of components that can be analyzed. Further, to
stain luminal components, the networks require permeabilization, which

has created additional problems. Immunostaining in most systems is difficult but the fragile nature of the *in vitro* nets makes them particularly sensitive to osmotic disruption during fixation and labeling. Still, by carefully controlling staining conditions it has been possible to establish that the networks do contain expected ER components such as the binding protein, BiP, and protein disulfide isomerase. The networks do not stain for Golgi markers. Interestingly, however, the networks do stain for Golgi markers after brefeldin A treatment of the cells before network formation.[16] Utilizing immunostaining techniques, the networks can be employed to catalog the components that will fuse or associate with the ER.

Membrane dynamics (diffusion and flow) in the ER can be readily analyzed in the *in vitro* systems. Preliminary results support earlier *in vivo* observations of rapid mixing between ERs of fused cells. With the *in vitro* system it is possible to differentiate between the fusion components of mixing and that of lateral diffusion. Another possible mechanism for mixing, membrane flow, is induced by the movements of membrane strands along microtubules.

Translation and translocation of newly synthesized proteins across the ER membrane is thought to produce a major channel or pore in the membrane. Although it is difficult to follow the movements of univalent cations across the ER membrane the strands are particularly sensitive to swelling,[3] which means the recovery from swelling could be used to measure the rate of ion movements across the ER. Further, the spatial localization of newly transported proteins as well as the distribution and dynamics of the ribosome–mRNA complexes on the ER surface can tell us much about the details of the translocation process.

There are many concerns about the dynamics of Ca^{2+} movement in the cytoplasm, which are thought to be controlled by the ER. Questions about the propagation of release and the focal concentration of Ca^{2+} transporters and release sites can all be addressed in the *in vitro* networks. Absolute levels of Ca^{2+} are difficult to measure because of the small luminal volume and only relative changes in Ca^{2+} concentration have been measured to date. Still, there are many issues that can be addressed.

Because of the relative ease with which the networks can be produced, they provide a useful alternative to *in vivo* studies of ER functions. There are potential difficulties resulting from the loss of components in preparation of the components for reconstitution. This, however, can be an advantage if the lost components and lost activities can be recovered with other fractions.

---

[16] J. Glickman and M. P. Sheetz, unpublished observations, (1989).

[10] Cell-Free Formation of Immature Secretory Granules and Constitutive Secretory Vesicles from Trans-Golgi Network

By SHARON A. TOOZE and WIELAND B. HUTTNER

Introduction

In eukaryotic cells, secretory proteins reach the cell surface by either the constitutive or the regulated pathway. The constitutive secretory pathway is common to all cells, whereas the regulated secretory pathway is present only in certain cells such as exocrine cells, endocrine cells, and neurons.[1] Sorting of secretory proteins to these two pathways occurs in the trans-Golgi network (TGN), where secretory proteins are packaged into either constitutive secretory vesicles (CSVs) or immature secretory granules (ISGs).[2-4]

To study the formation of CSVs and ISGs from the TGN and the sorting events occurring, these processes have been reconstituted in a cell-free system.[4] This chapter focuses on the experimental details of this cell-free system; the outline of the experimental approach is as follows.

1. Pulse-labeling of intact cells with radioactive sulfate to label marker proteins for the regulated and constitutive secretory pathway selectively in the TGN
2. Preparation of a postnuclear supernatant from the labeled cells
3. Incubation of the postnuclear supernatant under conditions that allow the formation of ISGs and CSVs containing the labeled marker proteins
4. Separation of ISGs and CSVs from the TGN
5. Separation of ISGs and CSVs

These five major steps are described in detail in Sections 1–5. In each, we first give the rationale of the experimental step, followed by a standard protocol and a discussion of important points therein. In addition, we describe possible modifications of the standard protocol that may be introduced to investigate the mechanism of formation of ISGs and CSVs in the cell-free system. In Section 6, we describe controls to validate the reconsti-

---

[1] T. L. Burgess, and R. B. Kelly, *Annu. Rev. Cell Biol.* **3**, 243 (1987).
[2] L. Orci, M. Ravazzola, M. Amherdt, A. Perrelet, S. K. Powell, D. L. Quinn, and H.-P. H. Moore, *Cell* **51**, 1039 (1987).
[3] J. Tooze, S. A. Tooze, and S. D. Fuller, *J. Cell Biol.* **105**, 1215 (1987).
[4] S. A. Tooze and W. B. Huttner, *Cell* **60**, 837 (1990).

tution of ISG and CSV formation in the cell-free system. Finally, Section 7 summarizes our current knowledge about the formation of ISGs and CSVs in the cell-free system.

1. Pulse-Labeling of Constitutive and Regulated Secretory Proteins in Trans-Golgi Network with Radioactive Sulfate

Rationale

In reconstituting the formation of vesicles in a cell-free system, it is important to ensure that the marker protein used is derived from the appropriate donor compartment. One approach to achieve this is to label this protein selectively in the donor compartment. In the case of the formation of ISGs and CSVs from the TGN, we use pulse-labeling with radioactive sulfate. Protein sulfation of both tyrosine and carbohydrate residues are TGN-specific posttranslational modifications,[5,6] and many of the proteins transported to the cell surface via the constitutive and regulated secretory pathway undergo these modifications.

To study the formation of ISGs and CSVs, any cell secreting proteins via the regulated pathway in addition to the constitutive pathway can be used. We chose the rat neuroendocrine cell line PC12 because the sorting of secretory proteins to the regulated pathway is efficient.[7] A further advantage of these cells is that they express easily detectable levels of sulfated marker molecules for both the constitutive and regulated pathway.[4] These are a heparan sulfate proteoglycan (hsPG) for the constitutive pathway and secretogranins I and II for the regulated pathway. With respect to the secretogranins,[8] we confine our analysis to secretogranin II (SgII) because the analysis of secretogranin I is complicated by its overlapping electrophoretic mobility with some of the hsPG.[4]

Standard Protocol: Steps 1–4

In the following standard protocol, all the indicated amounts and volumes are for one cell-free reaction. Depending on the number of cell-free reactions to be performed, these amounts and volumes need to be multiplied accordingly.

[5] P. A. Baeuerle and W. B. Huttner, *J. Cell Biol.* **105**, 2655 (1987).
[6] J. H. Kimura, L. S. Lohmander, and V. C. Hascall, *J. Cell. Biochem.* **26**, 261 (1984).
[7] H.-H. Gerdes, P. Rosa, E. Phillips, P. A. Baeuerle, R. Frank, P. Argos, and W. B. Huttner, *J. Biol. Chem.* **264**, 12009 (1989).
[8] W. B. Huttner, H.-H. Gerdes, and P. Rosa, *Trends Biochem. Sci.* **16**, 27 (1991).

Step 1. Take two 15-cm dishes of PC12 cells (clone 251)[9] grown to ~80% confluency at 37° and 10% CO_2 in Dulbecco's modified Eagle's medium (DMEM) supplemented with 10% (v/v) horse serum and 5% (v/v) fetal calf serum.

Step 2. Remove the growth medium and wash the attached cells once with sulfate-free medium. This medium consists of DMEM supplemented with 10 mM N-2-hydroxyethylpiperazine-N'-2-ethanesulfonic acid (HEPES)–NaOH, pH 7.2, in which magnesium sulfate is replaced with magnesium chloride and in which methionine and cysteine are present at 1% (w/v) of the normal concentration; the sulfate-free medium in addition contains 1% (v/v) horse serum and 0.5% (v/v) fetal calf serum, both dialyzed against phosphate-buffered saline (PBS). Incubate the cells for 15 min in 7.5 ml/dish of fresh sulfate-free medium on a rocking platform placed in the incubator. Replace the medium with 7.5 ml of fresh sulfate-free medium and continue the incubation for another 15 min on the rocking platform in the incubator.

Step 3. Add 500 μl of carrier-free [^{35}S]sulfate (SJS.1, 25 mCi/3 ml; Amersham, Arlington Heights, IL) to each dish, giving a final [^{35}S]sulfate concentration of ≈0.5 mCi/ml. Incubate for 4.5 min with rocking.

Step 4. Remove the dishes from the incubator and immediately place on ice. Rapidly remove the labeling medium and add 10 ml/dish of ice-cold TBSS (Tris-buffered saline plus sulfate: 137 mM NaCl, 4.5 mM KCl, 0.7 mM Na_2HPO_4, 1.6 mM Na_2SO_4, 25 mM Tris-HCl, pH 7.4).

Comments on Standard Protocol: Steps 1–4

Step 1: If PC12 cells are grown to more than ~80% confluency, they tend to form clumps of cells that detach from the dish in the subsequent manipulations. We use PC12 cells up to passage 18.

Step 2: This incubation serves to deplete the intracellular pool of inorganic sulfate. The reduced concentration of methionine and cysteine is used because this increases the amount of radioactive sulfate incorporation.[10,11] The incubation with rocking allows the labeling of cells in a "minimal" volume without drying, reducing the amount of radioactive sulfate required.

Step 3: Sulfate uptake, synthesis of phosphoadenosine phosphosulfate (PAPS), and PAPS translocation are known to take ~2 min[5]; the effective labeling time therefore is ~2.5 min. It is crucial to limit the labeling time:

[9] R. Heumann, V. Kachel, and H. Thoenen, *Exp. Cell Res.* **145**, 179 (1983).
[10] P. A. Baeuerle and W. B. Huttner, *Biochem. Biophys. Res. Commun.* **141**, 870 (1986).
[11] W. B. Huttner and P. A. Baeuerle, *Mod. Cell Biol.* **6**, 97 (1988).

after longer labeling, a portion of the labeled marker proteins will have left the TGN at the end of the pulse and will already be present in ISGs and CSVs.

Step 4: The immediate cooling of the dish and the replacement of the warm labeling medium with cold TBSS serve to achieve rapid cooling of the cells, thus preventing any transport of the marker proteins out of the TGN.

Modifications of Standard Protocol

The above protocol is applicable to other cell types and should also be applicable to tissues, provided that the label can be added to the tissue in an efficient manner (e.g., perfusion). The protocol also allows the exposure of cells, prior to labeling, to conditions that may affect the formation of ISGs and CSVs. However, any manipulation interfering with the sulfation machinery or resulting in a reduction in the level of marker proteins in the TGN would obviously be problematic. Finally, labels other than radioactive sulfate can be used, provided that these are incorporated into the marker molecule and that the labeling conditions are selective for the donor compartment of interest.

2. Preparation of Postnuclear Supernatant

Rationale

Two types of cell-free membrane traffic systems can be distinguished: those based on perforated/semiintact cells and those using cell homogenates and fractions derived from these homogenates. We chose the latter type of system, primarily because the physical separation of ISGs and CSVs from the TGN is a central step in our assay, and we anticipated that this separation would be easier in this type of system. Our cell-free system uses a postnuclear supernatant prepared from PC12 cells in concentrated form to prevent the dilution of components required for vesicle formation.

Continuation of Standard Protocol: Steps 5–11

Step 5. This and all subsequent steps are performed at 4°. Remove TBSS and wash PC12 cells one more time with TBSS and once with TBSS plus protease inhibitors [TBSS/PI: 0.5 mM phenyl methylsulfonyl fluoride (PMSF) and 10 μg/ml Trasylol in HBSS]. Add 10 ml of TBSS/PI per dish.

Step 6. Scrape cells off each dish in the 10 ml of TBSS/PI using a cut silicone stopper attached to a disposable pipette. Pool the cell suspension from the two dishes. *Note:* Because the total number of 15-cm dishes used in an experiment usually is more than two, multiple pools, each one containing the cells from two dishes, should be prepared and separately subjected to homogenization in the cell cracker (see step 10 below).

Step 7. Pellet the cells by centrifugation for 5 min at 700 g. Resuspend cells in 1 ml of homogenization buffer plus sulfate containing protease inhibitors [HBS/PI: 0.25 M sucrose, 1 mM ethylenediaminetetraacetic acid (EDTA), 1 mM magnesium acetate, 1.6 mM Na_2SO_4, 10 mM HEPES–KOH, pH 7.2, PI as above] and dilute with another 4 ml of HBS/PI.

Step 8. Pellet the cells by centrifugation for 5 min at 1700 g. Resuspend the cell pellet (~200 μl) in 1 ml of HBS/PI by trituration using a 1-ml Pipetman (four to six times up and down).

Step 9. Pass the cell suspension through a 22-gauge needle attached to a 1-ml syringe until a single-cell suspension, as monitored by phase-contrast microscopy (Fig. 1, top), is obtained (usually five or six times).

Step 10. Homogenize the cell suspension using a cell cracker[12] [European Molecular Biology Laboratory (EMBL) workshop] with an 18-μm clearance and two 1-ml syringes attached, until >95% of the cells are broken and the nuclei appear free of debris (Fig. 1, bottom; usually four or five passes back and forth). Remove the homogenate and rinse the cell cracker with 1 ml of HBS/PI; pool the rinse with the initial homogenate, yielding a total volume of typically 1.6 ml.

Step 11. Pellet the nuclei and unbroken cells by centrifugation at 1000 g for 10 min. Collect the resulting postnuclear supernatant (PNS; ~1.4 ml, typically 3–4 mg protein/ml) and keep on ice. *If applicable:* Pool the PNS from separate homogenizations at this step. This is the starting material for the cell-free reaction (see Section 3 below).

Comments on Standard Protocol: Steps 6–10

Step 6: After this step approximately 20% of the cells are permeable to trypan blue. An increase in the damage to the plasma membrane due to the scraping of the cells should be avoided as this may result in a loss of cytosol in the subsequent washes.

Step 7: This step serves to exchange the high-salt TBSS with the low-salt HBS. The removal of most of the salt is crucial to avoid the aggregation of membranes at 37° resulting in pelleting of the membranes during velocity sucrose gradient centrifugation (see Section 4 below). In addition, the HBS causes slight swelling of the cells, which facilitates their subsequent homogenization.

---

[12] W. E. Balch, W. G. Dunphy, W. A. Braell, and J. E. Rothman, *Cell* **39**, 405 (1984).

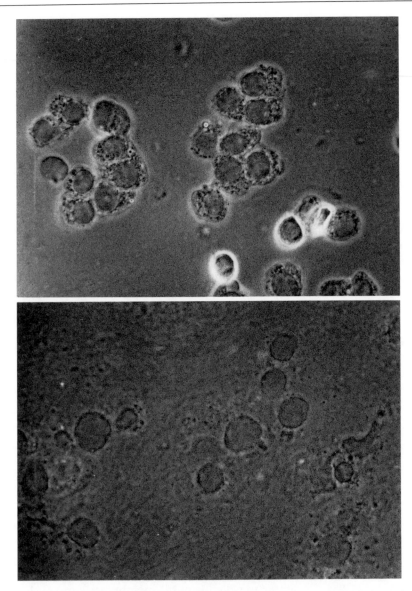

FIG. 1. Phase-contrast micrographs showing the appearance of the PC12 cell suspension (top, after step 9) and the PC12 cell homogenate (bottom, after step 10). Due to Brownian motion, the subcellular particles in the homogenate (bottom) moved during the exposure and are not seen as punctate structures.

Step 9: The preparation of a single-cell suspension is a prerequisite for the efficient and reliable homogenization of the cells.

Step 10: These conditions of homogenization result in the TGN being recovered as membrane structures significantly larger than the ISGs and CSVs. This size difference is the basis for the subsequent separation of ISGs and CSVs from the TGN by velocity sucrose gradient centrifugation (see Section 4 below).

Modifications of Standard Protocol

The major extension to this experimental section of the cell-free system would be the isolation, by physical or immunological techniques, of the TGN from the postnuclear supernatant prior to the cell-free reaction. Such an isolation would have to be carried out under conditions in which the TGN retains its vesicle budding and sorting capacity. The use of an isolated TGN in the cell-free reaction would also obviously require the replenishment of any factor required for vesicle formation that may have been removed in the course of isolation of the TGN.

3. Cell-Free Formation of Immature Secretory Granules and Constitutive Secretory Vesicles

Rationale

The PNS is incubated at 37° in a low ionic strength medium in the presence of ATP and an ATP-regenerating system. The low ionic strength medium is used because incubation of the PNS at 37° in the presence of salt causes the clumping of membranes, which interferes with the subsequent separation of ISGs and CSVs from the TGN. ATP is maintained during the reaction because the formation of CSVs and ISGs, like most other cell-free membrane traffic reactions, requires ATP hydrolysis.

Continuation of Standard Protocol: Steps 12 and 13

Step 12. Mix, by briefly vortexing, 1.25 ml of the PNS at 4° with 25 μl of 1 mM PAPS, 25 μl of 10 mM magnesium acetate, and 50 μl of ATP and an ATP-regenerating system [16.7 μl of 100 mM ATP, 16.7 μl of 800 mM creatine phosphate, 16.7 μl of 4 mg/ml creatine phosphokinase in 50% (w/v) glycerol].[13]

Step 13. Incubate sample at 37° without further agitation for 60 min. For the subsequent quantitation of cell-free ISG and CSV formation (see

[13] J. Davey, S. M. Hurtley, and G. Warren, *Cell* **43**, 643 (1985).

step 17 below), it is necessary to carry out another reaction for the same time at 0°. Samples are then placed on ice and subjected to velocity centrifugation (see Section 4 below).

Comments on Standard Protocol: Steps 12 and 13

Step 12: The PNS contains radioactive PAPS because of the pulse-labeling of intact PC12 cells with [^{35}S]sulfate. Therefore, unlabeled PAPS is added to inhibit further [^{35}S]sulfate incorporation during the cell-free reaction.

Step 13: If the cell-free reaction is carried out for various time periods, the samples after incubation at 37° are transferred to, and kept on, ice until the last time point, and then subjected to velocity centrifugation.

Modifications of Standard Protocol

Instead of ATP and the ATP-regenerating system, 50 μl of 10 mg/ml of hexokinase in 250 mMD-glucose[13] can be added to the PNS to investigate the ATP requirements of the cell-free formation of ISGs and CSVs.[4] Also, guanine nucleotides and guanine nucleotide analogs can be added (maximum volume added: 12.5 μl) to investigate the requirement for GTP hydrolysis.[14]

4. Separation of Immature Secretory Granules and Constitutive Secretory Vesicles from Trans-Golgi Network

Rationale

Most cell-free vesicle fusion systems are based on component A (in the donor vesicle) meeting component B (in the acceptor compartment) and thereby allowing a biochemical reaction to occur that is subsequently quantitated. Cell-free vesicle formation systems cannot use this approach. One feasible approach to monitor the cell-free formation of vesicles is to physically separate them from the donor compartment (e.g., see Ref. 15).

In the present cell-free system, the formation of ISGs and CSVs is assayed by monitoring the packaging of sulfate-labeled hsPG and SgII into CSVs and ISGs, respectively, which are then physically separated from the donor compartment, the TGN. Because the TGN of PC12 cells is recovered as relatively large membranous structures under the present homogenization conditions, velocity sucrose gradient centrifugation is used to achieve this separation.

[14] S. A. Tooze, U. Weiss, and W. B. Huttner, *Nature (London)* **347,** 207 (1990).
[15] M. K. Bennett, A. Wandinger-Ness, and K. Simons, *EMBO J.* **7,** 4075 (1988).

Continuation of Standard Protocol: Steps 14–17

Step 14. Prepare a linear sucrose gradient from 5.5 ml of 0.3 M sucrose and 6 ml of 1.2 M sucrose, both in 10 mM HEPES–KOH, pH 7.2. Load the sample after the cell-free reaction (typically 1.3 ml) onto the gradient at 4°.

Step 15. Centrifuge the gradient at 25,000 rpm in a Beckman (Palo Alto, CA) SW40 rotor for 15 min (after having reached the final speed) at 4°, with the brake applied at the end of the run.

Step 16. Collect 1-ml fractions from the top of the gradient. Subject aliquots from each fraction to sodium dodecyl sulfate-polyacrylamide gel electrophoresis (SDS-PAGE) followed by fluorography.

Step 17. Determine the efficiency of cell-free ISG and CSV formation as follows. Determine the amount of ^{35}S-labeled SgII and ^{35}S-labeled hsPG in each fraction of the velocity gradient after SDS-PAGE and fluorography by either scintillation counting or densitometric scanning.[4] For ISGs, then calculate the "total ^{35}S-labeled SgII in the TGN" by adding up the ^{35}S-labeled SgII in fractions 9–12, which contain the bulk of the TGN, of a sample kept at 4°. Calculate the "[35]S-labeled SgII in cell-free formed ISGs" by subtracting the ^{35}S-labeled SgII values in fractions 1–4 of the sample kept at 4° from the corresponding ^{35}S-labeled SgII values of samples incubated at 37° (fractions 1–4 contain the bulk of the cell-free formed ISGs). Express ^{35}S-labeled SgII in cell-free formed ISGs as the percentage of total ^{35}S-labeled SgII in the TGN. For CSVs, proceed accordingly with the ^{35}S-labeled hsPG (fractions 1–4 also contain the bulk of the cell-free formed CSVs).

Comments on Standard Protocol: Steps 15 and 16

Step 15: A 15-min centrifugation gives optimal separation of ISGs and CSVs from the TGN under the present conditions. With shorter centrifugation times, the TGN has not yet sufficiently entered the gradient. With longer centrifugation times, the ISGs enter the gradient more deeply and approach the position of the TGN.

Step 16: To reduce the exposure time for fluorography, 300-μl aliquots of the fractions are subjected to acetone precipitation prior to SDS-PAGE. Aliquots of the fractions can also be (1) subjected to protease protection assays to determine membrane integrity, and (2) assayed for sialyltransferase activity to validate the retention of a TGN marker enzyme (see Section 6 below). Although the primary function of the velocity gradient is to separate ISGs and CSVs from the TGN, the separation of ISGs and CSVs from each other can often be detected after velocity centrifugation.

Typical Results

When the PNS is kept at 0°, both sulfate-labeled SgII and hsPG are found in the bottom half of the velocity gradient in the fractions containing TGN (peak in fractions 9 + 10; see Ref. 4). When the PNS is incubated at 37° for 60 min, about half of both the sulfate-labeled SgII and hsPG are found in the top half of the gradient in the position of post-TGN vesicles.[4]

5. Separation of Immature Secretory Granules and Constitutive Secretory Vesicles

Rationale

Secretory granules are known to have a greater buoyant density in sucrose gradients than CSVs. We therefore physically separate the cell-free formed ISGs and CSVs using equilibrium sucrose gradient centrifugation.

Continuation of Standard Protocol: Steps 18–20

Step 18. Prepare a linear sucrose gradient from 5 ml of 0.5 M sucrose and 5 ml of 2 M sucrose, both in 10 mM HEPES–KOH, pH 7.2. Pool from the velocity gradient the remainder of fractions 1–4, which contain the bulk of ISGs and CSVs, and load the pooled material (typically ~3 ml) onto the gradient at 4°.

Step 19. Centrifuge the gradient at 25,000 rpm in a Beckman SW40 rotor for 5.5 hr at 4°, with the brake applied at the end of the run.

Step 20. Collect 1-ml fractions from the top of the gradient. Subject aliquots from each fraction to SDS-PAGE followed by fluorography to monitor the separation of the ISGs and CSVs formed during the cell-free reaction.

Comments on Standard Protocol: Step 19

Step 19: A 5.5-hr centrifugation has been found to be sufficient for ISGs and CSVs to reach their buoyant density in sucrose.

Modifications of Standard Protocol

Separation of ISGs and CSVs from each other can also be obtained by using a sucrose gradient prepared from 5 ml of 0.6 M sucrose and 5 ml of 1.6 M sucrose, both in 10 mM HEPES–KOH, pH 7.2.[16]

---

[16] A. Regnier-Vigouroux, S. A. Tooze, and W. B. Huttner, *EMBO J.* **10**, 3589 (1991).

Typical Results

Equilibrium centrifugation of the post-TGN vesicles formed during the cell-free reaction resolves these into two distinct populations: CSVs characterized by the presence of the ^{35}S-labeled hsPG (peak at a density of 1.098 g/ml) and ISGs characterized by the presence of ^{35}S-labeled SgII (peak at a density of 1.131 g/ml) (see Ref. 4). This indicates that during the cell-free reaction these two proteins are packaged into different secretory vesicles.

6. Controls

Rationale

In the cell-free formation of ISGs and CSVs, it is important to control for the possible fragmentation of the TGN by ascertaining (1) the membrane integrity of the formed vesicles, and (2) the retention of TGN markers in the TGN at the end of the reaction. The first is achieved by carrying out protease protection assays, the second by assaying the distribution of sialyltransferase activity across the velocity gradient.

Protease Protection

1. Pool aliquots of fractions 2–4 from the velocity gradient and incubate with 1 mM dibucaine (from a 100 mM stock in water) for 5 min at 4°. Dibucaine is known to stabilize membranes.[17] Split the pooled material into three samples.

2. Supplement the samples with one-tenth the volume containing either water, 1 mg/ml proteinase K, or 1 mg/ml proteinase K plus 3% (w/v) Triton X-100. Incubate at 4° for 30 min.

3. Stop the digestion by the addition of 2 mM (final concentration) phenylmethylsulfonyl fluoride to inactivate the proteinase K. Add sample buffer and boil immediately.

4. Analyze the samples by SDS-PAGE followed by fluorography.

Typically, the amount of ^{35}S-labeled SgII and hsPG is unaffected by the addition of proteinase K in the absence of Triton X-100, whereas these two marker proteins are degraded by the protease in the presence of the detergent.

---

[17] G. Scheele, R. Jacoby, and T. Carne, *J. Cell Biol.* **87**, 611 (1980).

Sialyltransferase Assay

1. Dilute the fractions from the velocity gradient 10-fold with 1 mM MgCl$_2$ and centrifuge at 4° for 60 min at 150,000 g.
2. Discard the supernatant and resuspend the pellet in 100 μl of 0.1% (w/v) Triton X-100.
3. Assay the sialyltransferase activity in the resuspended pellets as described[18] using asialofetuin as the exogenous acceptor.

Typically, the distribution of sialyltransferase activity across the velocity gradient is unaffected by the *in vitro* incubation at 37°.[4]

7. Requirements for Cell-Free Formation of Immature Secretory Granules and Constitutive Secretory Vesicles

ATP

The comparison of the efficiency of the cell-free ISG formation in the presence of ATP and an ATP-regenerating system or in the presence of hexokinase and glucose to deplete ATP (Table I) shows that this process is dependent on ATP.[4] Similarly, the formation of CSVs is ATP dependent,[4] consistent with observations on the formation of post-Golgi secretory vesicles obtained in other cell-free systems (e.g., see Ref. 15).

GTP Hydrolysis

As shown in Table I, addition of 1 μM GTPγS or 100 μM GMP-PNP (which is less effective than GTPγS on a molar basis) inhibits the formation of both ISGs and CSVs.[14] The inhibition by nonhydrolyzable GTP analogs is specific with respect to the guanine moiety because the addition of 50 μM ATPγS has no inhibitory effect.

Conclusion

The cell-free system described above faithfully reconstitutes the formation of ISGs and CSVs from the TGN, including the sorting events that occur during these processes. The cell-free formed ISGs and CSVs are indistinguishable from those formed *in vivo* by several criteria, and TGN markers are retained in the TGN during the reaction.[4] The use of this

[18] A. W. Brändli, G. C. Hansson, E. Rodriquez-Boulan, and K. Simons, *J. Biol. Chem.* **263**, 16283 (1988).

TABLE I

Effects of Various Nucleotides on the Cell-Free
Formation of Immature Secretory Granules and
Constitutive Secretory Vesicles[a]

Condition	Efficiency of vesicle formation (% of control)	
	ISGs	CSVs
ATP	100	100
Hexokinase/glucose	7	5
ATP + GTPγS (1 μM)	39	38
ATP + GMP-PNP (100 μM)	53	55
ATP + ATPγS (50 μM)	96	115

[a] Cell-free reactions were performed according to the standard protocol with the indicated modifications. Some of the values shown are from Ref. 14.

cell-free system has revealed that the formation of ISGs and CSVs requires ATP and GTP hydrolysis. The latter requirement suggests an involvement of GTP-binding proteins in the formation of ISGs and CSVs.[14] This cell-free system should be useful to identify other factors involved in the formation of ISGs and CSVs, and serves as a basis for future modifications, e.g., reactions using purified TGN membranes as the donor compartment.

Section II

Reconstitution Using Semiintact and Perforated Cells

[11] Nuclear Protein Import Using Digitonin-Permeabilized Cells

By STEPHEN A. ADAM, RACHEL STERNE-MARR, and LARRY GERACE

Molecular traffic between the cytoplasm and nucleus occurs through the nuclear pore complex, an elaborate supramolecular structure with a mass of approximately 125×10^6 D that spans the nuclear envelope[1] (for reviews see Refs. 2 and 3). The pore complex contains an aqueous channel with a diameter of about 10 nm, which allows rapid nonselective diffusion of ions, metabolites, and other small molecules between the cytoplasm and nucleus.[4] Most proteins and RNAs are too large to diffuse across this aqueous channel at physiologically significant rates, and instead utilize mediated mechanisms involving specific signals to cross the pore complex. Mediated import of proteins into the nucleus is specified by short, basic amino acid sequences within the proteins called nuclear location sequences (NLS).[2]

Cell-free systems promise to be important tools to study the biochemistry of nuclear protein import. For an *in vitro* nuclear import system to be considered physiologically relevant, it should result in the concentration of proteins in the nucleus from the surrounding medium in a fashion dependent on NLS, ATP, and physiological temperature, properties characteristic of nuclear protein import *in vivo*.[5-7] Moreover, it is essential that proteins enter the nucleus through the nuclear pore complex. Most methods of cell lysis, particularly those involving mechanical disruption, lead to ruptures in the nuclear envelope that permit large macromolecules to enter the nucleus by passive diffusion (e.g., see Ref. 5). Nuclear accumulation of NLS-containing proteins does not necessarily indicate that nuclei are intact, because NLS-binding proteins have been found to occur in intranuclear structures[8,9] and could serve to trap NLS-containing proteins that cross the nuclear envelope by a nonphysiological route. An important

[1] R. Reichelt, A. Holzenburg, E. L. Buhle, M. Jarnik, A. Engel, and U. Aebi, *J. Cell Biol.* **110**, 883 (1990).

[2] L. Gerace and B. Burke, *Annu. Rev. Cell Biol.* **4**, 335 (1988).

[3] D. S. Goldfarb, *Curr. Opinions Cell Biol.* **1**, 441 (1989).

[4] P. L. Paine, L. C. Moore, and S. B. Horowitz, *Nature (London)* **254**, 109 (1975).

[5] D. D. Newmeyer, D. R. Finlay, and D. J. Forbes, *J. Cell Biol.* **103**, 2091 (1986).

[6] D. D. Newmeyer and D. J. Forbes, *Cell* **52**, 641 (1988).

[7] W. D. Richardson, A. D. Mills, S. M. Dilworth, R. A. Laskey, and C. Dingwall, *Cell* **52**, 655 (1988).

[8] S. A. Adam, T. J. Lobl, M. A. Mitchell, and L. Gerace, *Nature (London)* **337**, 276 (1989).

[9] U. T. Meier and G. Blobel, *J. Cell Biol.* **111**, 2235 (1990).

criterion for "intactness" of nuclei in cell-free preparations is exclusion of nonnuclear proteins that are too large to diffuse into nuclei *in vivo* (i.e., globular proteins larger than ~ 40kDa).[4] In addition, a useful indication of nuclear entry via a pore complex route *in vitro* is inhibition of nuclear accumulation with reagents that bind to specific O-linked glycoproteins of the pore complex and inhibit nuclear import *in vivo*. These include monoclonal antibodies[10,11] and the lectin wheat germ agglutinin.[12-14]

Two cell-free nuclear import systems characterized up to now adequately satisfy the criteria considered above.[5,15] One of these systems is based on *Xenopus* egg extracts, which have strong nuclear assembly activity. The egg extracts can "reseal" the ruptured nuclear envelopes of isolated rat liver nuclei and also can assemble intact nuclei around exogenous chromatin or chromosomes to produce transport-competent nuclei.[5] This system is powerful for studying *de novo* assembly of the nuclear envelope,[16] but is limited because a source of *Xenopus* eggs must be maintained and the user is restricted to use of *Xenopus* egg cytosol to support import. In addition, while cytosol is required in this nuclear transport system, it is difficult to distinguish cytosolic factors essential for nuclear resealing or assembly from those directly involved in transport across the pore complex.[17]

We have developed an *in vitro* system for nuclear import that does not rely on nuclear assembly. This system utilizes digitonin-permeabilized cultured cells grown on glass coverslips.[15] Treatment of cells with low concentrations of digitonin selectively permeabilizes the plasma membrane due to its relatively high cholesterol content, while most other intracellular membranes, including the nuclear envelope (which is cholesterol poor), remain intact, and the cytoskeleton is largely undisturbed. After permeabilization, soluble proteins of the cell diffuse out through the digitonin-induced holes in the plasma membrane. When the permeabilized cells are supplemented with exogenous cytosol and ATP, the nuclei of these cells rapidly accumulate a fluorescent protein containing a nuclear location sequence in a manner that satisfies all the criteria established for authentic nuclear import. Transport in permeabilized cells absolutely re-

[10] M.-C. Dabauvalle, R. Benavente, and N. Chaly, *Chromosoma* **97,** 193 (1988).
[11] C. M. Featherstone, M. K. Darby, and L. Gerace, *J. Cell Biol.* **107,** 1289 (1988).
[12] M.-C. Dabauvalle, B. Schulz, U. Scheer, and R. Peters, *Exp. Cell Res.* **174,** 291 (1988).
[13] D. R. Finlay, D. D. Newmeyer, T. M. Price, and D. J. Forbes, *J. Cell Biol.* **104,** 189 (1987).
[14] Y. Yoneda, N. Imamoto-Sonobe, M. Yamaizumi, and T. Uchida, *Exp. Cell Res.* **173,** 586 (1987).
[15] S. A. Adam, R. E. Sterne-Marr, and L. Gerace, *J. Cell Biol.* **111,** 807 (1990).
[16] D. R. Findlay and D. J. Forbes, *Cell* **60,** 17 (1990).
[17] D. D. Newmeyer and D. J. Forbes, *J. Cell Biol.* **110,** 547 (1990).

quires soluble factors provided by exogenously added cytosol. This permeabilized cell system has a number of advantages, including the ease with which it can be set up and the wide variety of sources that can be used for permeabilized cells and cytosol.

Preparation of Cytosol Fraction

Choice of Cell Type

A valuable feature of this system is the ability to use cytosol prepared from a wide variety of vertebrate cells. Among other things, this could be useful for analyzing regulation of transport during cell growth and development (e.g., see Ref. 18). Cell types from which we have successfully obtained import-competent cytosol are indicated in Table I. When cultured cells are used to prepare cytosol extracts, the cells must be in log-phase growth for optimal activity. Stationary cultures, as well as cell types that do not rapidly divide, do not yield cytosol with transport activity comparable to that obtained from dividing cells. In addition, we have found that cells propogated in suspension culture using spinner flasks yield cytosol that is more active than the same cells grown in monolayer culture. This may be due to the fact that suspension cells form a more compact pellet and yield more highly concentrated cytosol on homogenization (see below).

For most of our routine work we use rabbit reticulocyte lysate obtained from commercial distributors (Promega Biotech, Milwaukee, WI) as a cytosol source. These extracts have consistently higher import activity than cultured cell extracts, possibly due to the higher protein concentration found in reticulocyte lysate (~ 80 mg/ml) compared to cultured cell cytosol (less than 40 mg/ml). We have found that cytosol prepared by hypotonic lysis of red blood cells (RBCs) is equivalent in activity to reticulocyte extracts. Red blood cells can easily be obtained in large quantities from appropriate sources of fresh blood and provide a convenient source of material for biochemical fractionation of soluble import components.[19] While mammalian RBCs are nondividing and lack nuclei, the cytosolic components required for nuclear import apparently persist in a stable form in RBCs from earlier developmental stages.

We also have found that extracts prepared from *Xenopus laevis* oocytes can support import in this system. In contrast to the *in vitro* nuclear import system involving *Xenopus* egg extracts where membranes are required for

[18] C. M. Feldherr and D. Akin, *J. Cell Biol.* **111,** 1 (1990).
[19] S. A. Adam and L. Gerace, *Cell* **66,** 837 (1991).

TABLE I
SOURCES OF CYTOSOL AND PERMEABLE CELLS

Cytosol	Permeable cells
Rabbit reticulocyte	HeLa (human)
HTC (rat hepatoma)	NRK (normal rat kidney)
Rat erythrocyte	BRL (buffalo rat liver)
Bovine erythrocyte	
Xenopus laevis oocyte	
Rat liver	

nuclear resealing,[5] nuclear import in our system is supported by high-speed supernatants of oocyte lysates that are devoid of membranes. It is likely that oocytes from other organisms also would be able to provide import-competent cytosol. Although we have not extensively investigated the possibility of using yeast cytosol for transport in permeabilized mammalian cells, our preliminary attempts to prepare active cytosol from *Saccharomyces cerevisiae* have been unsuccessful. We feel that this may be due to the presence of a nonspecific inhibitory factor, and that with appropriate conditions it may be possible to obtain transport-competent cytosol from yeast.

Finally, we have found that high-speed supernatants prepared from concentrated homogenates of rat liver also support *in vitro* nuclear import. Like lysates of RBCs, cytosol prepared from this and other tissue sources may prove useful for isolating factors involved in protein import.

Homogenization of Cells

For preparation of cytosol to support *in vitro* nuclear import, we have extensively used two lines of cultured mammalian cells grown in suspension: HeLa cells and a rat hepatoma cell line (HTC). The homogenization conditions described below also can be used for other cultured cell lines. Suspension cultures of HeLa cells or HTC cells are grown in Joklik's modified minimum essential medium with 10% (v/v) fetal bovine serum (FBS), 20 mM N-2-hydroxyethylpiperazine-N'-2-ethanesulfonic acid (HEPES), pH 7.2, penicillin/streptomycin. Cultures are maintained at 37° in 1-liter microcarrier flasks (Bellco Glass, Inc., Vineland, NJ), mixing at 30–50 rpm. Exponentially growing cultures are collected by centrifugation at 250 g for 10 min at 4° in a J6B centrifuge equipped with a JS5.2 rotor (Beckman, Palo Alto, CA) and washed at least two times with cold phosphate-buffered saline (PBS; 10 mM sodium phosphate, pH 7.4,

140 mM sodium chloride) by resuspension and centrifugation. The cells are then resuspended in 10 mM HEPES, pH 7.3, 110 mM potassium acetate, 2 mM magnesium acetate, 2 mM dithiothreitol (DTT), and pelleted. The cell pellet is gently resuspended in 1.5 vol of lysis buffer [5 mM HEPES, pH 7.3, 10 mM potassium acetate, 2 mM magnesium acetate, 2 mM DTT, 20 μM cytochalasin B, 1 mM phenylmethylsulfonyl fluoride (PMSF), 1 μg/ml each aprotinin, leupeptin, and pepstatin] and swelled for 10 min on ice. Because the cytosol is sensitive to inactivation by oxidation, it is important to keep reducing agents present at all times. The cells are lysed by five strokes in a tight-fitting stainless steel or glass Dounce (Wheaton Industries, Millville, NJ) homogenizer. The resulting homogenate is centrifuged at 1500 g in a JS5.2 rotor for 15 min at 4° to remove nuclei and cell debris. The supernatant is then sequentially centrifuged at 15,000 g for 20 min at 4° in a Beckman JA20 rotor followed by 100,000 g for 30 min at 4° in a Beckman 70.1 Ti rotor. The final supernatant is dialyzed for 2–3 hr with a collodion membrane apparatus (molecular weight cutoff 25,000; Schleicher & Schuell, Inc., Keene, NH) against multiple changes of import buffer [20 mM HEPES, pH 7.3, 110 mM potassium acetate, 5 mM sodium acetate, 2 mM magnesium acetate, 0.5 mM ethylene glycol-bis (β-aminoethyl ether)-N,N,N',N'-tetraacetic acid (EGTA), 2 mM DTT, 1 μg/ml each of aprotinin, leupeptin, and pepstatin] and frozen in aliquots in liquid nitrogen prior to storage at −80°.

Red blood cells from many different mammalian species can be used to prepare active cytosol for transport, according to the following procedure that we use for rat erythrocytes. Blood is collected from several rats after CO_2 asphyxiation by cardiac puncture into acid–citrate–dextrose (ACD; 75 mM trisodium citrate, 38 mM citric acid, 136 mM dextrose, pH 5.0) in the proportion 1 ml ACD/7 ml blood. A 250-g rat typically yields from 3 to 8 ml blood. Prior to lysis, the blood is processed to separate erythrocytes from platelets and leucocytes. Immediately after being collected, the blood is centrifuged at 500 g for 20 min at room temperature in a Beckman J6B centrifuge equipped with a JS5.2 rotor. The supernatant and light-colored "buffy coat" are aspirated and an equal volume of 2% (w/v) gelatin (250 bloom; Sigma Chemical Co., St. Louis, MO) in PBS is added and gently mixed. The red cells are allowed to sediment undisturbed at 37° for 30 min. If the red cells do not sediment, a larger volume of the gelatin/PBS solution may be added. The cloudy upper layer containing platelets and white blood cells is aspirated and the RBCs are washed four times by resuspension in PBS and centrifugation at 1600 g in the JS5.2 rotor. The washed cells are then lysed by adding to the pellet 1–2 vol of ice-cold magnesium lysis buffer (5 mM HEPES, pH 7.3, 2 mM magnesium acetate,

1 mM EGTA, 2 mM DTT, 0.1 mM PMSF, 1 μg/ml each aprotinin, leupeptin, and pepstatin) and vortexing on a high setting for 30 sec. The lysate is allowed to set on ice for 5 min before the cell debris is removed by centrifugation at 6000 g for 20 min at 4° in a Beckman JS5.2 rotor. The clear supernatant is removed, avoiding the cloudy layer of red cell ghosts filling the lower half of the tube. This supernatant is centrifuged at 100,000 g in a Beckman 70.1 Ti rotor and dialyzed into transport buffer as described above for reticulocyte lysate. After dialysis, it may be necessary to recentrifuge the extract at 15,000 g to remove material that may precipitate during dialysis. Occasionally, some red cell ghosts may remain and can be removed by filtration of the extract through a 0.45-μm syringe filter. To prepare commercially obtained rabbit reticulocyte lysate for transport studies, the lysate is dialyzed into transport buffer as described above and is centrifuged at 100,000 g prior to freezing.

For the preparation of oocyte extract, ovaries from *X. laevis* are first dissected into PBS. The ovary is cleaned of extraneous tissue, blotted to remove excess PBS, and placed in a glass Dounce homogenizer with 1 vol of 20 mM HEPES, pH 7.3, 110 mM potassium acetate, 2 mM magnesium acetate, 1 mM DTT, and 1 μg/ml each of aprotinin, leupeptin, and pepstatin. The oocytes are crushed by two strokes of a loose-fitting pestle. The crude lysate is centrifuged at 2000 g for 10 min and the supernatant is removed with a Pasteur pipette, taking care to avoid the lipid layer floating on top and the diffuse cloudy layer near the bottom of the tube. This supernatant is then centrifuged in a Beckman 70.1 Ti rotor at 100,000 g for 45 min. Again the clear supernatant is removed and saved, avoiding the top and bottom layers. This supernatant is then dialyzed against import buffer as described above.

Preparation of Fluorescent Transport Substrates

Short synthetic peptides containing an NLS are capable of directing the mediated nuclear import of a protein to which they have been chemically coupled.[20,21] Transport of such synthetic substrates closely resembles the NLS-directed import of native nuclear proteins. As a matter of convenience, we have used synthetic substrates consisting of fluorescent reporter proteins conjugated with NLS-containing peptides for most of our transport studies. A reporter protein chosen to study mediated nuclear import must be too large to diffuse through the pore complex passively (i.e., larger than 40–60 kDa in the case of a globular protein).[4] We commonly use the

---

[20] R. E. Lanford, P. Kanda, and R. C. Kennedy, *Mol. Cell. Biol.* **8**, 2722 (1986).
[21] D. S. Goldfarb, *Cell Biol. Int. Rep.* **12**, 809 (1988).

104-kDa naturally fluorescent protein allophycocyanin (APC; Calbiochem, San Diego, CA) for this purpose. Allophycocyanin has a high fluorescence output and is resistant to photobleaching. This protein is chemically coupled to a synthetic peptide containing the well-characterized NLS of the simian virus 40 (SV40) large T antigen (PKKKRKVE)[22,23] through an N-terminal cysteine on the peptides. Synthetic peptides containing the SV40 large T antigen wild-type nuclear location signal (CGGGPK[128]KKRKVED) or an import-deficient mutant sequence (CGGGPK[128]NKRKVED) are obtained from Multiple Peptide Systems (San Diego, CA). If the peptide preparations are not largely homogeneous, they are purified by high performance liquid chromatography (HPLC). Prior to conjugation, the peptide should be reduced to maximize its ability to couple through the cysteine sulfhydryl. To accomplish this, 1 mg of peptide is resuspended in 0.1–0.2 ml 50 mM HEPES, pH 7.0 and incubated with 5 mg DTT for 1 hr at room temperature. A 1% (v/v) concentration of glacial acetic acid is added to the reduced peptides and the mixture is separated by chromatography on a 20 × 1 cm Sephadex G-10 column. The column is eluted with 1% (v/v) acetic acid and 0.5-ml fractions are collected. The peptide elution profile is monitored by the use of the Ellman reaction. This involves addition of a 10-μl aliquot of each fraction to a test tube containing 1 ml 0.1 M sodium phosphate buffer, pH 7.4, 5 mM ethylenediaminetetraacetic acid (EDTA), followed by 100 μl 1 mM dithiobisnitrobenzoic acid (Ellman's reagent) in methanol. Fractions containing free sulfhydryl residues yield a bright yellow color. Fractions containing peptide (eluting first off the column) are pooled and the concentration of free cysteine-containing peptide is determined by measuring absorbance at 412 nm after carrying out the Ellman reaction as described above, using glutathione to prepare a standard curve.

To activate the APC for peptide conjugation, the protein is first dissolved in 0.1 M sodium phosphate, pH 8.0, at 2 mg/ml and dialyzed extensively against this same buffer. The bifunctional cross-linker sulfo-SMCC (Pierce Chemical Co., Rockford, IL) is dissolved in dimethyl sulfoxide (DMSO) and a 20-fold molar excess is immediately added to the APC solution and incubated for 30 min at room temperature. The activated APC, conjugated at primary amino groups with the cross-linker, is immediately separated from the unreacted sulfo-SMCC by desalting on a Sephadex G-25 column equilibrated in 0.1 M sodium phosphate, pH 7.0. The desalting column should be at least 20 times the volume of the sample to be desalted. Next, a 50-fold molar excess of peptides containing reduced

[22] D. Kalderon, B. L. Roberts, W. D. Richardson, and A. E. Smith, *Cell* **39**, 499 (1984).
[23] R. E. Lanford and J. S. Butel, *Cell* **37**, 801 (1984).

amino-terminal cysteine residue is mixed with the activated APC. After overnight incubation at 4°, the APC–peptide conjugate is separated from free peptide by desalting on Sephadex G-25. The blue color of the APC makes selection of appropriate fractions easy. The number of peptides conjugated to the protein is estimated by comparing the mobility of the APC–peptide conjugates to the mobility of activated APC on sodium dodecyl sulfate-polyacrylamide gels. (It should be noted that APC is a hexamer containing two different subunits.) The conditions described above usually yield four to eight peptides per APC molecule. The conjugates are then dialyzed against 10 mM HEPES, pH 7.3, 110 mM potassium acetate. The concentration of the allophycocyanin conjugates is determined by a commercial dye-binding assay [Pierce BCA (Pierce Chemical Co.) or Bio-Rad protein assay (Bio-Rad, Richmond, CA)]. They can be stored for short periods at 4°, or can be frozen in small aliquots in liquid nitrogen and stored for extensive periods at −80°.

In addition to APC, we also have successfully used peptide conjugates of rhodamine isothiocyanate-labeled bovine serum albumin (BSA) for transport studies. Highly purified crystallized BSA (e.g., Sigma Chemical Co.) should be used for this work so that the labeled BSA is not contaminated with other minor labeled proteins. Purified IgG conjugated to NLS-peptides also has been used for transport studies,[20] and IgG can be readily isolated by adsorption to immobilized protein A or protein G. Fluorescent conjugates of BSA or IgG are prepared as described below, or can be obtained commercially. Whatever carrier protein is used as a reporter, the protein must be labeled with a fluorochrome prior to conjugation with peptides to avoid inactivation of the NLS by modification of essential amino acid residues. In general, rhodamine conjugates of proteins are preferable to fluorescein conjugates for these transport studies, due to their greater resistance to photobleaching.

Xenopus nucleoplasmin is a nuclear protein with a naturally occurring NLS[24] and is widely used for analysis of mediated nuclear import. Furthermore, nucleoplasmin is easily purified to homogeneity in sufficient quantities for these studies.[25] In our hands rhodamine-labeled nucleoplasmin exhibits high nonspecific binding to the permeabilized cells. However, we have obtained efficient import of 20-nm gold particles coated with nucleoplasmin (e.g., see Ref. 26) into nuclei of the digitonin-permeabilized cells.

For labeling a reporter protein with fluorochrome, a protein solution of 1–2 mg/ml must first be dialyzed extensively in 0.1 M sodium carbonate,

[24] J. Robbins, S. Dilworth, R. Laskey, and C. Dingwall, *Cell* **64,** 615 (1991).

[25] C. Dingwall, S. Sharnick, and R. Laskey, *Cell* **30,** 449 (1982).

[26] C. Feldherr, E. Kallenbach, and N. Schultz, *J. Cell Biol.* **99,** 2216 (1984).

pH 9.0. The fluorescein isothiocyanate (FITC) or tetramethylrhodamine isothiocyanate (TRITC) is dissolved in dimethyl sulfoxide at 1 mg/ml immediately prior to use. For each milligram of protein to be labeled, 25 μl of the fluorochrome is added slowly while mixing. The mixture is incubated overnight in the dark at 4°. To stop the reaction, Tris-hydrochloride, pH 8.0, is added to 50 mM and incubated for 1 hr in the dark. The conjugated protein is separated from free fluorochrome by chromatography through Sephadex G-25 or G-50 equilibrated in 0.1 M sodium phosphate, pH 8.0. The labeled protein should then be concentrated to approximately 2 mg/ml and coupled to the peptide as described above for APC.

Permeabilization of Cells and Import Assay

Human (HeLa) cells and normal rat kidney (NRK) cells are grown on plastic petri dishes in Dulbecco's modified Eagle's medium containing 10% (v/v) FBS and penicillin/streptomycin. Cultures are maintained in a humidified incubator at 37° with a 5% CO_2 atmosphere. Twenty-four to 48 hr before use in a transport assay, cells are removed from the plastic dishes by trypsinization and replated on 18 × 18 mm glass coverslips in six-well multiwell plates. To permeabilize the cells, coverslips are rinsed in ice-cold import buffer and immersed in ice-cold import buffer containing 40 μg/ml digitonin (Calbiochem; diluted from a 40-mg/ml stock solution in dimethyl sulfoxide) contained in six-well plastic tissue culture dishes (Corning Glass Works, Corning, NY) or a Coplin jar. The cells are incubated for 5 min, after which the digitonin-containing buffer is removed by aspiration and replaced with cold import buffer. Cells permeabilized by this treatment retain about 50% of the proteins present in unpermeabilized cells, and appear similar to unpermeabilized cells in phase-contrast microscopy.[15] In applying this digitonin permeabilization scheme to a cell type other than the ones we have tested (Table I), it is important to conduct a titration with different digitonin concentrations to determine a detergent concentration that permeabilizes the plasma membrane to macromolecules at the same time as retaining the structural integrity of the nucleus envelope (see below).

To carry out nuclear import reaction with the permeabilized cells, cover slips are drained and blotted to remove excess buffer, and inverted over 40–50 μl of complete import mixture on a sheet of Parafilm in a humidified plastic box. The complete import mixture contains the following components: 50% (v/v) cytosol, 100 nM APC–peptide conjugate, 20 mM HEPES, pH 7.3, 110 mM potassium acetate, 5 mM sodium acetate, 2mM magnesium acetate, 2 mM DTT, 0.5 mM EGTA, 1 mM magnesium ATP, 5 mM creatine phosphate (Calbiochem), 20 units/ml creatine phosphokinase (Calbiochem), 1 μg/ml each of aprotinin, leupeptin, and pepstatin. The entire box is then floated in a water bath of 30°, usually for 30 min. At the end of the assay, each coverslip is rinsed in import buffer and mounted on a glass microscope slide in a small amount of import buffer and the coverslip edges sealed with nail polish. It is important to avoid the use of nonphysiological media for mounting coverslips, because an intact nuclear envelope is required to maintain accumulated nuclear fluorescence in unfixed samples.[15] If many samples are to be observed or if they cannot be examined immediately, the rinsed cells may be fixed by immersion in import buffer containing 4% (w/v) formaldehyde prior to mounting. Samples are observed by phase-contrast and epifluorescence microscope with a Zeiss Axiophot microscope equipped with ×40 or ×63 planapochromat objectives. When reticulocyte lysate is used as a source of cytosol, the final concentration of protein in the import mixture is about 40 mg/ml. We have found that cytosols from some other sources are inhibitory at high concentrations, and therefore cytosols should be titrated to determine concentrations that support maximal transport.

The results of a typical import reaction are shown in Fig. 1. The nuclei of cells incubated with APC conjugated with the wild-type NLS of the SV40 T antigen accumulate high levels of fluorescent reporter (Fig. 1, wild-type panels), while nuclei incubated with APC conjugated with the transport-defective mutant peptide show no accumulation of the reporter (Fig. 1, mutant panels). In an import system consisting of rabbit reticulocyte lysate and NRK cells, we obtain an average of 10- to 20-fold accumulation of the reporter protein in the nucleus with respect to the surrounding medium after 30 min.[15] In most experiments, ∼95% of the nuclei on the cover slip show significant nuclear accumulation of the transport probe. However, there is extensive heterogeneity from nucleus to nucleus in the actual level of nuclear accumulation. This can be seen from quantitive analysis of nuclear import among a population of nuclei (Fig. 1, bottom). This heterogeneity may be due to cell cycle differences between different nuclei in a population of exponentially growing cells.

It must be stressed that when the system is initially set up, appropriate controls must be done to authenticate import. First, it should be demonstrated that the nuclei of permeabilized cells are intact and exclude large proteins lacking NLSs. This feature constrains the import of large proteins to a mediated pathway. To determine nuclear integrity, we incubate unfixed permeabilized cells with anti-DNA antibodies in 0.2% (w/v) gelatin in import buffer for 15 min at room temperature, rinse the coverslips, and then incubate the coverslips with a secondary antibody (diluted in gelatin–import buffer) to detect the anti-DNA antibodies. If the permeabilized cell nuclei are intact they completely exclude anti-DNA antibodies,

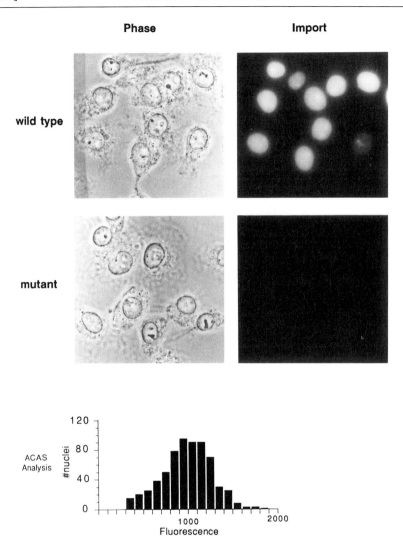

FIG. 1. Results of a typical nuclear import reaction. *Top:* Phase-contrast (left) and fluores-cence (right) images of NRK cells that were permeabilized with digitonin and incubated for 30 min in complete transport mix containing APC conjugated with either the wild-type SV40 T antigen NLS or with the mutant, nonfunctional NLS (see text). APC containing the wild-type NLS is strongly concentrated in the nuclei, while APC with the mutant NLS is not accumulated. *Bottom:* Quantitation of nuclear accumulation of the wild-type NLS-APC conjugate by ACAS analysis (see text).

– Triton   + Triton

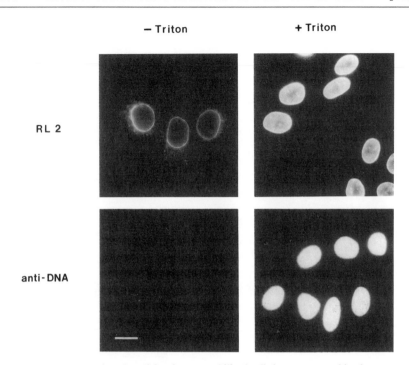

RL 2

anti-DNA

FIG. 2. Demonstration that digitonin-permeabilized cells have a permeable plasma membrane and retain an intact nuclear envelope. NRK cells were permeabilized with digitonin and in some cases were subsequently incubated with Triton X-100 to permeabilize the nuclear envelope completely (right-hand side). Afterward, samples were incubated with a monoclonal antibody that recognizes the cytoplasmic face of the pore complex (RL2) or with anti-DNA antibodies, and the bound antibodies were detected with fluorescent secondary antibodies (see text). RL2 binds to the nuclear envelope of permeabilized cells with or without Triton treatment, while the anti-DNA antibody binds to nuclei only if cells have been treated with Triton after the digitonin treatment. (From Ref. 15. Reproduced with permission of The Rockefeller University Press.)

while if they are not intact they become strongly labelled (Fig. 2, bottom). Incubation of permeabilized cells with antibodies to insoluble cytoplasmic antigens can demonstrate that the plasma membrane of digitonin-treated cells is permeable to large macromolecules. For example, proteins exposed on the cytoplasmic surface of the pore complex are readily decorated with antibodies in the permeabilized cells (Fig. 2, top). Another control involves demonstrating that reporter proteins coupled with peptides containing the nontransported mutant T antigen NLS sequence are not accumulated in the nucleus (see above). A related experiment involves analysis of the import capability of the carrier protein alone that has been modified with the cross-linking reagent, because some carrier proteins may exhibit high nonspecific binding to the permeabilized cells and/or nuclear surface. Other physiological controls involve the demonstration that no nuclear accumulation is obtained when cytosol is depleted of ATP by hexokinase/ glucose treatment or when the transport system is incubated at 0°.[5,15] Finally, nuclear import in vertebrate cells should be inhibited by addition of wheat germ agglutinin to the reaction at 50 µg/ml,[13,15] which would argue that proteins enter the nuclei of permeabilized cells through the pore complexes.

Quantifying Import

The amount of import into individual nuclei can be quantified by several methods. One procedure involves performing densitometry on photographic negatives. A field of cells is photographed with Kodak (Rochester, NY) TMax film (ASA400) and the film processed. A standard curve of fluorescence intensity values is obtained by preparing a dilution series of the fluorescent APC from about a 1.5- to 50-fold dilution of the stock solution. Eight microliters of each dilution is pipetted onto a slide and mounted under an 18×18 mm glass cover slip. This dilution series is then photographed at the same exposure time as the cells. The negatives are scanned with a scanning laser densitometer (LKB Instruments, Inc., Gaithersburg, MD). Comparison to values generated from the standard curve can be used to determine the APC concentration in each nucleus relative to the initial concentration in the transport medium. Alternatively, fluorescent images can be directly recorded from the microscope using a video camera, and image-processing software can be used to analyze the data for quantitation.[17] We have performed quantitation in this manner with the ACAS (anchored cell analysis and sorting) 470 interactive laser cytometer (Meridian Instruments, Okemos, MI) (Fig. 1, bottom). Many image analysis systems are available to carry out the simple analysis required here.

It also should be possible to use ^{125}I-labeled reporter proteins and γ counting to quantitate this assay. However, in this case it would be necessary to perform additional controls to determine what percentage of radioactivity becomes associated with permeabilized cells during the import assay due to adsorption to extranuclear components, as opposed to actual nuclear import. Inhibition of nuclear import by wheat germ agglutinin could be useful for this purpose.

Summary and Prospects

The nuclear import system described here is simple, efficient, and can be tailored to the needs of the individual laboratory. A wide range of different cell types can be used as the source of permeabilized cells or cytosol. The system is particularly useful for purifying and characterizing cytosolic factors that have a role in nuclear import,[19] in part because it does not involve the complications of import assays that rely on nuclear assembly. Use of a fluorescent reporter protein to measure nuclear import makes it possible to visualize directly nuclear accumulation of the reporter and to quantitate the concentration it achieves in the nucleus relative to the surrounding medium, as well as to determine variations in nuclear import among a population of different cells.

While this permeabilized cell system has been developed to analyze nuclear protein import, we believe that it also will be applicable to analysis of RNA export from the nucleus. Virally infected cells may provide an especially useful model for such studies, because large amounts of a relatively small number of transcripts are produced after infection of mammalian cells by certain viruses.

In addition to being useful for studying nucleocytoplasmic transport, digitonin-permeabilized cells also should be useful for studying a number of transport phenomena related to the secretory pathway. Because digitonin-permeabilized cells retain a largely intact cytoarchitecture, morphological visualization of the movement of transported proteins between different membrane compartments would be greatly facilitated. Transport of a viral membrane protein from the ER to the Golgi has been successfully visualized by this approach.[27]

[27] W. E. Balch, personal communication.

[12] Transport of Protein between Endoplasmic Reticulum and Golgi Compartments in Semiintact Cells

By R. Schwaninger, H. Plutner, H. W. Davidson, S. Pind, and W. E. Balch

Semiintact (perforated) cells are a population of cells that have lost a portion of their plasma membrane as a result of physical perforation.[1-11]

[1] C. J. M. Beckers, D. S. Keller, and W. E. Balch, Cell **50,** 523 (1987).

Perforated cells can be formed from both yeast spheroplasts[10,11] and mammalian cells.[1,7] Although the soluble cytoplasmic contents are lost, semiintact cells retain intracellular organelles of the secretory pathway including the endoplasmic reticulum (ER) and the Golgi apparatus. These organelles efficiently reconstitute vesicular transport between individual compartments and are accessible to exogenous factors including inhibitors, and enzymes and antibodies, allowing for the identification of components involved in intracellular transport.

The assays described below measure transport of protein *in vitro* from the ER through the cis- and medial-Golgi compartments. They are based on the vesicular stomatitis virus (VSV) G protein, a viral glycoprotein that is transported to the cell surface in infected cells through the same pathway as endogenous plasma membrane proteins. Transport is measured by following the maturation of the two asparagine-linked oligosaccharide chains on VSV G protein during transit through secretory compartments. The cis-Golgi-localized α-1,2-mannosidase I (Mann I) trims the high-mannose ($Man_9GlcNAc_2$) oligosaccharide (the ER form) to the $Man_5GlcNAc_2$ form, which is uniquely sensitive to cleavage of the oligosaccharide chains by the enzyme endoglycosidase D (endo D) (Fig. 1). Thus, appearance of the endo D-sensitive form is a measure of transport of VSV G protein from the ER to the cis-Golgi compartment. The $Man_5GlcNAc_2$ form of VSV G protein is further processed by *N*-acetylglucosamine transferase I (GlcNAc Tr I) and α-1,2-mannosidase II (Mann II) to an endoglycosidase H (endo H)-resistant form ($GlcNAc_1Man_3GlcNAc_2$) in the cis- and medial-Golgi compartments (Fig. 1). Oligosaccharides preceding processing by GlcNAc Tr I and Mann II are sensitive to the enzyme endoglycosidase H (endo H). All subsequent oligosaccharides containing additional GlcNAc, Gal, and sialic to form the complex structures are endo H resistant (Fig. 1). Each oligosaccharide processing intermediate confers on VSV G protein unique electrophoretic

[2] C. J. M. Beckers and W. E. Balch, *J. Cell Biol.* **108**, 1245 (1989).

[3] H. Plutner, R. Schwaninger, S. Pind, and W. E. Balch, *EMBO J.* **9**, 2375 (1990).

[4] R. Schwaninger, C. J. M. Beckers, and W. E. Balch, *J. Biol. Chem.* **266**, 3055 (1991).

[5] H. Plutner, A. D. Cox, S. Pind, R. Khosravi-Far, J. R. Bourne, R. Schwaninger, C. J. Der, and W. E. Balch, *J. Cell Biol.* **115**, 31 (1991).

[6] K. Simons and H. Virta, *EMBO J.* **6**, 2241 (1987).

[7] I. de Curtis and K. Simons, *Proc. Natl. Acad. Sci. USA* **85**, 8052 (1988).

[8] I. de Curtis and K. Simons, *Cell* **58**, 719 (1989).

[9] B. Podbilewicz and I. Mellman, *EMBO J.* **9**, 3477 (1990).

[10] H. Ruohola, A. K. Kabcenell, and S. Ferro-Novick, *J. Cell Biol.* **107**, 1465 (1988).

[11] D. Baker, L. Hicke, M. Rexach, M. Schleyer, and R. Schekman, *Cell* **54**, 335 (1988).

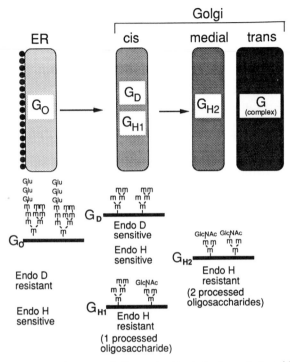

FIG. 1. Cleavage of oligosaccharide chains by endoglycosidases.

mobility during sodium dodecyl sulfate-polyacrylamide gel electrophoresis (SDS-PAGE) to provide a direct and simple quantitative assay for transport.[1,4]

In principle, the assay conditions described below are applicable to any marker protein that traverses the secretory pathway and that contains appropriate posttranslational modifications to quantitate specific transfer between ER and Golgi compartments.

I. Reconstitution of Endoplasmic Reticulum-to-Cis-Golgi Transport: Acquisition of Endo D Sensitivity

Preparation of Vesicular Stomatitis Virus-Infected, Labeled Semiintact Cells

To assay transport between the ER and the cis-Golgi compartment, cells are infected with VSV strain ts045, a temperature-sensitive VSV mutant. The ts045 G protein is retained within the ER at 39.5°, the restrictive temperature. This transport inhibition is reversible at 32°, the permissive temperature. Because the endo D-sensitive $Man_5GlcNAc_2$-containing protein is only a transient intermediate in the exocytic pathway in wild-type cells, transport assays use the clone 15B mutant cell line of Chinese hamster ovary (CHO) cells.[12] These cells lack GlcNAc Tr I, so that all VSV G protein that has reached or progressed beyond the cis-Golgi compartment remains endo D sensitive. The cumulative appearance of the endo D-sensitive form of VSV G protein in perforated 15B cells provides a direct measurement of ER-to-cis-Golgi transport.

The CHO 15B cells are maintained in monolayer culture using 10-cm dishes in α-minimal essential medium (α-MEM: Earle's salts, with glutamine and nucleosides), supplemented with 100 IU/ml penicillin, 100 mg/ml streptomycin, and 8% (v/v) fetal calf serum. The cells are passaged such that at the time of infection they form a complete monolayer while still maintaining a well-spread morphology. On the morning of use cells are infected with virus at the permissive temperature (see procedure 2), and then viral proteins are labeled *in vivo* at the restrictive temperature (see procedure 3). Virus stocks are propagated and stored as described in procedure 1.

Procedure 1: Propagation of Virus

Materials

Vesicular stomatitis virus (VSV)
G-MEM (minimal essential medium, Glasgow), 10% (v/v) tryptose phosphate broth (TPB)
TD buffer: 138 mM NaCl, 5 mM KCl, 25 mM Tris base, 0.4 mM Na_2HPO_4, pH 7.2
Baby hamster kidney (BHK) cells: Grow in monolayer on 10-cm dishes in G-MEM, 5% (v/v) fetal calf serum, 10% (v/v) TPB. For infection of some cell lines, it is necessary for the virus to be adapted to that cell line to achieve a good infection for transport studies. In this case, the cell line of interest is used for viral propagation instead of BHK cells

Methods

1. Wash BHK cells with TD.
2. Add per 10-cm dish: 10 ml G-MEM, 10% TPB, containing approximately 10^6 plaque-forming units (pfu) of VSV stock (multiplicity of infection = 0.1).
3. Incubate 36–48 hr (until cells begin to round up) at 32° (for ts045) or at 37° (for wild-type VSV).
4. Remove virus containing supernatant.

[12] C. Gottlieb, J. Baenziger, and S. Kornfeld, *J. Biol. Chem.* **250**, 3303 (1975).

5. Centrifuge the supernatant at 800 g for 10 min at 4° to remove cells. Generally, the yield of virus should be $1-2 \times 10^9$ pfu/ml.

6. Freeze the cell-free supernatant in 100 μl-aliquots in liquid nitrogen.

7. Store at $-80°$. Virus stocks may be refrozen twice without significant loss of titer.

Procedure 2: Infection of CHO Cells with Vesicular Stomatitis Virus

Materials

Rocking platform (md. T-415-110; Lab Industries, Berkeley, CA) in a 32° CO_2 incubator

Infection medium: α-modified Eagle's medium (α-MEM), 25 mM N-2-hydroxyethylpiperazine-N'-2-ethanesulfonic acid (HEPES)–KOH, pH 7.2

Actinomycin D in ethanol, 1 mg/ml; store at $-20°$

VSV stock solution (approximately 2×10^9 pfu/ml; see procedure 1)

Postinfection medium: α-MEM, 8% (v/v) fetal calf serum

Method

1. Thaw VSV at 32°.

2. Prepare infection cocktail: 1 ml infection medium, prewarmed to 32°, 5 μl actinomycin D, 100 μl of virus stock solution.

3. Remove the medium from a 10-cm tissue culture dish with cells that have just reached confluency.

4. Add the infection cocktail to the dish and rock to distribute evenly.

5. Incubate with constant rocking of the dish for 45 min at 32°.

6. Add 5 ml postinfection medium to the tissue culture dish.

7. Incubate 4.0 to 4.5 hr at 32° (for ts045-infected cells) or at 37° (for wild-type VSV-infected cells).

Procedure 3: Radioactive Labeling of Vesicular Stomatitis Virus-Infected CHO Cells

Materials

Water bath (40°) with a perforated stainless steel platform: The platform should be situated just below the surface of the water. The depth of the water (a few millimeters above the platform) is just sufficient to immerse the base of the tissue culture dish without the dish floating in the water when the lid is removed

Prewarmed (40°) methionine-deficient labeling medium (M 7270; Sigma, St. Louis, MO) supplemented with leucine, lysine, and 25 mM HEPES–KOH, pH 7.2

[^{35}S]Methionine (Translabel; ICN Biomedicals, Costa Mesa, CA)
Methionine, 0.25 M

Method

1. Place the dish in the 40° water bath and aspirate the medium.
2. Wash the cells three times with 3-ml portions of 40° prewarmed labeling medium.
3. Incubate for 5 min with 5 ml of 40° labeling medium.
4. Replace the medium with another 1.5 ml labeling medium and add 100 μCi of [^{35}S]methionine. Rock the plate gently to ensure even coverage of the cells. It is essential that the temperature of the cells does not drop below 39° to ensure that labeled ts045 G protein remains in the ER.
5. Incubate for 10 min at 40°, rocking the dish briefly at about 2-min intervals to prevent the cells from drying out. Maintain contact with the water bath while rocking.
6. Add 30 μl of 0.25 M methionine and incubate for 2 min at 40°.
7. Terminate labeling by rapidly and sequentially aspirating the medium, immediately adding perforation buffer (see below), and transferring cells to an ice-water bath as described in procedure 4 below.

Procedure 4: Preparation of Semiintact Cells by Hypotonic Swelling

The hypotonic swelling method described below results in perforation of 90–95% of wild-type or 15B CHO cells when cells are scraped from the dish using a rubber policeman. The swelling procedure can be adjusted for different cell types by changing the swelling time, composition of the swelling buffer, or the degree of adherence to the tissue culture plate. In the case of strongly adherent cells, the use of hypotonic swelling buffer is unnecessary to achieve efficient perforation during scraping. In contrast, some cell lines are not perforated efficiently even with the hypotonic swelling method due to their weak adherence to tissue culture plates. An alternative approach in this case is to attach the cells more firmly to the culture dish by pretreating with poly(L-lysine) (procedure 5 below).

In addition to scraping, perforation can be performed using nitrocellulose as described in Refs. 6–9. In the latter technique, adherent cells are covered with a sheet of nitrocellulose and then peeled off, tearing pieces of plasma membrane from the surface of cells. This approach is particularly useful for cell lines such as polarized epithelial cells grown on membrane filters, in which it is desired to study transport with cells remaining attached to the filter.[9] Alternatively, perforated cells remaining attached to the dish after nitrocellulose stripping may be removed by gentle scraping with a rubber policeman for assay in suspension.

In all cases, the degree of perforation can be readily quantitated by treatment of cells with the membrane-impermeant chromatin-binding dye, trypan blue.

Materials

Ice–water bath with a stainless steel platform situated just below the surface of the water

Rubber policeman (Cat. No. 36300-0014; Macalaster-Bicknell, Inc.)

Perforation buffer: 50 mM

HEPES–KOH, pH 7.2, 90 mM potassium acetate

Swelling buffer: 10 mM HEPES–KOH, pH 7.2, 18 mM potassium acetate

Trypan blue (1%, v/v) in water

Method

1. Wash the cells three times with cold perforation buffer.
2. Overlay the cells with 5 ml swelling buffer.
3. Let stand for 10 min on ice.
4. Aspirate the swelling buffer.
5. Add 3 ml cold perforation buffer.
6. Scrape the cells immediately with the rubber policeman, using rapid, firm strokes.
7. Transfer the cells with a Pasteur pipette into a 15-ml polypropylene centrifuge tube.
8. Pellet for 3 min at 800 g at 4°.
9. Wash the pellet with 3 ml perforation buffer.
10. Pellet for 3 min at 800 g at 4°.
11. Resuspend the pellet in 200–300 μl perforation buffer.
12. Check perforation index by mixing 10 μl cell suspension and 1 μl of the 1% trypan blue on a cover slip. Using a ×20 objective (phase contrast), trypan blue-stained nuclei (blue in color) identify the perforated cells.

Procedure 5: Preparation of Poly(L-lysine)-Coated Plates

Materials

Poly(L-lysine) stock solution (5 mg/ml, filter sterilized; store at 4°

Poly(L-lysine), 5 μg/ml H$_2$O

Sterile TD buffer (see procedure 1)

Method

1. Wash the culture dish with 10 ml sterile H_2O.
2. Add 2.5 ml poly(L-lysine) (5 μg/ml) to the culture dish.
3. Incubate for 60 min at 32°.
4. Wash the dish three times with sterile TD buffer.
5. Plate the cells in regular medium on a pretreated dish.
6. Incubate for 1 to 2 days, until the cells reach confluency.
7. Perforate the cells using isotonic or hypotonic conditions as need dictates.

Procedure 6: In Vitro Transport from Endoplasmic Reticulum to Cis-Golgi Compartment: Quantitation of Appearance of Endo D-Sensitive Form

Freshly prepared semiintact cells can be stored on ice for 4 to 6 hr, although the transport efficiency is best if the cells are used within 2 hr. Freezing of perforated cells results in inactivation of transport.

Transport is reconstituted by incubation of semiintact cells in the presence of cytosol (a 100,000 g supernatant of CHO cell homogenate; see procedure 6A) as described below (procedures 6b and 6c).

Procedure 6a: Preparation of a Cytosol Fraction (100,000 g Supernatant)

Materials

Ball-bearing homogenizer (see Ref. 13) or glass Dounce
Forty 15-cm dishes with confluent CHO cells (wild type or 15B)
TEA buffer: 125 mM potassium acetate, 10 mM triethanolamine (TEA)-HCl, pH 7.2)
25/125 HEPES/KOAc (25 mM HEPES–KOH, pH 7.2, 125 mM potassium acetate)
Sephadex G-25 column equilibrated with 25/125 HEPES/KOAc

Method

1. Scrape confluent cells gently with a rubber policeman in 2–3 ml of growth medium.
2. Pellet the cells 5 min at 800 g at 4°.
3. Resuspend the cells in 15 ml TEA buffer.
4. Pellet the cells for 5 min at 800 g at 4°.
5. Determine the volume of the cell pellet and add 4 vol of 25/125 HEPES/KOAc.

6. Homogenize the cells with the ball-bearing homogenizer using sufficient strokes to give 80–90% breakage. Homogenization should not result in extensive fragmentation of nuclei.
7. Centrifuge for 60 min at 100,000 g at 4°.
8. Remove the white lipid layer at the top of the tube by aspiration.
9. Remove the supernatant (cytosol), taking care not to disturb the pellet.
10. To desalt [to remove low ($M_r < 3000$) components], load the high-speed supernatant on a Sephadex G-25 column (five-fold bed volume over cytosol volume) equilibrated with 25/125 HEPES/ KOAc.
11. Collect the excluded (void) fraction and pool.
12. Concentrate using ultrafiltration to at least 10 mg/ml.
13. Freeze the cytosol in 100-μl aliquots in liquid nitrogen and store at $-70°$. Cytosol is thawed immediately before use by a brief incubation at 32° and placed on ice. Generally, unused cytosol is discarded at the end of each experiment as it is unstable to refreezing.

Procedure 6b: In Vitro Reconstitution of Endoplasmic Reticulum-to-Cis-Golgi Transport in Semiintact Cells

Materials

Semiintact cells
Cytosol
ATP-regenerating system (20× stock solution): 20 mM ATP (sodium form), 100 mM creatine phosphate (CP), 100 U creatine phosphate kinase (CPK)/ml, store at $-70°$
100 mM magnesium acetate pH 7
Potassium acetate, 1 M
HEPES–KOH (1 M), pH 7.2
Stock Ca^{2+}/EGTA buffer (10X): 50 mM EGTA, 18 mM CaCl$_2$, 20 mM HEPES–KOH, pH 7.2 (final pH adjusted to pH 7.2 with KOH)

Method

1. The transport cocktail (Table I) is prepared in 1.5-ml microcentrifuge tubes on ice [we recommend Sarstedt (No. 72.690, Princeton, NJ) as some brands contain residual chemicals or surfactants that inhibit transport]. Mix in order: H$_2$O, salts, ATP, and cytosol. Semiintact cells should always be added last, immediately prior to incubation. It is essential to avoid exposing the semiintact cells to excessive variation in ionic strength and salt. When inhibitors or other factors (i.e., antibodies or purified proteins) prepared in buffers different from those described are added to an incubation, nonspecific effects of the buffer need to be tested.

TABLE I
TRANSPORT COCKTAIL

Solution	Volume (μl)	Final concentration
Semiintact cells	5 μl	~25 μg protein (11 mM potassium acetate, 6 mM HEPES)
Cytosol	2.5–5 μl	25–50 μg protein (16–32 mM potassium acetate, 3–6 mM HEPES)
ATP-regenerating system	2 μl	1 mM ATP, 5 mM CP, 0.2 IU CPK
100 mM Magnesium acetate	1 μl	2.5 mM Magnesium acetate
1 M Potassium acetate	2 μl	50 mM Potassium acetate
1 M HEPES–KOH, pH 7.2	1 μl	25 mM HEPES
10 × Ca^{2+}/EGTA buffer	4 μl	5 mM EGTA, 1.8 mM Ca^{2+} (100 nM free Ca^{2+})
H$_2$O	To 40 (final)	

2. Vortex the transport cocktail gently for 3 sec before incubation to ensure mixing.

3. Incubate the cocktail for 90 min at 32° in the case of VSV ts045-infected cells or at 37° for wild-type VSV-infected cells.

4. Terminate by transfer to ice.

Procedure 6c: Detection and Quantitation of Transport

To distinguish between the Man$_9$GlcNAc$_2$ ER form and Man$_5$GlcNAc$_2$ cis-Golgi form, the cells are pelleted, lysed, and digested with endo D. The endo D-digested glycoprotein (Fig. 2, lower band) has a faster mobility on SDS-PAGE than the undigested Man$_9$GlcNAc$_2$ form (Fig. 2, upper band). The fraction converted to the band with faster electrophoretic mobility can be quantitated using densitometry. Endo D can be prepared from *Diplococcus pneumoniae* as described in Ref. 14, or may be purchased from Boehringer Mannheim (Indianapolis, IN; Cat. No. 752 991). Because endo D is sensitive to low concentrations of SDS and may be used only in the presence of nonionic detergents, many proteins contain Man$_5$GlcNAc$_2$ oligosaccharides, which are insensitive to endo D in the native structure. The absence of apparent Man$_5$GlcNAc-containing oligosaccharides in other glycoproteins based on endo D resistance needs to be confirmed using more direct oligosaccharide analyses. In wild-type cells the

[13] W. E. Balch and J. E. Rothman, *Arch. Biochem. Biophys.* **240**, 413 (1985).
[14] L. R. Glasgow, J. C. Paulson, and R. L. Hill, *J. Biol. Chem.* **252**, 8615 (1977).

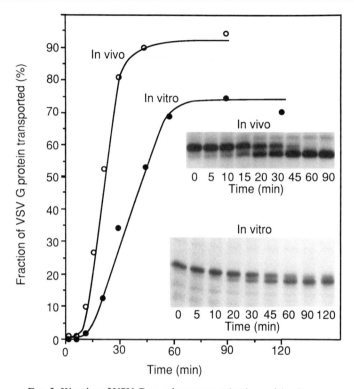

FIG. 2. Kinetics of VSV G protein transport *in vivo* and *in vitro*.

Man$_5$GlcNAc$_2$ form of VSV G and other glycoproteins is a transient intermediate; thus endo D is generally applicable only to studies conducted with 15B cells and the equivalent lec 1 mutant.

Materials

Endo D buffer: 50 mM phosphate buffer, pH 6.5, 5 mM ethylenedia-minetetraacetic acid (EDTA), 0.2% (v/v) Triton X-100
Endo D

Method

1. Pellet cells by centrifugation for 20 sec in a microcentrifuge at 15,000 g at 4°.
2. Resuspend the pellet in 40 μl endo D buffer and 5 mU of endo D.
3. Incubate overnight at 37°.
4. The fraction of VSV G protein that has been transported to the cis-Golgi is determined using SDS-PAGE, fluorography, and densitometry.

Samples are analyzed using 6.75% (w/v) SDS-polyacrylamide gels according to the method of Laemmli.[15] The composition of the separating gel is critical and is prepared by mixing 21 ml of H_2O, 9 ml of acrylamide [30% (w/v) acrylamide, 0.8% (w/v) bis acrylamide], 10 ml of a 4× separating gel buffer [containing 1.5 M Tris-HCl (pH 8.8) and 0.4% (w/v) SDS], 200 μl 10% (w/v) ammonium persulfate, and 20 μl N,N,N',N'-tetramethylethylenediamine (TEMED). Samples are normally separated using 15 × 20 cm plates (20 0.5-cm gel lanes/plate) at 35-mA constant current for 1 hr, followed by 45-mA constant current for 5 hr (or until tracking dye reaches 0.5 cm from the bottom of the gel).

5. Acrylamide gels are treated with a fluorographic enhancement solution [125 mM salicylic acid (sodium salt, pH 7.0), 30% methanol] for 20 min, dried, and exposed to Kodak (Rochester, NY) XAR-5 film at −80°. Generally, a 12- to 14-hr exposure is sufficient. The fraction of VSV G protein transported (the G_D form) is determined by densitometry of the autoradiogram. In a standard reaction, transport is generally 50–80% efficient. The rate and extent of transport and cytosol dependence can vary between different preparations of semiintact cells.

II. Successive Transport from Endoplasmic Reticulum to the Cis- and Medial-Golgi Compartments

Successive transport of VSV G protein between the ER and the cis- and medial-Golgi compartments can be measured in wild-type semiintact cells through the appearance of two endoglycosidase H (endo H)-resistant forms that can be detected using SDS-gel electrophoresis. These two forms represent VSV G protein with one (G_{H1}) or both (G_{H2}) oligosaccharides processed to the endo H-resistant form as a consequence of transport to the cis- and medial-Golgi compartments, respectively[4] (Fig. 1). Note that some strains of VSV have spontaneously deleted one of the oligosaccharide attachment sites. It is therefore important in these studies to verify that a strain harboring both oligosaccharides is utilized.

For these analyses we use wild-type VSV G protein, which is more efficiently transported to the medial-Golgi compartment than is ts045 G protein. To distinguish between the two endo H-sensitive forms, the cells are lysed and digested with endo H. Endo H, like endo D, cleaves sensitive oligosaccharide side chains from the protein backbone. The ER form of VSV G protein with no endo H-resistant side chains has the fastest mobility on SDS-PAGE (G_{H0}; Fig. 3, upper), the cis-Golgi form contains one

[15] U. K. Laemmli, *Nature (London)* **227**, 680 (1970).

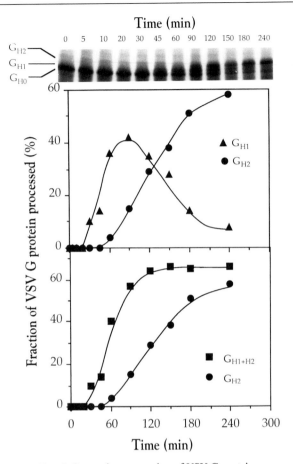

FIG. 3. Successive conversion of VSV G protein.

endo H-resistant and one endo H-sensitive oligosaccharide side chain (G_{H1}; Fig. 3, upper), and the medial-Golgi form with both oligosaccharide side chains processed to endo H resistance has the slowest mobility (G_{H2}; Fig. 3, upper). Figure 3 (lower) illustrates the successive conversion of VSV G protein from the G_{H0} (endo H-sensitive) form at $t = 0$ min to the intermediate G_{H1} form (after a lag period of ~ 20 min), followed by the appearance of the G_{H2} form. The G_{H2} form appears after a lag period of ~ 45 min and reaches a maximum after $90-120$ min of incubation.

The rates and efficiency of VSV G protein conversion to the endo H-resistant structures vary between cell lines and cytosols used to promote transport *in vitro*.

Procedure 8: Transport and Postincubation with Endo H

Wild-type VSV-infected cells are labeled for 3 min at $37°$ with 200 μCi [^{35}S]methionine as described in procedure 3. Because wild-type VSV G protein rapidly exits the ER, appearance of the G_{H1} form is less sensitive to reagents that detect early acting components required for export of ts045 from the ER.[4] The transport assay using wild-type semiintact cells is the same as described in procedure 6, with the following changes.

1. To achieve GlcNAc addition to the oligosaccharide, uridine diphosphate-GlcNAc (UDPGlcNAc) is added to the transport cocktail. A stock solution of 10 mM UDPGlcNAc in water is stored in 50-ml aliquots at $-80°$. Two microliters of this stock solution is added to a 40-μl assay to make a final concentration of 0.5 mM UDPGlcNAc. The stability of UDPGlcNAc in the incubation may be increased by the addition of 1 mM UTP.

2. To enhance the appearance of the G_{H1}, the total potassium acetate in the buffer is reduced to 35 mM. Cells are incubated for 90–120 min and transport terminated by transfer to ice. Appearance of the endo H-resistant oligosaccharide forms of VSV G are determined as described below.

Materials

Endo H buffer: 0.3% (w/v) SDS, 0.1 M sodium acetate, pH 5.6, 20 μl/ml 2-mercaptoethanol (added just prior to use)
Sodium acetate (0.1 M), pH 5.6
Endo H (Cat. No. 1088 734; Boehringer Mannheim)

Method

1. Pellet cells by centrifugation for 15 sec in a microcentrifuge at 15,000 g at $4°$.
2. Resuspend the pellet in 20 μl concentrated endo H buffer.
3. Boil immediately (to prevent proteolysis of denatured G protein) for 5 min.
4. Cool the samples.
5. Add 40 μl of 0.1 M NaOAc (pH 5.6) and 3 mU endo H.
6. Mix by vortexing.
7. Incubate overnight at $37°$.
8. The percentage of G protein in each of the bands (G_{H0}, G_{H1}, G_{H2}) is determined by densitometry. The fraction of VSV G protein that has been transported from the ER to the cis-Golgi is expressed as the total G_{H1} plus G_{H2} observed at each time point. Transport to the medial-Golgi compartment is expressed as the fraction found in the G_{H2} form.

Comments

Preparation of cells that actively transport VSV G protein has been most thoroughly characterized using CHO cells and CHO cytosol. However, there is considerable variability in transport using different cell lines or conditions. Where indicated in the procedures, measurement of transport using different marker proteins, or the use of different cell lines or cytosol preparations, will need to be optimized to obtain maximal efficiencies of transport. Extensive fragmentation and lysis of the ER (particularly in the case of strongly adherent cells) generally result in reduced transport. In particular, perforation conditions can lead to the release of soluble marker proteins from the ER during preparation of semiintact cells. In contrast, poor perforation leads to transport that is efficient, but cytosol independent. In the latter case, cytosol dependence can be enhanced by gently homogenizing the semiintact cells with a loose-fitting pestle in a glass Dounce (10–20 strokes) prior to washing.

[13] Reconstitution of Transport from Endoplasmic Reticulum to Golgi Complex Using Endoplasmic Reticulum-Enriched Membrane Fraction from Yeast

By LINDA J. WUESTEHUBE and RANDY W. SCHEKMAN

Protein transport from the endoplasmic reticulum (ER) to the Golgi apparatus may be monitored *in vitro* with gently lysed semiintact cells[1–3] or with crude homogenates fractionated by differential centrifugation.[3,4] A complete resolution of this reaction will require the purification of donor ER and acceptor Golgi membranes. Toward this end we have developed an *in vitro* assay that utilizes an ER-enriched membrane fraction to reconstitute ER-to-Golgi transport. This new assay allows a detailed analysis of the specific contributions of both membrane and soluble components to the transport process not possible in semiintact cells and crude homogenates.

[1] C. J. M. Beckers, D. S. Keller, and W. E. Balch, *Cell* **50,** 523 (1987).
[2] D. Baker, L. Hicke, M. Rexach, M. Schleyer, and R. Schekman, *Cell* **54,** 335 (1988).
[3] H. Ruohola, A. K. Kabcenell, and S. Ferro-Novick, *J. Cell Biol.* **107,** 1465 (1988).
[4] W. E. Balch, K. R. Wagner, and D. S. Keller, *J. Cell Biol.* **104,** 749 (1987).

Preparation of Subcellular Fractions

Three subcellular fractions are required to reconstitute ER-to-Golgi transport: sucrose-purified membranes, cytosol, and a high-speed pellet fraction. The protein concentration of all these fractions is determined in the presence of 1% (w/v) sodium dodecyl sulfate (SDS) by the procedure of Lowry et al.[5] Fractions are divided into small aliquots, fast-frozen in liquid nitrogen, and stored at $-80°$. No significant loss in transport activity is detected in subcellular fractions stored for as long as 1 year.

Sucrose Density Gradient-Purified Membrane Preparation

Sucrose density gradient-purified membranes are prepared from osmotically lysed yeast spheroplasts. The method of lysis is important in the preparation of transport-competent membranes. We found the yield and specific activity of membranes prepared from osmotically lysed spheroplasts to be substantially higher than membranes prepared from glass bead-lysed cells.

Formation of Yeast Spheroplasts. Wild-type yeast cultures are grown at $30°$ in YP medium [2% (w/v) Bacto-peptone, 1% (w/v) yeast extract] containing 2% (w/v) glucose to a density of $2-4$ OD_{600} U/ml (1 OD_{600} U = 1 × 10^7 cells). We recommend using protease-deficient strains such as RSY445 (*MATα ura3-52 trpl leu2 his4 pep4::URA3*) or RSY607 (*MATα, leu2-3,112 ura3-52 pep4::URA3*). Cells are harvested by centrifugation (1000 g, 5 min, 24°), resuspended to 100 OD_{600} U/ml in 10 mM dithiothreitol (DTT), 100 mM Tris-HCl, pH 9.4, and incubated for at least 10 min at 24° with gentle agitation. Cells are collected by centrifugation (1000 g, 5 min, 24°) and resuspended to 100 OD_{600} U/ml in 0.7 M sorbitol, 1.5% (w/v) Bacto-peptone, 0.75% (w/v) yeast extract, 0.5% (w/v) glucose, 10 mM Tris-HCl, pH 7.4. Lyticase[6] (fraction II, ~60,000 U/ml, $10-20$ U/OD_{600} U cells) is added and the cell suspension is incubated at $30°$ until the OD_{600} of a 1:200 dilution in distilled H_2O is less than 5% of the initial value. The effectiveness of the lyticase treatment is an important parameter in determining the degree of lysis achieved. Therefore, the OD_{600} of a 1:200 dilution in distilled H_2O is monitored periodically during the incubation and additional lyticase is added if necessary. The total time of incubation with lyticase is usually no more than 1 hr. Lyticase-treated cells (spheroplasts) are collected by centrifugation in an HB-4 (Sorvall

[5] O. H. Lowry, N. J. Rosebrough, A. L. Farr, and R. J. Randall, *J. Biol. Chem.* **193**, 265 (1951).
[6] J. Scott and R. Schekman, *J. Bacteriol.* **142**, 414 (1980).

DuPont, Wilmington, DE) swinging bucket rotor (10,500 g, 10 min, 4°) through a cushion of 0.8 M sucrose, 1.5% (v/v) Ficoll 400, 20 mM N-2-hydroxyethylpiperazine-N'-2-ethanesulfonic acid (HEPES), pH 7.4 (10–20 ml spheroplast suspension/15-ml cushion in a 40-ml polycarbonate tube). Alternatively, for large-scale preparations spheroplasts may be collected by centrifugation in a GSA fixed-angle rotor (1000 g, 5 min, 4°) and washed once or twice in 0.7 M sorbitol, 20 mM HEPES, pH 7.4. Spheroplasts may either be lysed directly or resuspended to 1000 OD_{600} U/ml in 0.7 M sorbitol, 20 mM HEPES, pH 7.4 and slowly frozen at −80° in insulated containers for long-term storage prior to lysis.

Osmotic Lysis and Homogenization. Spheroplasts are resuspended to 100 OD_{600} U/ml in lysis buffer [0.1 M sorbitol, 20 mM HEPES, pH 7.4, 50 mM potassium acetate, 2 mM ethylenediaminetetraacetic acid (EDTA), 1 mM DTT, 1 mM phenylmethylsulfonyl fluoride, 0.7 μg/ml pepstatin, 0.5 μg/ml leupeptin] at 4° and homogenized with a motor-driven Potter–Elvehjem homogenizer (4–10 strokes). Care should be taken to ensure that the lysate is kept at temperatures ≤ 4° at all times. The degree of lysis may be assessed by visualization of the lysate under a phase-contrast microscope. Intact spheroplasts, with characteristic bright halos resulting from the difference in refractive index between the spheroplasts and the surrounding buffer, are easily distinguished from the fragments of lysed spheroplasts. Greater than 95% lysis usually is obtained with this method.

Differential Centrifugation and Sucrose Density Gradient Fractionation. The lysate is centrifuged (1,000 g, 10 min, 4°) and the pellet and supernatant fractions are collected separately. Increased yields of membranes may be achieved by resuspending the low-speed pellet in lysis buffer to at least one-half the original volume of the lysate and repeating the Potter-Elvehjem homogenization and low speed centrifugation. The supernatant from the rehomogenized low-speed pellet then is combined with the previous supernatant fraction. Up to 30% additional membrane protein may be obtained by rehomogenization of the low-speed pellet. Rehomogenizations of subsequent low-speed pellets usually are not worthwhile.

The membrane pellet is collected from the combined low-speed supernatants by centrifugation (27,000 g, 10 min, 4°) and resuspended to 5000 OD_{600} U cell equivalent/ml in lysis buffer with gentle Dounce (Wheaton, Millville, NJ) homogenization (five strokes). Up to 0.5 ml of the membrane suspension is overlayed onto a 2.0-ml sucrose density step gradient (1.0 ml each of 1.5 M sucrose and 1.2 M sucrose in lysis buffer) and centrifuged in the SW50.1 swinging bucket rotor (100,000 g, 1 hr, 4°). The membranes at the 1.2/1.5 M interface are collected and washed twice in reaction buffer (20 mM HEPES, pH 6.8, 150 mM potassium acetate,

5 mM magnesium acetate, 250 mM sorbitol) by centrifugation (27,000 g, 10 min, 4°). Sucrose-purified membranes are resuspended to approximately 5 mg/ml in reaction buffer with gentle Dounce homogenization. Approximately 1.5 mg sucrose-purified membranes per 1000 OD_{600} U cells is obtained with this procedure.

Cytosol Preparation

The cytosol fraction is prepared using glass bead breakage of whole cells. Wild-type yeast cultures are grown at 30° in YP medium containing 2% (w/v) glucose to late log phase (10–18 OD_{600} U/ml). Cells are harvested by centrifugation (3000 g, 5 min, 4°), washed twice in reaction buffer, and resuspended to 2000 OD_{600} U/ml in reaction buffer containing 1 mM DTT, 1 mM ATP, 100 μM GTP, 0.5 mM phenylmethylsulfonyl fluoride. Approximately 1 g acid-washed glass beads (425–600 μm) per 1000 OD_{600} U of cells is added and the cells are lysed in 10 cycles by 30-sec periods of agitation on a VWR Vortexer 2 (San Francisco, CA) at full speed, spaced by 1-min intervals on ice. For large preparations ($\geq$40,000 OD_{600} U cells), a 100-ml bead beater chamber (Biospecs Products, Bartlesville, OK) may be used. The extent of lysis is monitored by phase-contrast microscopy as described above. Lysis efficiency is generally 50–60% using this procedure. The lysate is clarified by centrifugation (27,000 g, 10 min, 4°) and the supernatant fraction further clarified by centrifugation in a Beckman (Palo Alto, CA) TLA 100.3 rotor (300,000 g, 1 hr, 4°). The high-speed supernatant (cytosol) is collected, carefully avoiding the pellet and the lipid layer on top of the centrifuge tube. If desired, small molecules may be removed by filtration on a Sephadex G-25 column equilibrated in reaction buffer. Approximately 15 mg cytosol (7–10 mg protein/ml) is obtained per 1000 OD_{600} U cells.

High-Speed Pellet Preparation

The high-speed pellet (HSP) fraction may be prepared either from osmotically lysed spheroplasts or from glass bead-lysed cells. The specific transport activity of HSP fractions prepared from osmotically lysed spheroplasts may be somewhat higher than HSP fractions prepared from bead-lysed cells, but the yield is comparable regardless of the method of lysis used. Because the HSP and membrane fractions have different sedimentation properties, it is convenient to prepare these fractions from the same preparation of osmotically lysed spheroplasts. It also is possible to prepare the HSP fraction from the glass bead-lysed cells used in the preparation of cytosol; however, this method precludes treatment of the lysate with

RNase, which we have found to be a critical step in obtaining high specific activity HSP fractions.

Yeast cells are either glass bead lysed (as described for cytosol preparation) or spheroplasts are osmotically lysed and homogenized (as described for membrane preparation) in lysis buffer at a density of $100-300$ OD_{600} U/ml. The lysate is clarified by centrifugation (27,000 g, 10 min, 4°) and the supernatant is collected. RNase A (from a 10 mg/ml stock solution) is added to a final concentration of 100 μg/ml and the lysate is incubated with gentle agitation either at 30° for 40 min or on ice for 1 hr. In large preparations of $\geq$ 4000 OD_{600} U cells, a visible white precipitate will form during this incubation. The RNase-treated lysate is clarified by centrifugation (27,000 g, 10 min, 4°) and the supernatant is collected. If the supernatant is cloudy, then the RNase treatment and centrifugation is repeated. The HSP fraction is collected by centrifugation in a Beckman Ti45 rotor (100,000 g, 1 hr, 4°) and then is washed into reaction buffer by centrifugation in a Beckman TLA 100.3 rotor (300,000 g, 20 min, 4°). To collect the HSP fraction, the supernatant is removed by aspiration and the clear pellet is scraped from the bottom of the centrifuge tube with a metal spatula and transferred to a Dounce homogenizer on ice. Residual amounts of HSP fraction remaining in the bottom of the tube are recovered more easily if allowed to incubate on ice in a small volume of reaction buffer for $5-10$ min prior to resuspension. The HSP fraction is resuspended to approximately 20 mg/ml in reaction buffer with gentle Dounce homogenization. Approximately 1.5 mg HSP fraction/1000 OD_{600} U cells are obtained. It should be noted that the transport activity of the HSP fraction is resistant to several cycles of freezing and thawing but is not stable at 4°. Therefore, it is best to work quickly when preparing this fraction and to freeze it as soon as possible.

Enriched Endoplasmic Reticulum Transport Assay

Design of Endoplasmic Reticulum to Golgi Transport Assay

Transport of prepro-α-factor, precursor of the secreted mating pheromone α-factor, has been reconstituted $in\ vitro$ using ER-enriched membranes from yeast (Fig. 1). We chose prepro-α-factor as a substrate because it is translocated efficiently postranslationally into yeast membranes[7-9] and its transport $in\ vivo$[10-12] and $in\ vitro$[2,3] has been well characterized. The

[7] J. Rothblatt and D. Meyer, $EMBO\ J.$ **5,** 1543 (1986).
[8] M. G. Waters and G. Blobel, $J.\ Cell\ Biol.$ **102,** 1543 (1986).
[9] W. Hansen, P. D. Garcia, and P. Walter, $Cell$ **45,** 397 (1986).
[10] D. Julius, R. Schekman, and J. Thorner, $Cell$ **36,** 309 (1984).

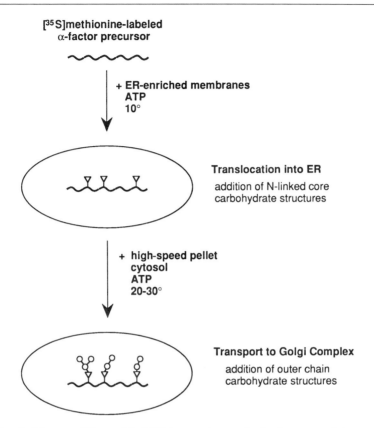

FIG. 1. Diagram of the enriched ER transport assay. In the first stage of the reaction, α-factor precursor labeled with [³⁵S]methionine is translocated into the endoplasmic reticulum, where it acquires N-linked core carbohydrate structures (▽). In the second stage, ER-enriched membranes containing core-glycosylated α-factor precursor are incubated with a high-speed pellet (HSP) fraction, cytosol, and energy at 20–30°. Arrival of the α-factor precursor in the Golgi complex is marked by the addition of outer chain carbohydrate structures (○). Transport efficiency is measured as the fraction of core-glycosylated α-factor precursor that receives outer chain modifications.

in vitro assay requires sucrose-purified membranes as a source of translocation- and transport-competent ER, cytosol, and a high-speed pellet fraction prepared as described above. The reaction in addition requires *in vitro* translated [³⁵S]methionine-labeled α-factor precursor, GDP-

¹¹ R. S. Fuller, R. E. Sterne, and J. Thorner, *Annu. Rev. Physiol.* **50,** 345 (1988).
¹² A. J. Brake, R. S. Fuller, and J. Thorner, *in* "Neuropeptides and Their Receptors" (T. W. Schwartz, L. M. Hilstead, and J. F. Rehfeld, eds.), p. 197. Munksgaard, Copenhagen, 1990.

mannose, and an ATP-regenerating system. Preparation of *in vitro* translated [^{35}S]methionine-labeled α-factor precursor is described in detail by Baker and Schekman.[13]

Transport is determined *in vitro* by measuring the amount of [^{35}S]methionine-labeled core-glycosylated α-factor precursor originating in the ER that has received outer chain carbohydrate in the Golgi apparatus. On translocation into the ER, the α-factor precursor acquires N-linked core carbohydrate and can be precipitated with concanavalin A, a lectin that recognizes mannose-containing oligosaccharides. α-Factor precursor that is transported to the Golgi apparatus acquires outer chain carbohydrate[14,15] and is identified by precipitation with anti-(α1→6)-mannose serum. This antiserum was produced in response to *mnn1 mnn2* yeast strains[16] as described by Ballou.[17]

Reagents

Sucrose-purified membranes: 2–20 μg protein (~ 3–5 mg/ml)
Cytosol: 100–120 μg protein (~ 10–15 mg/ml)
High-speed pellet: 20–40 μg protein (~ 20 mg/ml)
[^{35}S]Methionine-labeled prepro-α-factor: ~ 100,000 cpm/μl
ATP mixture (10×): 10 m*M* ATP (equine muscle disodium salt; Sigma, St. Louis, MO), 400 m*M* creatine phosphate (disodium salt; Boehringer Mannheim, Indianapolis, IN), 2 mg/ml creatine phosphokinase (type 1 from rabbit muscle; Sigma) in reaction buffer
GDP-mannose (10 m*M*) (type 1 sodium salt from yeast; Sigma) in reaction buffer
Reaction buffer: 20 m*M* HEPES, pH 6.8, 150 m*M* potassium acetate, 5 m*M* magnesium acetate, 250 mM sorbitol
Concanavalian A-Sepharose (Sigma)
Antiserum specific for (α1→6)-linked mannose
Protein A-Sepharose CL-4B (Pharmacia, Piscataway, NJ)

Translocation (First Stage)

Sucrose-purified membranes (2–20 μg) are mixed with 3–6 μl [^{35}S]methionine-labeled prepro-α-factor, 1.2 μl 10× ATP mixture, and

[13] D. Baker and R. Schekman, *Methods Cell Biol.* **31**, 127 (1989).
[14] B. Esmon, P. Novick, and R. Schekman, *Cell* **25**, 451 (1981).
[15] C. E. Ballou, *in* "The Molecular Biology of the Yeast *Saccharomyces*" (J. N. Strathern, E. Jones, and J. Broach, eds.), pp. 335–360. Cold Spring Harbor Lab., Cold Spring Harbor, New York, 1982.
[16] W. C. Rashke, K. A. Kern, C. Antalis, and C. E. Ballou, *J. Biol. Chem.* **248**, 4660 (1973).
[17] C. Ballou, *J. Biol. Chem.* **245**, 1197 (1970).

reaction buffer in a final volume of 12 μl. The reaction mixture is incubated at 10° for 30 min and then is diluted at least 1 : 5 in reaction buffer. Membranes are collected by centrifugation (27,000 g, 10 min, 4°), washed once, and resuspended to 1 mg/ml in reaction buffer. Up to eighteen 12-μl reactions usually are combined in a single 1.5-ml microfuge tube for this first-stage reaction and then aliquoted into separate tubes for the second-stage reaction.

Transport (Second Stage)

Membranes (2–20 μg) containing translocated [^{35}S]methionine-labeled pro-α-factor from the first-stage reaction are mixed with cytosol (100–120 μg), HSP (20–40 μg), 5 μl 10× ATP mixture, 5 μl 10 mM GDP-mannose, and reaction buffer in a final volume of 50 μl. After incubation for 2 hr at 20°, the reaction is terminated by addition of 50 μl 2% (w/v) SDS and heated for 5 min at 95°. The amount of cytosol and HSP added should be titrated for each preparation to obtain optimal transport efficiency. As a control for background, 5 μl 100 U/ml apyrase (grade VII; Sigma) may be added in place of 10× ATP mixture.

Concanavalin A Precipitation and Immunoprecipitation

The total amount of core-glycosylated [^{35}S]methionine-labeled pro-α-factor is quantified by precipitation with concanavalin A-Sepharose. The terminated reaction mixture (up to 100 μl), 30 μl 20% (v/v) concanavalin A-Sepharose, and 1 ml 500 mM NaCl, 1% (v/v) Triton X-100, 20 mM Tris-HCl, pH 7.5 are combined in a 1.5-ml microfuge tube and incubated with continuous rotation for 2 hr at room temperature or overnight at 4°. The precipitates are collected by centrifugation (12,000 g, 1 min, 20°) and washed once with 1 ml of 150 mM NaCl, 1% (v/v) Triton X-100, 0.1% (w/v) SDS, 15 mM Tris-HCl, pH 7.5, twice with 1 ml of 2 M urea, 200 mM NaCl, 1% (v/v) Triton X-100, 100 mM Tris-HCl, pH 7.5, once with 1 ml of 500 mM NaCl, 1% (v/v) Triton X-100, 20 mM Tris-HCl, pH 7.5, and once with 1 ml of 50 mM NaCl, 10 mM Tris-HCl, pH 7.5.

The amount of core-glycosylated [^{35}S]methionine-labeled pro-α-factor that has received outer chain modifications is quantified by precipitation with antiserum specific for (α1→6)-linked mannose and protein A-Sepharose. The amount of antiserum added to precipitate all of the outer chain-modified pro-α-factor in a reaction should be titrated for each batch of antiserum. Generally, 5 μl 20% (v/v) protein A-Sepharose is added per 1 μl of antiserum. The terminated reaction mixture (up to 100 μl), an optimal amount (X μl) of antiserum specific for (α1→6)-linked mannose, 5X μl 20% (v/v) protein A-Sepharose, and 1 ml 150 mM NaCl, 1% (v/v)

Triton X-100, 15 mM Tris-HCl, pH 7.5, are combined in a 1.5-ml micro-fuge tube. All incubations and washes are carried out as described above. The washed precipitates are analyzed by scintillation counting. Samples are resuspended in 150 μl 2% (w/v) SDS, heated at 95° for 5 min, and then transferred to scintillation vials. Five milliliters of scintillant (Universol; ICN Biomedicals, Costa Mesa, CA) is added and samples are counted in a scintillation counter.

Determination of Transport Efficiency

The efficiency of transport is expressed as the percentage of core-glycosylated [35S]methionine-labeled pro-α-factor that has received outer chain modifications. Transport efficiency is determined for each reaction by the following equation: counts per minute (cpm) precipitated with antiserum specific for (α1→6)-linked mannose divided by cpm precipitated with concanavalin A-Sepharose times 100.

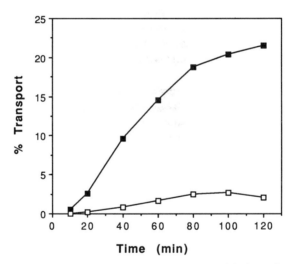

FIG. 2. Time course of ER–Golgi transport *in vitro*. ER-enriched membranes containing core-glycosylated 35S-labeled α-factor precursor were combined with high-speed pellet (HSP), cytosol, and either an ATP-regenerating system (■) or 10 U/ml apyrase (□) on ice. The reaction mixtures were incubated at 20° and aliquots were terminated at the indicated times. Reaction products were precipitated with concanavalin A-Sepharose (core-glycosylated α-factor precursor) or with anti-(α1→6)-mannose serum and protein A-Sepharose (outer chain glycosylated α-factor precursor, indicative of arrival in the Golgi complex) and processed for scintillation counting. Percentage transport equals anti-(α1→6)-mannose-precipitable cpm divided by concanavalin A-precipitable cpm times 100.

Features of Enriched Endoplasmic Reticulum Transport Assay

The enriched ER transport assay shares many basic features with previously developed *in vitro* assays that use semiintact yeast cells prepared by freeze-thaw[2] and osmotic[3] lysis. Both semiintact and enriched ER assays utilize [35S]methionine-labeled prepro-α-factor as a substrate to measure transport from the ER to the Golgi apparatus (Fig. 1). The *in vitro* assays have properties expected of intercompartmental protein transport: transport requires energy, is stimulated by addition of cytosolic protein, and is time and temperature dependent (Figs. 2 and 3).

Unlike the *in vitro* assay with freeze-thaw lysed semiintact cells, the enriched ER transport assay shows a marked dependence on the addition of a HSP fraction for transport activity. Sucrose density gradient purification of membranes appears to be a critical step in obtaining a substantial HSP dependence in this assay. Stimulation of the transport reaction by the HSP fraction is increased from ~2- to 3-fold to > 20-fold after density gradient purification of membranes (Fig. 4).

The HSP and cytosol fractions provide different limiting factors because these fractions act synergistically rather than additively in transport (Fig. 3). Proteins are required because HSP or cytosol fractions treated

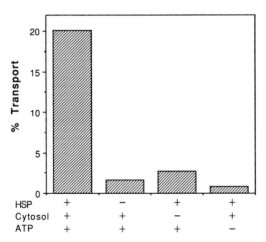

FIG. 3. The high-speed pellet (HSP) and cytosol fractions act synergistically to promote ER–Golgi transport. Endoplasmic reticulum-enriched membranes containing core-glycosylated 35S-labeled α-factor precursor were prepared and aliquoted to tubes containing HSP, cytosol, and either an ATP-regenerating system of 10 U/ml apyrase as indicated. Samples were incubated for 2 hr at 20°. Reaction products were precipitated and analyzed as described in the legend to Fig. 2.

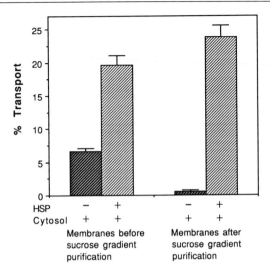

FIG. 4. Sucrose density gradient-purified membranes show an increased dependence on the high-speed pellet (HSP) fraction for ER–Golgi transport *in vitro*. ^{35}S-Labeled α-factor precursor was translocated into ER membranes isolated by differential centrifugation and/or sedimentation on sucrose density gradients as described in the text. Transport efficiency was measured after incubation of these membranes with cytosol and an ATP-regenerating system in the presence or absence of HSP as indicated.

with heat or with trypsin fail to stimulate transport while control treatments with trypsin inhibitor have little effect (L. Wuestehube, unpublished observations, 1989). The factors associated with the HSP are unique to this subcellular fraction. Addition of low-speed pellet (1000 *g*, 10 min, 4°) or medium-speed pellet (10,000 *g*, 10 min, 4°) fractions does not replace the HSP in the transport assay (Fig. 5).

Applications and Future Directions

The development of an enriched ER *in vitro* assay represents a methodological advance in defining the molecular requirements for ER-to-Golgi transport. We have used this assay to demonstrate a direct role in transport for Ypt1p,[18] a small GTP-binding protein that is required for transport of secretory proteins from the ER to the Golgi apparatus *in vivo*.[19,20] Antibody fragments directed against Ypt1p inhibit transport and

[18] D. Baker, L. J. Wuestehube, R. Schekman, D. Botstein, and N. Segev, *Proc. Natl. Acad. Sci. USA* **87**, 355 (1990).

[19] N. Segev, J. Mulholland, and D. Botstein, *Cell* **52**, 915 (1988).

[20] H. D. Schmitt, M. Puzicha, and D. Gallwitz, *Cell* **53**, 635 (1988).

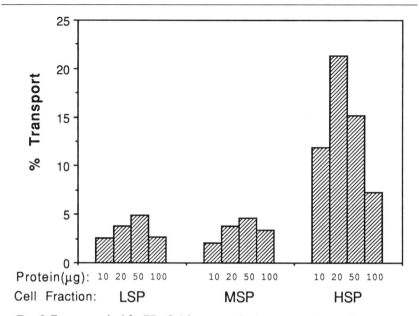

FIG. 5. Factors required for ER–Golgi transport *in vitro* are associated uniquely with the HSP fraction. Endoplasmic reticulum-enriched membranes containing core-glycosylated ^{35}S-labeled α-factor precursor were prepared and aliquoted to tubes containing cytosol, ATP-regenerating system, and increasing amounts of cell fractions as indicated. Cell fractions were isolated from homogenized yeast spheroplasts by sequential sedimentation at 4° at 1000 *g* for 10 min (LSP), 10,000 *g* for 10 min (MSP), and 100,000 *g* for 1 hr (HSP). Transport efficiency was measured after incubation at 20° for 2 hr.

subcellular fractions prepared from the *ypt1-1* mutant strain are defective in the enriched ER transport assay. Further analysis has shown that Ypt1p acts after the formation of ER-derived transport vesicles.[21,22] The function of other gene products could similarly be determined by inhibiting transport with specific antibodies or by reproducing a conditional defect *in vitro* for other temperature-sensitive mutant strains in this assay.

The enriched ER transport assay offers more flexibility than assays using semiintact cells in that membrane fractions from different sources can readily be mixed. We have taken advantage of this to demonstrate the biochemical interaction of Sec12p and Sar1p, two proteins that are required for transport *in vivo*[23,24] and whose genes display genetic interactions.[24] Transport *in vitro* is inhibited by the addition of membranes

[21] M. Rexach and R. Schekman, *J. Cell Biol.* **114,** 219 (1991).
[22] N. Segev, *Science* **252,** 1553 (1991).
[23] P. Novick, C. Field, and R. Schekman, *Cell* **21,** 205 (1980).
[24] A. Nakano and M. Muramatsu, *J. Cell Biol.* **109,** 2677 (1989).

containing elevated levels of Sec12p, and is restored by the addition of purified Sar1p.[25] Independent studies in which a temperature-sensitive defect in the *sec12-4* mutant is restored by the addition of bacterially expressed Sar1p[26] confirm these results. The enriched ER transport assay thus authentically reproduces *in vivo* protein transport and can be used to detect specific, physiologically relevant biochemical interactions.

Finally, it is possible to dissect the enriched ER transport assay into specific subreactions. The formation of ER-derived transport vesicles from sucrose-purified membranes can be determined by measuring [^{35}S]methionine-labeled core-glycosylated α-factor precursor released into a protease-protected compartment that sediments after centrifugation at 100,000 *g* at 4° for 1 hr.[27] It is therefore possible to distinguish the specific biochemical requirements for vesicle formation from the set of reactions that compose transport from the ER to the Golgi apparatus.

The development of the enriched ER transport assay represents an advance toward the reconstitution of ER-to-Golgi transport in a fully resolved system. This assay provides an efficient means to identify membrane and soluble components essential for intracellular membrane transport. The use of this assay in analyzing the biochemical requirements of transport *in vitro* will aid in defining the molecular details of transport *in vivo*.

Acknowledgments

We would like to thank Dr. Jürg Sommer for helpful discussions throughout the course of this work. We are grateful to Dr. Linda Silveira for advice on RNase treatment of yeast cell lysates and to Kristina Whitfield for critically reading the manuscript. This work was supported by grants from the National Institutes of Health (National Institute of General Medical Sciences) and the Howard Hughes Medical Institute. L. Wuestehube was supported by Fellowship DRG-991 of the Damon Runyon-Walter Winchell Cancer Fund and by a fellowship from the American Cancer Society, California Division.

[25] C. d'Enfert, L. J. Wuestehube, T. Lila, and R. Schekman, *J. Cell Biol.* **114**, 663 (1991).
[26] T. Oka, S. Nishikawa, and A. Nakano, *J. Cell Biol.* **114**, 671 (1991).
[27] L. J. Wuestehube and R. Schekman, *J. Cell Biol.* **111**, 325a (1990).

[14] Reconstitution of Endoplasmic Reticulum to Golgi Transport in Yeast: *In Vitro* Assay to Characterize Secretory Mutants and Functional Transport Vesicles

By Mary E. Groesch, Guendalina Rossi, and Susan Ferro-Novick

We describe here an *in vitro* assay that reconstitutes vesicular transport between the endoplasmic reticulum (ER) and Golgi complex in the yeast *Saccharomyces cerevisiae*.[1] In this assay, we follow the processing of a precursor of the secreted pheromone α-factor. Prepro-α-factor is synthesized as a 19-kDa precursor molecule that contains four tandem repeats of the mature protein.[2] In the lumen of the ER, the signal sequence is cleaved and three N-linked core oligosaccharides are added to the protein, forming a 26-kDa species. After transport to the Golgi complex, outer chain carbohydrate is added to this proprotein, yielding a high molecular weight species that migrates as a heterogeneous smear on sodium dodecyl sulfate (SDS) polyacrylamide gels.[3-5] Pro-α-factor eventually is processed to a 13-amino acid peptide in a late compartment of the Golgi complex and is secreted.[5,6]

The *in vitro* transport assay is composed of four elements: a donor compartment, acceptor membranes, soluble transport factors, and an ATP-regenerating system. The donor compartment is provided by permeabilized yeast cells (PYCs) that have radiolabeled pro-α-factor translocated into the lumen of the ER (26 kDa). When the donor PYCs are incubated with a yeast lysate (S3 fraction; supernatant of a 3000 g spin) in the presence of ATP, the 26-kDa form of pro-α-factor is converted to the high molecular weight species. This species is enclosed within sealed membranes and contains outer chain carbohydrate. The S3 fraction, which is added exogenously to the PYCs, supplies both soluble factors and the acceptor Golgi membranes that are necessary for transport. When this fraction is subfractionated into a high-speed supernatant (HSS) and pellet (HSP), the HSS supplies the necessary soluble transport factors and the HSP provides the functional acceptor compartment. Thus, the donor, acceptor, and soluble components required for transport can be separated in this assay.[1] The ability to do this has enabled us to determine the

[1] H. Ruohola, A. K. Kabcenell, and S. Ferro-Novick, *J. Cell Biol.* **107**, 1456 (1988).

[2] J. Kurjan and I. Herskowitz, *Cell* **30**, 933 (1982).

[3] D. Julius, L. Blair, A. Brake, G. Sprague, and J. Thorner, *Cell* **32**, 839 (1983).

[4] D. Julius, A. Brake, L. Blair, R. Kunisawa, and J. Thorner, *Cell* **37**, 1075 (1984).

[5] D. Julius, R. Schekman, and J. Thorner, *Cell* **36**, 309 (1984).

[6] R. S. Fuller, R. E. Sterne, and J. Thorner, *Annu. Rev. Physiol.* **50**, 345 (1988).

defective fraction in mutants blocked in transport from the ER through the Golgi complex.[7,8]

We have separated the *in vitro* transport assay into two stages: vesicle formation and vesicle utilization.[9] Vesicles are formed and released in the first stage of the reaction when the donor PYCs are incubated with soluble factors and ATP. The donor PYCs are separated from the released vesicles by a brief centrifugation step. The vesicles can then fuse with the acceptor Golgi membranes that are added in the second stage of the reaction. In this assay, the functional acceptor compartment is supplied only by the exogenously added crude membrane fraction, and not by Golgi membranes retained within the PYCs.[1,9] Our studies have shown that when vesicles containing the 26-kDa species are incubated with acceptor membranes, cytosol, and an ATP-regenerating system, pro-α-factor is processed to a 28-kDa species before it is converted to the high molecular weight reaction product. We have proposed that 28-kDa pro-α-factor is formed in an early subcompartment of the Golgi complex prior to its conversion to the high molecular weight species.[7,9] Our efforts have been directed toward identifying protein components of the transport vesicles. We found that the *BOS1* gene product, which is required for ER-to-Golgi transport in yeast, comigrates with the transport vesicles on sucrose gradients. In addition, Bos1p is released from the donor PYCs under the same conditions in which the pro-α-factor containing vesicles are released, but a resident ER protein is retained.[10] Thus, release is specific. These findings imply that Bos1p is a constituent of the transport vesicles that mediate ER-to-Golgi transport.

Preparation of Reagents for *in Vitro* Transport Assay

Preparation of Yeast Spheroplasts

The spheroplast protocol described below is used for the preparation of the PYCs and the fractions for the *in vitro* transport assay.

Media and Solutions

YP medium: 10 g/liter yeast extract, 20 g/liter peptone; the solution is autoclaved and stored at room temperature

---

[7] R. Bacon, A. Salminen, H. Ruohola, P. Novick, and S. Ferro-Novick, *J. Cell Biol.* **109,** (1989).
[8] G. Rossi, Y. Jiang, A. P. Newman, and S. Ferro-Novick, *Nature (London)* **351,** 158 (1991).
[9] M. E. Groesch, H. Ruohola, R. Bacon, and S. Ferro-Novick, *J. Cell Biol.* **111,** 45 (1990).
[10] A. Newman, M. E. Groesch, and S. Ferro-Novick, *EMBO J.* in press (1992).

YPD medium: YP medium with 20 g/liter of glucose; the glucose is added from a 40% (w/v) autoclaved stock solution

Low glucose medium: $1 \times$ YP medium with 1 g/liter glucose

Spheroplasting medium: $1 \times$ YP medium, 1 g/liter glucose, 1.4 M sorbitol, 50 mM KP$_i$, pH 7.5, 50 mM 2-mercaptoethanol; the solution is titered to pH 7.5 by the addition of KOH, and approximately 30 min before use, zymolyase 100T (*Arthrobacter luteus;* ICN ImmunoBiologicals, Costa Mesa, CA) is added to a final concentration of 0.67 U/OD$_{599}$ units of cells (1 mg/50 ml medium). The zymolyase is dissolved during a 30-min incubation at room temperature and the medium is cleared of particulate material by centrifugation at 2000 g (7 min at 4°)

Recovery medium: $1 \times$ YP medium, 1 g/liter glucose, 1 M sorbitol

General Comments. Either 50-ml conical tubes or 250-ml conical bottles are expedient for harvesting cells. For convenience, we generally prepare spheroplasts in multiples of 75 OD$_{599}$ units of cells. This quantity of cells requires 50 ml each of low glucose and recovery media and 25 ml of spheroplast medium.

Procedure. Cells are grown overnight in YPD medium to a density of 2–4 OD$_{599}$ units/ml ($\sim 10^7$ cells/OD$_{599}$ unit). The cells are harvested by centrifugation at 2000 g for 7 min at room temperature, resuspended in low glucose medium, and incubated for 30 min at 25° with vigorous shaking (~ 200 rpm). The cells are pelleted and resuspended in spheroplast medium. This is followed by an incubation for 30 min at 37° with slow swirling (~ 90 rpm). To begin partial regeneration of the cell wall, the spheroplasts are resuspended gently in recovery medium and incubated for 90 min at 37° with mild agitation (~ 90 rpm). Metabolic activity resumes during this step.

Before pelleting the cells, it is useful to apportion the cell suspension according to its future use. As a guide, 75 OD$_{599}$ units of cells are used for preparing the PYCs. This will eventually provide enough material for 50 *in vitro* transport reactions. If the cells are used to generate cell lysate fractions, aliquots of 50 or 200 ml (75 or 300 OD$_{599}$ units) are convenient. Spheroplasts prepared from 75 OD$_{599}$ units of cells yield approximately 150 μl of yeast lysate. The cells can be stored on ice overnight as a packed pellet, or used immediately. The packed recovered spheroplasts retain their translocation activity for approximately 24 hr on ice.

Preparation of Permeabilized Yeast Cells

The donor PYCs used in our transport assay are prepared from regenerated yeast spheroplasts that have been lysed in the presence of osmotic support. The PYCs retain their overall structure in terms of organellar contents, but they are depleted of cytoplasm.

Materials and Solutions

Regenerated yeast spheroplasts

Spheroplast lysis buffer: 100 mM potassium acetate, 200 mM sorbitol, 20 mM 4-(2-hydroxyethyl)-1-piperazineethanesulfonic acid–KOH (HEPES–KOH), pH 7.2, 2 mM MgCl$_2$ or magnesium diacetate; can be stored at 4°

PYC buffer: 20 mM HEPES–KOH, pH 7.4, 250 mM sucrose, 2 mM dithiothreitol (DTT), 1 mM ethylene glycol-bis(β-aminoethyl ether)-N,N,N',N'-tetraacetic acid (EGTA)

Procedure. If the regenerated spheroplasts are stored overnight on ice, the residual supernatant is removed by aspiration. This is an important step, because the concentration of sorbitol is higher in the recovery medium than in the permeabilization buffer, and the cells will not lyse well unless the recovery medium is removed efficiently. The pellet of regenerated spheroplasts is resuspended in spheroplast lysis buffer, 5 ml/75 OD$_{599}$ unit cell equivalents. The slurry is pipetted up and down with moderate force (five times with a 5-ml pipette). This is generally sufficient to resuspend the pellet. The PYCs are collected by centrifugation at 3000 g for 5 min at 4°. After thoroughly removing the supernatant, the pellet is mixed with PYC buffer (50 μl/75 OD$_{599}$ unit cell equivalents). The cells are gently resuspended into a thick slurry that is used in the translocation step described below.

Morphological and Biochemical Analysis of Permeabilized Yeast Cells

Thin sections of spheroplasts and PYCs can be prepared for a morphological assessment of the permeabilization step. When viewed by electron microscopy, both preparations of cells retain their overall structure. The plasma membrane appears to be largely intact, and large organelles such as the nucleus and vacuole are readily apparent. At high magnification, ribosome-studded membranes are visible. However, when compared to a spheroplast, the nuclear envelope of a permeabilized yeast cell appears dilated and the cytosolic contents are depleted.[1]

Materials and Reagents

Spheroplasts
Permeabilized spheroplasts (PYCs)
PYC fixation buffer: 1% (w/v) paraformaldehyde, 0.8% (w/v) glutaralde-

hyde, 100 mM potassium acetate, 200 mM sorbitol, 2 mM MgCl$_2$, 150 mM cacodylate buffer, pH 7.2
Cacodylate buffer (1.5 M), pH 7.2
Spheroplast fixation buffer: 150 mM cacodylate buffer, pH 7.2 that is 1% (w/v) paraformaldehyde, 0.8% (w/v) glutaraldehyde, 100 mM potassium acetate, 1 M sorbitol, 2 mM MgCl$_2$
Postfixation buffer: 1% (w/v) osmium tetroxide, 100 mM cacodylate buffer, pH 7.2
NaCl, 150 mM
Uranyl acetate in 0.5% (w/v) NaCl
2% (w/v) Bacto-agar (Difco, Detroit, MI)
Spurr medium (Polysciences, Inc.)

Procedure. Spheroplasts and PYCs are incubated in their respective fixation buffers for 3.5 hr, washed three times with cacodylate buffer, and then incubated in postfix buffer for 1 hr at 4°. After three washes with 150 mM NaCl, the samples are stained with uranyl acetate for 2 hr at 4°. The stained cells are then washed with 150 mM NaCl and resuspended in Bacto-agar and formed into blocks. The agar blocks are dehydrated with ethanol and embedded in Spurr medium. Thin sections are stained with uranyl acetate and lead citrate and viewed in a Phillips 301 electron microscope at 80 kV.

Biochemical analysis of the spheroplasts and PYCs supports morphological observations. Antibodies to carboxypeptidase Y, a vacuolar enzyme, and hexokinase B, a cytoplasmic enzyme, have been used to determine the retention of these markers before and after permeabilization. The majority of the cytoplasmic marker is released from the permeabilized cells, while most of the vacuolar marker is retained.[1] A similar experiment using antibodies to Kex2, an enzyme of the Golgi complex, has revealed that the PYCs contain Golgi membranes.[9] These and other studies,[9] together with the observation that functional acceptor membranes must be supplied exogenously to the cells in the assay,[1] indicate that while the donor PYCs contain Golgi membranes, they are not competent as acceptor membranes *in vitro.*

Preparation of Transport Fractions: S3, High-Speed Supernatant, and High-Speed Pellet

The S3 fraction is the supernatant of a yeast lysate that has been spun at 3000 g. It contains both the cytoplasmic components and the functional acceptor membranes necessary for transport in the *in vitro* assay. This fraction can be subfractionated into a high-speed supernatant (HSS) and pellet (HSP) that contain, respectively, the soluble transport factors and acceptor membranes.[1]

Media and Reagents

HEPES–KOH (20 mM), pH 7.4; store at 4°

Yeast spheroplasts; store on ice as pellet, aspirate any supernatant before use

TB (transport buffer), 10×: 250 mM HEPES–KOH, pH 7.2, 1.15 M potassium acetate, 25 mM MgCl$_2$; store at 4°

Sorbitol (4 M); store at room temperature

Protease inhibitor cocktail (1000× PIC): 1 mg/ml each leupeptin, chymostatin, pepstatin, antipain, and aprotinin; store in small aliquots at −80°

Procedure. Regenerated spheroplasts are lysed in 20 mM HEPES–KOH, pH 7.4 (210 μl/75 OD$_{599}$ unit cell equivalents) and the slurry is homogenized 30 times with a 1-ml micropipettor. To preserve the activity of the S3 fraction, the cells are kept on ice and gently homogenized so that no bubbles form. The lysate is vortexed on a low setting (20 short bursts) and then spun in a 3-ml conical glass tube. A maximum of 300 OD$_{599}$ unit cell equivalents/tube is centrifuged at 3000 g for 5 min at 4°. The supernatant of this spin is the S3 fraction. The HSS and HSP fractions are prepared by centrifuging the S3 fraction at 120,000 g for 1 hr at 4°. The HSP is resuspended in a volume of 20 mM HEPES–KOH, pH 7.4 that is equivalent to the supernatant. Just before the S3, HSS, and HSP are used in the assay, transport buffer and PIC are added to 1× concentration, and sorbitol is added to a final concentration of 200 mM. The HSS can be prepared (with transport buffer and PIC) and stored for several months at −80° with no loss of activity.

Plasmid DNA and Preparation of Prepro-α-factor mRNA

In our assay, the marker protein prepro-α-factor is transcribed and translated *in vitro*. Prepro-α-factor is transcribed from pDJ100, a plasmid that contains the α-factor gene cloned behind the *SP6* promoter.[11] To do this, the plasmid is linearized with *Xba*I (Boehringer Mannheim, Indianapolis, IN) and then transcribed with *SP6* RNA polymerase using methods described elsewhere.[11-14] The RNA prepared from transcribing 30 μg

[11] W. Hansen, P. D. Garcia, and P. Walter, *Cell* **45**, 397 (1986).
[12] M. G. Waters and G. Blobel, *J. Cell Biol.* **102**, 1543 (1986).
[13] J. A. Rothblatt and D. I. Meyer, *Eur. Mol. Biol. Organ. J.* **5**, 1031 (1986).
[14] J. Sambrook, E. F. Fritsch, and T. Maniatis, "Molecular Cloning: A Laboratory Manual," 2nd Ed. Cold Spring Harbor Lab., Cold Spring Harbor, New York, 1989.

of pDJ100 is resuspended in 100 μl of diethyl pyrocarbonate (DEPC)-treated water (described below) and used in the in vitro translations described below.

Preparation of Yeast Lysate Used for Translations

Media and Reagents. All RNase-free solutions are prepared with DEPC-treated water and stored in new sterile plastic or baked glass containers; they are dispensed with baked glass pipettes or sterile micropipette tips.

Spheroplast medium: As described for the preparation of PYCs, except 15 mg of zymolyase 100T is added to 120 ml of solution

Recovery medium: 1 × YP medium, 10 g/liter glucose, 1.4 M sorbitol

Protease inhibitor cocktail (1000 × PIC, RNase free): 1 mg/ml each leupeptin, chymostatin, pepstatin, antipain, and aprotinin; store in small aliquots at −80°

Lysis buffer: 20 mM HEPES–KOH, pH 7.4, 100 mM potassium acetate, 2 mM DTT, 1 × protease inhibitor cocktail, 2 mM magnesium acetate, 0.5 mM phenylmethylsulfonyl fluoride (PMSF); add PIC and PMSF immediately before use, store on ice

DEPC-treated water: Diethyl pyrocarbonate in water, 1 ml/liter; heat the solution until steaming to disseminate the DEPC and then autoclave to destroy the DEPC

Gel-filtration buffer (RNase free): 20 mM HEPES–KOH, pH 7.4, 100 mM potassium acetate, 20% (v/v) glycerol, 2 mM magnesium diacetate, 2 mM DTT

Micrococcal nuclease from *Staphylococcus aureus* (Boehringer Mannheim), 15,000 U/mg: Prepare a stock solution of 30,000 U/ml in DEPC-treated water and store in small aliquots in RNase-free tubes at −80°

CaCl$_2$ (1 M, RNase free); store at room temperature

200 mM EGTA (200 mM, RNase free) adjusted with KOH to pH 8.0; store at room temperature

Dounce (7-ml) tissue grinder fitted with an "A" pestle (Wheaton Scientific, Millville, NJ), chilled on ice

Sephadex G-25 gel-filtration column: 4-ml bed volume; swell the beads in DEPC-treated water in a baked glass beaker, then equilibrate in column buffer and pour into a column that has been rinsed extensively with DEPC-treated water; attach new tubing to the column and collect fractions into new borosilicate glass tubes.

Procedure. Cells are grown overnight in YPD medium to a density of 4–4.5 OD$_{599}$ units/ml. Approximately 6000–8000 OD$_{599}$ units of cells is collected by centrifugation in 500-ml bottles at 2000 g for 10 min at room temperature. The cells are resuspended in 120 ml of spheroplast medium and incubated at 37° for ~40 min with gentle swirling. The efficiency of spheroplasting is monitored by measuring the A_{599} of cells in water. As the cell wall is removed, the spheroplasts lyse in water and the turbidity decreases. Spheroplasting is complete when there is a 95% reduction in the A_{599} value. The spheroplasts are harvested in a 250-ml conical bottle during a spin at ~3500 g (5 min) at 4°. The supernatant should be cloudy. The pellet is resuspended gently into 120 ml of recovery medium and incubated at 37° with slow swirling. The total recovery period, which includes the time required to resuspend the pellet after spheroplasting (~3 min), is 20 min.

It is important to stress that, in our hands, the length of the recovery period is the most critical factor for obtaining a concentrated, active lysate. With a longer recovery period (60 min) the cells do not lyse well, and the translation activity of the lysate is low. With no recovery period, the cells lyse well, and the lysate is concentrated, but there is no translation activity. We have not tested recovery periods between 20 and 60 min.

The regenerated spheroplasts are pelleted and resuspended in 7.5–9 ml of cold lysis buffer. This thick slurry is transferred to a chilled 7-ml Dounce tissue grinder and homogenized with 40 strokes using an "A" pestle. The lysate is transferred to two 50Ti centrifuge tubes and spun at 20,000 rpm (~26,500 g_{av}) for 15 min in a Beckman (Palo Alto, CA) 50Ti rotor at 4°. The supernatant is yellow and the pellet contains two or three distinct layers of membranes and unlysed cells. If there is a third layer (intact cells) at the bottom of the pellet, one can use fewer cells and/or improve the lysis by adjusting the recovery period and/or the volume of lysis buffer. The supernatants are pooled into a single 50Ti tube and centrifuged for an additional 30 min at 40,000 rpm (~106,000 g_{av}) in the same rotor. The resultant supernatant is yellow and is sometimes viscous. Occasionally, there is a visible band in the supernatant that is collected, while the loose pellet is discarded. The A_{260} of a 1:100 dilution of the supernatant is typically ~1.2–2.0. This extract is loaded onto a 40-ml Sephadex G-25 column that has been equilibrated in gel-filtration buffer and fractions of 25 drops (~1.2 ml) are collected. The excluded volume fractions with peak A_{260} absorbances are pooled (about five to seven fractions). Their A_{260} values are typically 0.9–1.2 when read at a 1:100 dilution, although absorbances have ranged from 0.75 to 1.7. The pooled fractions (~6–9 ml) are transferred to a new sterile plastic tube and PIC is added to a 1 × concentration along with $CaCl_2$ (1 mM) and micrococcal nuclease (150 U/ml). This mixture is incubated for 20 min at 20° and the digestion is terminated by the addition of EGTA (4 mM final concentration). The translation lysate is frozen in RNase-free microfuge tubes in 100-μl aliquots and stored at $-80°$. The lysate can be frozen and thawed at least two or three times without loss of activity.

Protein Translation and Translocation Assays

Translation of Prepro-α-factor

Materials and Reagents

All of the stock solutions used in the *in vitro* translation reaction are prepared with DEPC-treated water and stored in small aliquots at $-80°$ unless indicated otherwise.

DEPC-treated water; store at room temperature
RNasin ribonuclease inhibitor, 40,000 U/ml (Promega Corporation, Madison, WI)
PIC, 100×: 1:10 dilution of 1000 × PIC (see Preparation of Yeast Lysate Used for Translations, above) into DEPC-treated water
L-[^{35}S]Methionine (in aqueous solution, stabilized with 2-mercaptoethanol and pyridine 3,4-dicarboxylic acid; Amersham, Arlington Heights, IL), 15 mCi/ml
Creatine phosphokinase (CPK), 5 mg/ml
Yeast tRNA, 5 mg/ml
Compensation buffer, 10×: 120 mM HEPES–KOH, pH 7.4, 1.1 M potassium acetate, pH 7.4, 22 mM magnesium diacetate, 10 mM DTT, 752 mM sucrose
Amino acid mix (minus methionine), 1 mM, pH 7.4; store in small aliquots at $-20°$
ATP/GTP mix: 100 mM ATP, 10 mM GTP; titer to pH 7.4
Energy mix, 10×: 0.3 mM amino acids (minus methionine), $\frac{1}{10}$ vol ATP/GTP mix, 10 mM DTT, 200 mM creatine phosphate
Translation lysate
Prepro-α-factor RNA

Procedure. A translation cocktail is prepared for the appropriate number of reactions. The recipe that follows is for a single translation reaction: 2.65 μl DEPC-treated water, 0.1 μl RNasin, 0.25 μl 100 × PIC, 3 μl [^{35}S]methionine (45 μCi), 1 μl CPK, 1 μl yeast tRNA, 2.5 μl 10 × compensation buffer, 2.5 μl 10 × energy mix, 10 μl translation lysate, 2 μl RNA (appropriately diluted in DEPC-treated water and heated at 65° for 1 min). The cocktail is gently mixed in a sterile microfuge tube and incubated at 20° for 60 min. A 20° water bath can be made using a Styrofoam box or ice bucket filled with water that has been cooled to 20° by the addition of ice. The bath holds its temperature for several hours at room temperature.

Translocation of Prepro-α-factor into Permeabilized Yeast Cells

Materials and Reagents. After prepro-α-factor has been translated, RNase-free materials and methods are no longer necessary.

Creatine phosphate (CP), 1.5 M; store in small aliquots at $-80°$

Translated prepro-α-factor in translation cocktail; can be stored for short periods at $-80°$

PYC slurry in PYC buffer (see Preparation of Permeabilized Yeast Cells, above); permeabilize the cells just before use

TB (transport buffer), $10\times$: See Preparation of Transport Fractions, above)

TBPS (transport buffer + PIC + sorbitol), $1\times$: $1 \times$ TB, $1 \times$ PIC, 280 mM sorbitol

Procedure. A single translocation reaction consists of the following: 25 μl translated prepro-α-factor in translation mix, 0.34 μl CP, 3 μl PYC slurry (60 μg protein). The PYCs are mixed in with a micropipettor, and the solution is incubated at $20°$ for 30 min. At the end of the incubation, the cells are pelleted by centrifugation for 20 sec in a microfuge. The supernatant is aspirated and the PYCs are washed once with $1 \times$ TBPS. The wash and final resuspension volume is 94% of the volume of the translocation mix (cocktail + CP + PYCs), and resuspension is more efficient if about one-fourth the wash is used to resuspend the pellet initially and then the remainder is added. The cells are pelleted as above and resuspended in the same manner. The washed PYCs, with radiolabeled prepro-α-factor, are now referred to as donor PYCs. A 25-μl aliquot of the donor PYCs is the amount used in one *in vitro* transport assay. This corresponds to approximately 1.5 OD_{599} unit cell equivalents.

A small aliquot of the translocation mix can be removed following the incubation at $20°$ (before the cells are pelleted and washed) and prepared for SDS polyacrylamide gel electrophoresis (PAGE). After autoradiography, several radiolabeled species with apparent molecular masses between 19 and 26 kDa are seen. The 19-kDa species is soluble prepro-α-factor and is sensitive to digestion by trypsin. The larger species are resistant to trypsin digestion except in the presence of detergent. These species are forms of pro-α-factor that has been translocated into the lumen of the ER retained within the PYCs. Comparing the relative ratios of the trypsin-sensitive to trypsin-resistant species allows one to assess the efficiency of the translocation reaction.

In Vitro Transport Assay

Materials and Reagents

Donor PYCs in transport buffer
S3 fraction
HSS and HSP fractions
ATP-regeneration mix: 1 mM ATP, pH 7.4, 0.1 mM GTP, pH 7.4, 20 mM creatine phosphate, 0.2 mg/ml creatine phosphate kinase; mix just before use
50 mM GDP-mannose; store in small aliquots at $-80°$
Trypsin (10 mg/ml); store at $-20°$
Trypsin inhibitor (20 mg/ml); store at $-20°$
Concanavalin A (Con A)-Sepharose, diluted to 20% (v/v) in high salt wash
High salt wash: 20 mM Tris, pH 7.5, 1% (v/v) Triton X-100, 500 mM NaCl
Low salt wash: 10 mM Tris, pH 7.5, 50 mM NaCl
Urea wash: 100 mM Tris, pH 7.5, 200 mM NaCl, 2 M urea, 1% (v/v) Triton X-100
SDS wash: 15 mM Tris, pH 7.5, 150 mM NaCl, 1% (v/v) Triton X-100, 0.1% (w/v) SDS
2-Mercaptoethanol (BME) 1% (v/v) in water
Phosphate-buffered saline (PBS), 10×: 125 mM sodium phosphate, pH 7.5, 2 M NaCl
PBS–Triton–BSA: 1 × PBS that contains 1% (v/v) Triton X-100, and 1 mg/ml bovine serum albumin (BSA)
Protein A-Sepharose (Sigma Chemical Co. St. Louis, MO): Prepare a stock solution of 29 mg beads/ml PBS–Triton–BSA; swell for 30 min at room temperature followed by one wash and resuspension in PBS–Triton–BSA; we generally use 30 μl of this suspension per microliter of antibody
PBS–Triton wash: 1 × PBS containing 2% (v/v) Triton X-100

General Comments. The transport assay can be performed in one or two steps. The one-step assay involves the use of the S3 fraction as the source of soluble factors and functional acceptor Golgi membranes. In the two-step assay, vesicles are first formed and released from the donor PYCs, and then allowed to fuse with the acceptor membranes in a second incubation.

One-Step Assay Procedure. A single transport reaction contains 25 μl of donor PYCs (60 μg of protein), $\sim$25–50 μl S3 lysate (0.7–1.0 mg of protein), 1.5 μl GDP-mannose (0.92 mM), and 5.1 μl of the ATP-regeneration mix. The transport reaction is incubated at 20° for 60–90 min and then terminated by cooling on ice. The donor cells are pelleted in a microfuge (15-sec spin) and the resulting supernatant is removed and added to a second microfuge tube. After resuspending the pellet to the starting reaction volume in 1 × TBPS, both the supernatant and pellet fractions are treated with trypsin (0.47 mg/ml for 20 min on ice). Trypsin inhibitor is then added to a final concentration of 1.88 mg/ml (5 min on ice), and the samples are heated to 100° for 3 min in 1% (w/v) SDS. Because pro-α-factor is glycosylated, its release from the PYCs can be quantitated by precipitation with Con A-Sepharose. The digestion with trypsin ensures that only the pro-α-factor contained within intact membranes will be measured.[9] To quantitate pro-α-factor release, the boiled supernatant and pellet fractions are diluted with 1 ml of high salt wash and then spun for 15 min in a microfuge at 4°. The supernatant is incubated for 2 hr, at room temperature with shaking, in the presence of 90 μl of Con A-Sepharose [20% (v/v) slurry in high salt wash]. The beads are washed with 1 ml each of SDS wash, urea wash, high salt wash, and low salt wash and then boiled in 70 μl of sample buffer. Aliquots from the samples are counted in a scintillation counter and examined by SDS-PAGE.

The percentage release of pro-α-factor from the donor PYCs is calculated by dividing the Con A-precipitable counts released from the cells by the total Con A-precipitable counts (supernatant + pellet). The donor cells retain the 26-kDa species of pro-α-factor and lower molecular mass forms of pro-α-factor that contain one or two N-linked core oligosaccharide units. In addition to these forms of pro-α-factor, the supernatant contains a higher molecular mass species that migrates as a heterogeneous smear on SDS-PAGE. The decreased mobility of the pro-α-factor is due to the addition of outer chain mannose units that are added to yeast glycoproteins only in the Golgi complex.

The Golgi-specific modification of outer chain addition can be quantitated by immunoprecipitation with antibodies directed against outer chain carbohydrate. Aliquots (20 μl) of the samples treated with Con A-Sepharose (described above) are diluted 10-fold in PBS–Triton and incubated (2 hr at room temperature) in the presence of anti-outer chain antibody. The samples are treated subsequently with protein A-Sepharose (90 min at 4°) and washed twice with urea wash (1 ml each) followed by two washes (1 ml each) with 1% (v/v) BME. Pro-α-factor is solubilized from the beads by boiling in sample buffer. The ratio of outer chain-modified pro-α-factor to the total Con A-precipitable counts gives a measure of transport through the Golgi complex.

Two-Step Transport Assay Procedure. The transport assay can be performed in two steps. In the first step, vesicles are released from the donor

PYCs. This is accomplished by incubating donor cells in the presence of either an HSS or HSP (0.7–1.0 mg) for 10–60 min at 20°. We have shown that the components necessary for vesicle formation can be supplied by either the HSS or the HSP, but these fractions are not biochemically identical, because the HSP alone is not sufficient to complete the transport reaction.[9] In the second step of the assay, vesicles are targeted to the acceptor Golgi complex, where they bind and fuse to release their contents. These events require vesicles, ATP, cytosol, and acceptor membranes. The second incubation is performed at 20° for 1–1.5 hr and the samples are processed as described above.

Analysis of Yeast Secretory Mutants in Transport Assay

The genetic flexibility of yeast has facilitated the isolation of a large collection of mutants that are blocked at various stages of the secretory pathway.[15-18] In two such mutants, *ypt1* and *bet2,* traffic is disrupted between the ER and Golgi complex. As a consequence, precursors of exported proteins accumulate within the lumen of the ER at the restrictive growth temperature (37°). In the case of invertase, a portion of the protein that is synthesized is also secreted as an underglycosylated species.[8,17] The *in vitro* assay has enabled us to examine the defects in the *ypt1*[7] and *bet2*[8] mutants.

General Comments. Fractions used in this analysis are generated by the protocol described above with two exceptions. First, the steps involved in preparing regenerated spheroplasts are conducted at 25°. Second, the PYC, S3, HSS, and HSP fractions are prepared from regenerated spheroplasts that were frozen as a packed pellet at −80° and then lysed after thawing. The reasons for these changes are as follows. Normally, spheroplasting is performed at 37°, the optimum temperature for zymolyase, the enzyme used to remove the yeast cell wall. Cells prepared under these conditions lyse readily and yield concentrated S3, HSS, and HSP fractions. However, because the yeast secretory mutants are temperature sensitive for growth, fractions isolated from them are always prepared from cells that have been maintained at the permissive temperature of 25°. Because the cell wall is not removed efficiently at this temperature, fractions of insufficient protein concentration are obtained unless the freeze-thaw step is added to augment lysis. The fractions are then screened for temperature sensitive defects by performing the transport assay at higher temperatures (29°).

[15] P. Novick, S. Ferro, and R. Schekman, *Cell* **25,** 461 (1981).
[16] A. P. Newman and S. Ferro-Novick, *J. Cell Biol.* **105,** 1587 (1987).
[17] N. Segev, J. Mulholland, and D. Botstein, *Cell* **53,** 635 (1988).
[18] H. D. Schmitt, M. Puzicha, and D. Gallwitz, *Cell* **53,** 635 (1988).

Procedure. When the donor PYC and S3 fraction, both prepared from the *ypt1* mutant, are incubated in the presence of an ATP-regenerating system, the 26-kDa species fails to be processed to the high molecular weight Golgi form. Instead, the 28-kDa species is observed. This form of pro-α-factor is likely to reside in an early subcompartment of the Golgi complex.[7,9] The low extent of conversion of the 26- to the 28-kDa species suggests that in addition to disrupting transport through the Golgi apparatus, *ypt1* mutant fractions display a partial block in ER-to-Golgi transport. Because the function of donor, acceptor, and soluble factors can be assayed separately in our assay,[1] we determined the defective fraction in this mutant. We observed that transport proceeds normally when permeabilized *ypt1* mutant cells are assayed in the presence of a wild-type S3 fraction, indicating that the donor compartment is not deficient in this mutant. However, the later stages of transport are defective when the S3 fraction is obtained from this mutant and assayed in the presence of wild-type PYCs. We next ascertained whether the soluble or membrane-bound factors are defective for transport. To do this, we prepared HSS (soluble) and HSP (membrane) fractions from *ypt1* mutant cells. The HSS from the *ypt1* mutant in combination with the wild-type HSP effectively supports transport. However, when the HSP from the *ypt1* mutant is assayed in the presence of a wild-type HSS fraction, transport events subsequent to vesicle release fail to occur. These results indicate that the *ypt1* mutant is defective in a component that is normally provided by the HSP. We have proposed that this component is the Golgi complex. The same experiments were also performed with *bet2* mutant fractions. Like *ypt1*, *bet2* mutant fractions fail to support the later stages of transport.[8] We have shown that *BET2* encodes a factor that is needed for the biological activity and membrane attachment of Ypt1.[8] These *in vitro* findings are consistent with this proposal.

Separation of Vesicles from Cytosolic Components

Vesicles formed from the incubation of donor PYCs, ATP, and HSS can be separated from the cytosolic components in two ways.[9] After the donor PYCs are removed by centrifugation, the vesicles can be pelleted at 120,000 *g* for 1 hr at 4°. The supernatant is aspirated and the vesicles are resuspended in 1 × transport buffer. We have shown that under these conditions the vesicles remain intact. The other method used to separate vesicles from cytosolic components is gel filtration. Generally, the supernatant from 10 to 12 transport reactions is gel filtered over a 13-ml Sephacryl S-300 column equilibrated with TBPS. The vesicles elute in the void volume on this column and are cleanly separated from soluble marker proteins (BSA, cytochrome *c*, and *p*-nitrophenol). This method of separation is preferable if the vesicles are to be used in the second step of the transport assay, because we have found that they are more competent for completing the transport assay than when they are pelleted at high speed. In addition, we concentrate the vesicle fractions to approximate the ratio of vesicles to acceptor normally maintained in the reaction. This is achieved by centrifugation in a Centricon-10 concentrator (5000 g for 1 hr at 4°). Approximately 40–50% of the vesicles are recovered after concentration.

Characterization of Endoplasmic Reticulum-to-Golgi Transport Vesicles

We have used several criteria to establish that the vesicles formed in our assay constitute a distinct intermediate in ER-to-Golgi transport.[9] Kinetic studies have shown that the vesicles are released from the PYCs before pro-α-factor receives Golgi-specific modifications. In addition, vesicle formation and release requires donor membranes, cytosolic proteins, and ATP. These are requirements shown to be necessary for vesicle-mediated transport *in vitro* and *in vivo*.[15,19–21] The vesicles containing pro-α-factor are a homogeneous population of membranes that lack detectable amounts of the ER marker protein, hydroxymethylglutaryl (HMG)-CoA reductase. Because they contain only the 26-kDa and lower molecular mass forms of pro-α-factor, we can conclude that they have not fused with post-ER compartments retained within the donor cells. Together, these results suggest that pro-α-factor is released from the cells within carrier vesicles derived from donor ER membranes. We have shown that these vesicles are competent to fuse with acceptor Golgi membranes. This has enabled us to determine the requirements to complete the transport reaction.[9] These studies have demonstrated that either cytosol or a crude membrane fraction can provide the factor(s) required for vesicle release. However, soluble factors needed to complete this reaction are supplied only by cytosol.

We have also begun to characterize the intermediate transport vesicles with respect to their protein composition. Toward this end, we have used sucrose gradients to partially purify the vesicles to determine whether specific proteins known to be involved in ER-to-Golgi transport comigrate with them. Our results suggest that Bos1p is a component of the vesicles.[10] The *BOS1* and *BET1* genes encode proteins that are essential for ER-to-Golgi transport.[16,22] In addition, they genetically interact with each other, implying that their products function at a common step in the transport

[19] J. Jamieson and G. Palade, *J. Cell Biol.* **39**, 589 (1968).
[20] W. E. Balch, B. S. Glick, and J. E. Rothman. *Cell* **39**, 525 (1984).
[21] B. W. Wattenberg, W. E. Balch, and J. E. Rothman, *J. Biol. Chem.* **261**, 2202 (1986).
[22] J. Shim, A. P. Newman, and S. Ferro-Novick, *J. Cell Biol.* **113**, 55 (1991).

pathway in yeast.[22,23] We have shown also that Bos1p and Bet1p colocalize to the cytoplasmic surface of the ER membrane.[10]

Procedure. To address whether a specific protein, such as Bos1p, comigrates with ER-to-Golgi vesicles on sucrose gradients, parallel sets of transport assays are performed. In the first set (four reactions), vesicles are formed and released from donor PYCs that contain radiolabeled pro-α-factor as a marker protein in the lumen of the ER. The second set consists of 48–96 reactions and utilizes PYCs that do not contain a marker protein in the donor compartment. (In these experiments, four reactions are pooled in a single microfuge tube.) Membranes released from the donor cells are collected by centrifugation at 120,000 g for 1 hr at 4° and the pellets are resuspended in transport buffer and loaded on the top of linear sucrose gradients. Depending on the conditions employed, the membranes are separated on the basis of their densities (density gradients) or on the basis of their size (velocity gradients). Density equilibrium gradients [5 ml of 20–50% (w/w) sucrose in 50 mM NaCl, 10 mM Tris, pH 7.5] are spun for 21 hr at 120,000 g (4°) while velocity gradients [8–20% (w/w) sucrose in 250 mM sorbitol, 50 mM NaCl, 10 mM Tris, pH 7.5] are sedimented for 2 hr at 100,000 g (4°). The gradients are fractionated into 300-μl aliquots and the fractions analyzed as follows. Fractions from the gradient that contain labeled transport reactions are treated with trypsin and then precipitated with Con A-Sepharose as described above. The unlabeled fractions are analyzed on Western blots with the appropriate antibodies. The blots are quantitated and the location of Bos1p is compared with that of pro-α-factor, the marker for the transport vesicles.

The methods described here compose a multifaceted approach for studying ER-to-Golgi transport. Yeast is a model system for such studies, because of its genetic simplicity. With our *in vitro* assay, we have been able to define the functional defects in some of the mutants that are blocked in transport from the ER to the Golgi complex. This was possible because the donor, acceptor, and soluble factors required for transport are separable in this assay. We can now divide the assay into two stages: vesicle formation and release from the donor membrane, and the utilization of these vesicles for subsequent stages of transport. The ability to do this has allowed us to identify and isolate functional carrier vesicles that mediate ER-to-Golgi traffic. Finally, by combining the assay with more classical biochemical methods, we have begun to identify the protein components of the transport vesicles.

[23] A. P. Newman, J. Shim, and S. Ferro-Novick, *Mol. Cell Biol.* **10,** 3405 (1990).

[15] Transport from Late Endosomes to Trans-Golgi Network in Semiintact Cell Extracts

By YUKIKO GODA, THIERRY SOLDATI, and SUZANNE R. PFEFFER

Introduction

We describe here a complementation system that reconstitutes the transport of mannose 6-phosphate receptors (MPRs) from late endosomes to the trans-Golgi network (TGN).[1] Mannose 6-phosphate receptors carry newly synthesized, soluble lysosomal hydrolases from the TGN to late endosomes, and are then transported back to the TGN to complete a cycle of biosynthetic, lysosomal enzyme transport.[2] Our endosome-to-TGN transport assay relies on the unique localization of sialyltransferase to the trans-Golgi and TGN, and utilizes a mutant cell line in which glycoproteins are not sialylated [Chinese hamster ovary (CHO) clone 1021;[3] Fig. 1]. Radiolabeled MPRs, present in late endosomes in a mutant cell extract, acquire sialic acid residues when they are transported to the TGN of wild-type Golgi complexes present in reaction mixtures. Sialic acid acquisition by MPRs in this system reflects a vesicular transport process, because it is time, temperature, ATP, and cytosol dependent, and also requires GTP hydrolysis.[1,4] Furthermore, MPRs and sialyltransferase remain in sealed membrane compartments throughout the reaction, and nonspecific membrane fusion is ruled out by several criteria.[1]

Materials

[35S]Methionine/[35S]cysteine, 1 mCi (protein labeling mix; New England Nuclear, Boston, MA)

Cells: CHO clone 1021 cells[3] were obtained from J. E. Rothman (Sloan Kettering, NY) and are grown as monolayers in minimum essential medium containing 7.5% (v/v) fetal bovine serum to a density of 2×10^7 cells/10-cm dish

Cytosol: CHO cytosol (5–7 mg/ml) is obtained from CHO cells grown in suspension as described by Balch *et al.*[5] The cells are broken in a steel ball-bearing homogenizer; cytosol is desalted into 25 mM Tris-HCl, pH 8.0, 50 mM KCl using a column of P6DG (Bio-Rad, Rich-

[1] Y. Goda and S. R. Pfeffer, *Cell* **55**, 309 (1988).
[2] S. Kornfeld and I. Mellman, *Annu. Rev. Cell Biol.* **5**, 483 (1989).
[3] E. B. Briles, E. Li, and S. Kornfeld, *J. Biol. Chem.* **252**, 1107 (1977).
[4] Y. Goda and S. R. Pfeffer, *J. Cell Biol.* **112**, 823 (1991).
[5] W. E. Balch, W. G. Dunphy, W. A. Braell, and J. E. Rothman, *Cell* **39**, 405 (1984).

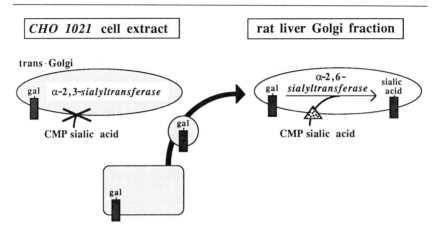

FIG. 1. Complementation scheme to detect transport of MPRs from late endosomes to the TGN. A semiintact cell extract prepared from ^{35}S-labeled CHO 1021 cells is incubated with wild-type Golgi membranes, cytosol, ATP, and an ATP-regenerating system at 37°. CHO 1021 cell MPRs (rectangles) possess galactose-terminating oligosaccharides and will acquire sialic acid if transported to the wild-type TGN. At the end of the reactions, receptors are isolated and sialic acid acquisition is monitored by slug lectin chromatography.

mond, CA) and is frozen in liquid nitrogen in small aliquots, as rapidly as possible

Golgi membranes: Golgi membranes are isolated from rat liver according to the procedure of Tabas and Kornfeld[6]; homogenization is carried out in 0.5 M sucrose/10 mM Tris-HCl, pH 7.4, containing 1 mM dithiothreitol (DTT), 5 mM ethylenediaminetetraacetic acid (EDTA), 10 μg/ml leupeptin, 1 μM pepstatin, and 0.5 mM phenyl-methylsulfonyl fluoride (PMSF) (4 ml/g liver). All other sucrose solutions are prepared in 10 mM Tris-HCl (pH 7.4) containing 1 mM DTT, 1 mM EDTA, and 50 μM ATP. The Golgi-enriched fraction at the 0.5/1.0 M sucrose interface is harvested in a minimal volume and is rapidly frozen in 50-μl aliquots. Golgi preparations with protein concentrations of 1.5 to 2.5 mg/ml are standardly used

Pentamannosyl 6-phosphate-Sepharose: Pentamannosyl 6-phosphate (PMP) fragments of *Hansenula holstii* phosphomannan are prepared according to Murray and Neville[7] and derivatized with the bifunctional reagent, p-(aminophenyl)ethylamine (PAPEA; Aldrich, Milwaukee, WI) as follows: PMP (90 mg) is added to 0.5 ml PAPEA and stirred in a sealed, 15-ml glass tube overnight. Absolute ethanol is

[6] I. Tabas and S. Kornfeld, *J. Biol. Chem.* **254,** 11655 (1979).
[7] G. J. Murray and D. M. Neville, *J. Biol. Chem.* **255,** 11942 (1980).

added (0.5 ml), followed by 12 mg NaBH$_4$ in 1 ml absolute ethanol. The mixture is stirred for 5 hr with venting. H$_2$O (4 ml) is added, and the reaction is transferred to ice. The pH is brought to 5.6 by the addition of glacial acetic acid. Ethanol is removed by blowing nitrogen onto the sample, and water is added to bring the total volume to 5 ml. The sample is applied to a column of Sephadex G-10 (2.5 × 60 cm) equilibrated with 1 M acetic acid, pH 5.0 with pyridine, and the sample is eluted in the same buffer. The void peak material, detected by absorbance is 285 nm, is pooled, lyophilized, and stored at −20°. The PAPEA–PMP (50 mg) is coupled to 5 g CNBr-Sepharose (Pharmacia, Piscataway, NJ) according to the manufacturer

Slug lectin Affi-Gel: *Limax flavus* agglutinin (Calbiochem, La Jolla, CA) is coupled to Affi-Gel 10 (Bio-Rad), at 1 mg lectin/ml Affi-Gel, as follows. Lyophilized lectin is reconstituted to 1 mg/ml in 0.05 M sodium phosphate, pH 7.5, 0.15 M NaCl, and dialyzed overnight into 10 mM sodium phosphate, pH 7.5, 0.15 M NaCl (with two changes). The lectin is mixed with a 1 : 1 slurry of Affi-Gel 10 in 0.1 M N-2-hydroxyethylpiperazine-N'-2-ethanesulfonic acid (HEPES), pH 7.2. Coupling is carried out overnight at 4°; the reaction is quenched by addition of 1 M Tris-HCl, pH 8, to 50 mM final concentration. The resin is washed and stored in ST (see buffers below). Coupling efficiency should be ≥ 95%

Buffers/Premixes

TD: 25 mM Tris-HCl, pH 7.4, 5.4 mM KCl, 137 mM NaCl, 0.3 mM Na$_2$HPO$_4$

PMP wash buffer: 50 mM HEPES, pH 7.5, 150 mM NaCl, 0.05% (v/v) Triton X-100, 5 mM β-glycerophosphate

PMP elute buffer: 50 mM HEPES, pH 7.5, 150 mM NaCl, 0.5% (w/v) 3-[(3-cholamidopropyl)dimethyl ammonio]-1-propane sulfonate (CHAPS), 5 mM β-glycerophosphate, 5 mM mannose 6-phosphate

PMP resin recycling buffer: 0.2 M citrate, pH 5.0, 2 M NaCl, 0.05% (v/v) Triton X-100

ST: 100 mM Tris-HCl, pH 7.5, 50 mM NaCl

STT: ST containing 3% (v/v) Triton X-100

STC: ST containing 1.0% (w/v) CHAPS

Slug lectin resin recycling buffer: STT + 10 mM sialic acid

Detergent stop mix (1.33×): 1.67 mM CDP, 187 mM NaCl, 67 mM TrisHCl, pH 7.2, 0.13% (w/v) sodium dodecyl sulfate (SDS), 2.2% (v/v) Triton X-100, 1.33% (w/v) deoxycholate, 6.7 mM β-glycerophosphate

Swelling buffer: 10 mM HEPES–KOH, pH 7.2, 15 mM KCl

Assay buffer (10×): 250 mM HEPES–KOH, pH 7.2, 15 mM magnesium diacetate, 1.15 M KCl

Low-salt assay buffer (10×): 250 mM HEPES–KOH, pH 7.2, 15 mM magnesium diacetate, 250 mM KCl

Protease inhibitors (100×): 3.4 U/ml aprotinin, 100 μg/ml leupeptin, 10 μM pepstatin

Procedure

1. Label one plate of CHO clone 1021 cells with 1 mCi [^{35}S]methionine/cysteine in methionine and cysteine-free medium for 3 hr, then chase in complete growth medium for 4 hr to permit MPRs to achieve their steady state distribution beyond the Golgi complex.

2. Transfer the culture dish to a wet paper towel-covered steel block, on ice. Wash three times with ice-cold swelling buffer, then allow the cells to swell for 10 min on ice.[8]

3. While the cells are swelling, prepare the following scrape mix:

Distilled H$_2$O, 420.5 μl
Assay buffer (10×), 80 μl
Protease inhibitors (100×), 10 μl
MgCl$_2$ (1 M), 2 μl
ATP, 40 mM (pH 7.0 with 1 M NaOH), 25 μl
Creatine phosphate (400 mM in H$_2$O), 37.5 μl
Creatine phosphokinase (1400 U/ml in H$_2$O), 15 μl
CMP sialic acid (10 mM in H$_2$O), 10 μl
CHO cytosol, 200 μl

4. Drain the buffer by holding the culture dish vertically for 30 sec and aspirating the collected liquid. Add the scrape mix and use a relatively inflexible rubber policeman to scrape cells from the plate with a firm, left-to-right motion, from top to bottom. The plates are scraped four times, turning the plate 90° between each scrape. Pipette the semiintact cell extract up and down three times in a 1-ml blue Eppendorf tip to yield a uniform suspension. [The washing procedure of Beckers et al.[8] is omitted because it significantly decreases the activity of the extract; control experiments suggest that a large fraction of organelles involved in transport (late endosomes) are released from the cell residue, and are lost during the washing step. Semiintact cells may be further disrupted by three passes through a 27-gauge needle.]

5. One plate of cells is sufficient for 5 to eight 200-μl assays or 10 to

[8] C. J. M. Beckers, D. S. Keller, and W. E. Balch, *Cell* **50**, 523 (1987).

sixteen 100-μl assays. Some volume is contributed by the drained, scraped cells, such that the final volume of the scrape mix will be slightly greater than 800 μl. For six reactions, use ~140 μl semiintact cell extract (with cytosol included), 5–10 μl (~ 10 μg) wild-type Golgi membranes, 6 μl low-salt assay buffer, and doubly distilled H$_2$O to a final volume of 200 μl. Incubations are generally carried out for 2 hr at 37° in 3-ml polycarbonate tubes (Sarstedt, Hayward, CA) and stopped on ice by addition of 600 μl 1.33 × detergent stop mix containing CDP to inhibit solubilized sialyltransferase. In reactions testing cytosol activity, cytosol may be omitted from the scrape mix.

6. Solubilized mixtures are clarified by centrifugation for 10 min at 95,000 rpm in a Beckman (Palo Alto, CA) TL-100 centrifuge. Mannose 6-phosphate receptors are isolated from preclarified reaction mixtures by incubation with 600 μl of a 1:1 slurry of pentamannosyl 6-phosphate-Sepharose resin (in PMP wash buffer) for at least 2 hr at 4° on a rotator.

7. Slurries are transferred to disposable column holders and flow-through fractions are collected. Columns are washed at room temperature with 10 ml PMP wash buffer, then eluted in 1.5–2 ml PMP elute buffer.

8. Sialic acid-containing MPRs are resolved from the remainder by incubation with 100 μl slug lectin Affi-Gel for 30 min at room temperature (or overnight at 4°). Flow-through fractions are collected. Columns are washed with 4 ml of STT followed by 0.75–1 ml STC. Mannose 6-phosphate receptors that have acquired sialic acid residues are eluted with 1 ml STC containing 10 mM sialic acid; eluted material is collected in 12 × 75 mm glass tubes for subsequent analysis.

9. The percentage of total MPR that had acquired sialic acid during the course of a reaction is determined by densitometry of autoradiograms after SDS-polyacrylamide gel electrophoresis (PAGE); 5–10% of the slug lectin Affi-Gel flow-through fractions are routinely analyzed to enable direct comparison with eluted fractions within the linear range for densitometry. Eluted fractions may be counted in a scintillation counter, but we usually run gels to ensure that the radioactivity is due to the 300-kDa MPR and not associated with contaminating low molecular weight components.

10. For polyacrylamide gel analysis, 25 μl of 5 mg/ml deoxycholate (pH 7.5) is added to the eluted fractions. Tubes are vortexed, followed by the addition of 750 μl of 10% (w/v) trichloroacetic acid (TCA). Incubate the samples on ice 15–30 min, then centrifuge them in a Beckman GPR centrifuge at 3500 rpm for 30 min at 4°. Invert the tubes onto a paper towel, and remove excess liquid from the tube walls using a cotton-tipped applicator. Add 25 μl SDS-gel sample buffer and 4 μl of 1N NaOH. Boil for 5 min, and load onto a 6% (w/v) polyacrylamide gel. Gels are treated with Entensify (Du Pont, Wilmington, DE) prior to drying to increase detection sensitivity. We generally include 2 μg bovine serum albumin (BSA) in each

sample prior to addition of TCA as an internal control for TCA precipitation efficiency.

11. PMP-Sepharose is recycled by washing once or twice in PMP resin recycling buffer, then three times in PMP wash buffer. Slug lectin Affi-Gel is recycled by washing once or twice in slug lectin recycling buffer and three or four times in ST. Both are stored at 4° and retain their binding capacity for at least 6 months.

Results: Late Endosome-to-Trans-Golgi Network Transport Does Not Require Calcium

We have shown previously that MPR transport from late endosomes to the TGN requires ATP, N-ethylmaleimide (NEM)-sensitive cytosolic factors, and GTP hydrolysis.[1,4] In addition, transport probably occurs via non-clathrin-coated vesicles.[9]

The transport of proteins between the ER and the Golgi complex requires calcium ions.[10,11] Beckers and Balch have shown that ER-to-Golgi transport is optimal between 0.01 and 0.1 μM Ca^{2+}. In addition, delivery of post-Golgi transport vesicles to the plasma membrane is a calcium-dependent process.[12] We tested whether calcium ions were also required for MPR transport from late endosomes to the TGN. Reactions were first carried out in 5 mM ethylene glycol-bis(β-aminoethyl ether)-N,N,N',N'-tetraacetic acid (EGTA) to chelate available calcium ions. Under these conditions, the concentration of free Ca^{2+} should have been less than 0.1 nM. As shown in Fig. 2, the transport reaction was unaffected in the presence of 5 mM EGTA.

We next tested the effect of additional Ca^{2+} on our transport reaction. If calcium was added in the presence of EGTA, transport was uninhibited up to the equivalent of ~0.1 μM Ca^{2+}, which is close to physiological calcium concentrations. Above this level, at ~3 mM Ca^{2+}/5 mM EGTA, transport was sharply inhibited. Thus, only supraphysiological levels of calcium ions inhibit endosome-to-TGN transport.

The lack of calcium ion dependence contrasts this transport step with the transport of proteins between the ER and the Golgi and also post-Golgi vesicle targeting. Further work will be needed to determine the precise role of calcium in specific vesicular transport steps, in addition to the calcium-sensitive target in endosome-to-TGN transport.

[9] R. K. Draper, Y. Goda, F. M. Brodsky, and S. R. Pfeffer, *Science* **248,** 1539 (1990).
[10] C. J. M. Beckers and W. E. Balch, *J. Cell Biol.* **108,** 1245 (1989).
[11] D. Baker, L. Wuestehube, R. Schekman, D. Botstein, and N. Segev, *Proc. Natl. Acad. Sci. USA* **87,** 355 (1990).
[12] I. De Curtis and K. Simons, *Proc. Natl. Acad. Sci. USA* **85,** 8052 (1988).

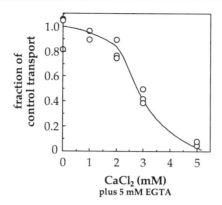

FIG. 2. Calcium ions inhibit late endosome-to-TGN transport at supraphysiological concentrations. Standard transport reactions were carried out in the presence of 5 mM EGTA and the indicated concentrations of $CaCl_2$ for 2 hr at 37°. In control reactions carried out in the absence of EGTA, 16.9% of MPRs acquired sialic acid.

[16] Microtubule-Mediated Golgi Capture by Semiintact Chinese Hamster Ovary Cells

By IRÈNE CORTHÉSY-THEULAZ and SUZANNE R. PFEFFER

Introduction

Interphase cells are believed to contain a single Golgi complex that is located near the microtubule organizing center (MTOC).[1] Disruption of cytoplasmic microtubules, by nocodazole for example, is sufficient to fragment the Golgi into "mini" Golgi stacks that distribute throughout the cytoplasm.[2] If nocodazole is removed, microtubules repolymerize, and the Golgi regains its perinuclear distribution, eventually reforming a single organelle. These experiments demonstrate that the mechanisms by which the Golgi achieves its localization are operational within interphase cells. Moreover, they indicate that Golgi localization should be amenable to analysis in semiintact cells.[3]

We have established a functional assay that detects the interaction of

[1] A. Rambourg and Y. Clermont, Eur. J. Cell Biol. **51,** 189 (1990).
[2] W. C. Ho, V. J. Allan, G. van Meer, E. G. Berger, and T. E. Kreis, Eur. J. Cell Biol. **48,** 250 (1989).
[3] C. J. M. Beckers, D. S. Keller, and W. E. Balch, Cell **50,** 523 (1987).

isolated Golgi complexes with the microtubule-based cytoskeleton.[4] Semiintact, but not intact, Chinese hamster ovary (CHO) cells are able to "capture" Golgi complexes by a process that requires ATP hydrolysis, cytosolic and peripheral membrane proteins, intact microtubules, and cytoplasmic dynein. Semiintact cells provide a native MTOC and other cytoplasmic components that may be important for physiologically relevant organelle–microtubule interactions.

Figure 1 outlines the "Golgi capture" assay. The system utilizes semiintact CHO clone 15B cells,[3] which retain microtubules emanating from an MTOC but lack the medial Golgi-specific enzyme, N-acetylglucosamine (GlcNAc) transferase I.[5] Semiintact cells are incubated with isolated, wild-type Golgi complexes at 37°, in the presence of a crude cytosol fraction, ATP, and an ATP-regenerating system. The association of wild-type Golgi complexes with the mutant CHO cells is then measured after differential centrifugation: semiintact cells are sedimented through a sucrose cushion, and cell-associated Golgi complexes are detected by assaying the pellet for GlcNAc transferase I enzyme activity.

Materials

Cells: Chinese hamster ovary (CHO) clone 15B cells are grown as monolayers in minimum essential medium containing 7.5% (v/v) fetal bovine serum to a density of $1-2 \times 10^7$ cells/10-cm dish. This represents about 90% confluency. Cell density is critical for experiments requiring cytosol dependency (see below)

Cytosol: CHO cytosol (5–7 mg/ml) is purified from CHO cells grown in suspension as described by Balch et al.[6] The cells are broken in a steel ball-bearing homogenizer[7]; cytosol is desalted into 25 mM Tris-HCl, pH 8.0, 50 mM KCl using a column of P6DG[8] and is frozen in liquid nitrogen in small aliquots, as soon as possible.

Golgi membranes: Rat liver Golgi complexes (1.5–2.5 mg/ml protein) are isolated by sucrose gradient flotation as described[9]; the band of Golgi at the 0.5/1.0 M sucrose interface is harvested in a minimal volume and is rapidly frozen in 200-μl aliquots. Homogenization is carried out using 4 ml buffer/g liver.

[4] Corthésy-Theulaz, A. Pauloin, and S. R. Pfeffer, *J. Cell Biol.* **118,** in press (1992).
[5] C. Gottlieb, J. Baenziger, and S. Kornfeld, *J. Biol. Chem.* **250,** 3303 (1975).
[6] W. E. Balch, W. G. Dunphy, W. A. Braell, and J. E. Rothman, *Cell* **39,** 405 (1984).
[7] W. E. Balch and J. E. Rothman, *Arch. Biochem. Biophys.* **240,** 413 (1985).
[8] M. R. Block, B. S. Glick, C. A. Wilcox, F. T. Wieland, and J. E. Rothman, *Proc. Natl. Acad. Sci. USA* **85,** 7852 (1988).
[9] I. Tabas and S. Kornfeld, *J. Biol. Chem.* **254,** 11655 (1979).

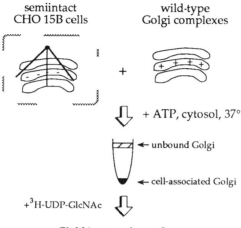

FIG. 1. Complementation scheme to detect Golgi capture by semiintact CHO cells. Semiintact CHO clone 15B cells, which lack the medial-Golgi enzyme, GlcNAc transferase I, are incubated at 37° with wild-type Golgi membranes, cytosol, ATP, and an ATP-regenerating system. At the end of the reaction, unbound Golgi are separated from cell-associated Golgi by differential centrifugation through a sucrose cushion. The presence of cell-associated Golgi complexes is then detected by assaying the pellet fractions for GlcNAc transferase I activity.

Buffers/Premixes

Swelling buffer: 10 mM N-2-hydroxyethylpiperazine-N'-2-ethanesulfonic acid (HEPES)–KOH, pH 7.2, 15 mM KCl

Capture buffer (10×): 250 mM HEPES–KOH, pH 7.2, 15 mM magnesium diacetate, 1.15 M KCl

Protease inhibitors (10×): 3.4 U/ml aprotinin, 100 μg/ml leupeptin, 10 μM pepstatin

Sucrose cushion: 1 M sucrose, 10 mM Tris-HCl, pH 7.4

Phosphate-buffered saline (PBS): 137 mM NaCl, 2.7 mM KCl, 8.8 mM Na$_2$HPO$_4$, 5.4 mM KH$_2$PO$_4$, pH 7.4

Concanavalin A (con A) elution buffer: 0.1 M α-methylmannoside, 10 mM Tris-HCl, pH 8.0

Con A storage buffer: 1 M NaCl, 0.1 M sodium acetate, pH 6.0, 1 mM MgCl$_2$, 1 mM MnCl$_2$, 1 mM CaCl$_2$

Con A wash buffer A: 0.1 M Sodium acetate, pH 4.5, 0.5 M NaCl

Con A wash buffer B: 0.1 M Tris-HCL, pH 8.5, 0.5 M NaCl

[Glucosamine-6-^{3}H(N)]:UDP[^{3}H]GlcNAc (New England Nuclear, Boston, MA)

Concanavalin A-Sepharose 4B (Sigma, St. Louis, MO)
Ovalbumin glycopeptide V: $Man_5(GlcNAc)_2$-asparagine (Biocarb, Lund,
    Sweden)

Method

1. Semiintact cells are prepared by osmotic swelling as described by
Beckers et al.[3] Briefly, a metal block is placed on ice, and covered with a
wet paper towel. Dishes are transferred to the towel-covered blocks, and are
washed three times with ice-cold swelling buffer. Cells are then swollen in
the same buffer for 10 min on ice. The plates are drained, and then held
vertically for 30 sec to facilitate aspiration of any trace of remaining buffer.

2. While the cells are swelling, a scraping mix is prepared. A 10-cm
dish is sufficient for six to eight 200-μl reactions. One hundred and twenty
microliters of scraping mix is needed for each assay; for six reactions,
prepare the following mix on ice:

Double distilled H_2O, 493 μl
Capture buffer (10×), 120 μl
Protease inhibitors (100×), 12 μl
$MgCl_2$ (1M), 2.4 μl
ATP, 40 mM (pH 7.0 with 1 M NaOH), 30 μl
Creatine phosphate (400 mM in H_2O), 45 μl
Creatine phosphokinase (1400 U/ml in H_2O), 18 μl

3. After aspirating the cells, the reaction mix is added directly to the
dishes and the cells are scraped with a trimmed, inflexible rubber policeman. Scraping is accomplished using a firm, squeaking, horizontal, left-to-
right motion, beginning at the top of the plate. The plates are scraped four
times, turning the plate 90° between each scrape. The dish is then tilted up,
and the cells are collected at the bottom edge by scraping. A uniform
suspension of semiintact cells is generated by pipetting up and down three
times through a 1-ml blue tip using a Gilson Pipetman; the extract is then
transferred to a tube on ice.

4. Capture reactions are carried out in 1.5-ml Eppendorf tubes, containing 20–40 μl CHO cytosol, 15–30 μl rat liver Golgi membranes,
120 μl cell extract, and enough water to bring the final volume to 200 μl.
Tubes are incubated in a 37° water bath for 60 min.

5. Reactions are transferred to tubes containing 1 ml of sucrose cushion, and are centrifuged in a horizontal microfuge (Beckman, Palo Alto,
CA) for 2 min at 12,000 g at 4°. Unbound Golgi complexes remain at the
top of the sucrose cushion; cell-associated Golgi complexes pellet with the
cells. The sucrose is carefully aspirated using a 21-gauge needle attached to
a vacuum aspirator. Tube walls are dried with a Kimwipe-covered cotton-
tipped applicator, and pellets are then frozen in liquid nitrogen.

6. The extent of Golgi capture is measured by assaying thawed cell pellets for GlcNAc transferase I activity.[10] Tubes containing cell pellets are brought to a total volume of 0.02 ml with sucrose cushion solution; assays are carried out in a final volume of 0.05 ml for 30 min at 37° in 0.1 M 2-(N-morpholino)ethanesulfonic acid (MES), pH 6.5, 10 mM MnCl$_2$, 0.2% (v/v) Triton X-100, 2 mM ATP, 0.5 mM UDPGlcNAc, 5 μCi UDP[^{3}H]GlcNAc, and 2–10 nM ovalbumin glycopeptide V [Man$_5$(GlcNAc)$_2$-asparagine]. Reactions are stopped by boiling for 5 min. One milliliter of PBS is added, and the samples are then clarified by centrifugation for 5 min at 12,000 g at 4°. Supernatants are applied to a 300-μl concanavalin A-Sepharose 4B column [poured in a Bio-Rad (Richmond, CA) 10-ml disposable column] that has been prewashed with 10 ml of PBS. Columns are washed with 15 ml PBS and then eluted with 2.5 ml elution buffer. Eluted fractions are mixed with 12 ml of scintillation fluid and counted in a scintillation counter.

7. Protocols are designed such that reactions are carried out in duplicate or triplicate to ensure reliability of results. The normal blank for the assay is a reaction lacking rat liver Golgi, which yields approximately 350 cpm.

8. Concanavalin A-Sepharose is recycled by consecutive washes in 50 ml each of wash buffers A, B, and PBS; it is then stored in Con A storage buffer at 4°. The resin can be used seven or eight times.

Results

Figure 2 shows the results of Golgi capture reactions carried out in the presence of increasing concentrations of rat liver Golgi membranes. Although a small amount of capture is observed on ice (triangles), significantly more capture is observed at 37° (squares). In addition, the reaction appears to be saturable for Golgi complexes under the conditions employed in this experiment. We have shown elsewhere that only semiintact cells capture Golgi complexes, and that this process is entirely ATP dependent.[4] As much as 30% of added Golgi complexes can be captured by semiintact CHO cells.

The semiintact cells lose their capacity for Golgi capture if washed prior to use. Thus, a significant level of recipient cell cytosolic proteins are present in semiintact cell extracts. This is likely to explain the low level of Golgi capture that is observed in the absence of added cytosol (Fig. 3). Nevertheless, capture can be stimulated by ~2.5-fold by supplementing reactions with a crude cytosol fraction. Under most circumstances, capture assays are carried out using ~1 mg/ml cytosolic proteins.

[10] W. G. Dunphy and J. E. Rothman, *J. Cell Biol.* **97**, 270 (1983).

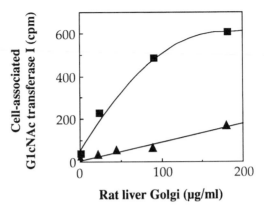

FIG. 2. The extent of Golgi capture is proportional to the amount of rat liver Golgi membranes added. Reactions were carried out at 37° (squares) or at 0° (triangles) with increasing concentrations of rat liver Golgi membranes.

When it is important to achieve maximal cytosol dependence, it is best to use a single 10-cm dish for 8–10 reactions, rather than 6–8 reactions. Cells are scraped in the same volume as before and then further diluted by addition of proportionally more capture buffer and ATP-regeneration reagents.

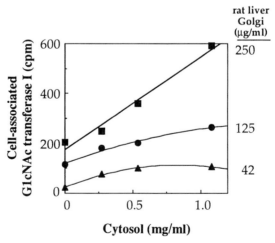

FIG. 3. Golgi capture is stimulated by the addition of cytosol. Golgi capture reactions were carried out for 60 min at 37° using the indicated concentrations of rat liver Golgi membranes, with increasing concentrations of cytosol.

Future Perspectives

The cell cycle-dependent redistribution of the Golgi complex under-scores the dynamic nature of organelle–microtubule interactions.[11] During mitosis, the Golgi vesiculates into smaller units that disperse throughout the cytoplasm.[12,13] At telophase, the vesicles coalesce and return to a pericentriolar location. Postmitotic Golgi vesicle movement requires polymerized microtubules,[11] follows microtubule tracks,[2] and is also energy dependent.[14] This feature has led many investigators to propose a role for cytoplasmic dynein in Golgi redistribution, because dynein has the capacity to drive organelle movements toward the minus ends of microtubules.

To date, there is no information regarding the proteins that link the Golgi to the microtubule-based cytoskeleton. We have shown elsewhere that cytoplasmic dynein is required for Golgi capture.[4] These results validate the prediction that cytoplasmic dynein can serve as a motor for the Golgi complex. Moreover, Golgi complexes bear peripherally associated proteins on their surfaces that mediate this process.[4] Perhaps it is these proteins that couple dynein to the Golgi complex. In summary, the Golgi capture assay provides the first functional assay that will permit the elucidation of the mechanism by which the Golgi complex is able to bind to, and move along, microtubules in a cell cycle-regulated manner.

[11] T. Kreis, *Cell Motil. Cytoskeleton* **15**, 67 (1990).
[12] J. M. Lucocq, J. G. Pryde, E. G. Berger, and G. Warren, *J. Cell Biol.* **104**, 865 (1987).
[13] J. M. Lucocq, E. G. Berger, and G. Warren, *J. Cell Biol.* **109**, 463 (1989).
[14] J. R. Turner and A. M. Tartakoff, *J. Cell Biol.* **109**, 2081 (1989).

[17] Regulated Exocytotic Fusion I: Chromaffin Cells and PC12 Cells

By RONALD W. HOLZ, MARY A. BITTNER, and RUTH A. SENTER

Introduction

An important advance in the study of regulated exocytosis came with the demonstration by Baker and Knight[1,2] of the feasibility of studying secretion from cells with leaky plasma membranes. High-voltage discharges caused dielectric breakdown of the plasma membrane of sus-

[1] P. F. Baker and D. E. Knight, *Nature (London)* **276**, 620 (1978).
[2] D. E. Knight and P. F. Baker, *J. Membr. Biol.* **68**, 107 (1982).

pended chromaffin cells and platelets, thus rendering the membranes permeable to small salts and nonelectrolytes. The studies demonstrated the central role of Ca^{2+} in stimulating exocytosis and the necessity of ATP as a cofactor. Subsequently, a variety of other techniques were developed to permeabilize selectively the plasma membrane, many of which were suitable for use with monolayer cells.[3-7] The focus of this chapter is the use of the detergent digitonin in the study of secretion from primary cultures of bovine chromaffin cells and the continuous cell line PC12 (rat adrenal pheochromocytoma), both of which store and secrete catecholamine. Digitonin[3,4] gives reproducible results, is easy to use, and most importantly permits exogenous proteins to enter the cells. At the end of the chapter we will compare some of the methods of permeabilization that are commonly used. The preparation and maintenance of adrenal medullary chromaffin cells *in vitro* have been reviewed.[8]

Methods for Digitonin-Permeabilized Cells

Cell density and digitonin concentration are critical factors in obtaining optimal permeabilization and secretion. In bovine chromaffin cell and PC12 cultures with densities of less than 150,000 cells/cm² it is difficult to define a digitonin concentration that supports vigorous Ca^{2+}-dependent secretion of catecholamine and does not cause significant release of catecholamine in the absence of Ca^{2+}. Optimal Ca^{2+}-dependent secretion with minimal release in the absence of Ca^{2+} occurs from primary cultures of bovine chromaffin cells at $2.5-7.5 \times 10^5$ cells/cm² with $10-20 \mu M$ digitonin. Optimal Ca^{2+}-dependent secretion with minimal catecholamine release in the absence of Ca^{2+} is obtained when PC12 cells are subcultured at 200,000 cells/cm² and permeabilized with $7.5 \mu M$ digitonin 4–6 days later (when densities are $4-6 \times 10^5$ cells/cm²).[9]

Knight and Baker found with electropermeabilized chromaffin cells that a solution in which glutamate is the primary anion supports excellent secretion. This is also true with digitonin-permeabilized cells. Typically, experiments are performed with KGEPM [139 mM potassium glutamate, 5 mM ethylene glycol-bis(β-aminoethyl ether)-N,N,N',N'-tetraacetic acid

[3] L. A. Dunn and R. W. Holz, *J. Biol. Chem.* **258**, 4989 (1983).
[4] S. P. Wilson and N. Kirshner, *J. Biol. Chem.* **258**, 4994 (1983).
[5] J. C. Brooks and S. Treml, *J. Neurochem.* **40**, 468 (1983).
[6] M.-F. Bader, D. Thierse, D. Aunis, G. Ahnert-Hilger, and M. Gratzl, *J. Biol. Chem.* **261**, 5777 (1986).
[7] J. M. Sontag, D. Aunis, and M. F. Bader, *Eur. J. Cell Biol.* **46**, 316 (1988).
[8] B. G. Livett, *Physiol. Rev.* **64**, 1103 (1984).
[9] S. C. Peppers and R. W. Holz, *J. Biol. Chem.* **261**, 14665 (1986).

(EGTA) and 20 mM piperazine-N,N'-bis(2-ethanesulfonic acid (PIPES), and 1 mM MgCl$_2$, pH 6.6] with 2 mM MgATP. Propionate and isethionate can substitute for glutamate; Na$^+$ can substitute for K$^+$. Secretion is also vigorous when 250 mM sucrose substitutes for potassium glutamate. The solution can contain 20 mM chloride without altering secretion. However, solutions with chloride as the dominant anion give a smaller secretory response. A pH of 6.6 is used to buffer Ca^{2+} concentrations as high as 30 μM. (Higher pH increases the effective affinity of the EGTA buffer for Ca^{2+} and reduces the free Ca^{2+} concentrations that can be buffered.) EGTA strongly binds numerous multivalent cations in addition to Ca^{2+},[10] including Ba^{2+} and Sr^{2+}. Magnesium ion is not strongly bound and the Ca/EGTA equilibrium is affected to only a small degree by the concentrations of Mg^{2+} and ATP that are present. In contrast, pH is critical for determining the free Ca^{2+} in the presence of Ca^{2+}/EGTA. A computer program is available to calculate the total Ca^{2+} necessary to attain a given free Ca^{2+} concentration in multicomponent solutions.[11] Alternatively, one cal calculate according to Portzehl $et\ al.$[12] effective association constants for Ca/EGTA, $K_{eff} = $ [CaEGTA]$_{total}$/([Ca^{2+}][EGTA]$_{total}$), where "total" refers to all forms of the Ca/EGTA and EGTA complexes. Logarithms of EGTA protonation constants from Martell and Smith[13] (adjusted for H$^+$ activities) are 9.59, 8.97, 2.80, and 2.12. Logarithms of the Ca^{2+}/EGTA association constants are 10.97 and 5.29. The effective association constants are 10$^{5.61}$ M^{-1} at pH 6.6 and 10$^{6.41}$ M^{-1} at pH 7.0. These constants give satisfactory estimates of free Ca^{2+} concentrations in solutions containing Ca^{2+} and millimolar total concentrations of Mg^{2+}, EGTA, and ATP. At pH 6.6, a total concentration of 1.45 mM Ca^{2+} with 5 mM EGTA gives a free Ca^{2+} concentration of 1 μM. Because EGTA releases protons on chelating divalent cations, it is important to verify pH in the incubation solutions. It should be noted that sometimes different published constants are used in calculations, which can result in estimates for free Ca^{2+} that may differ by as much as twofold from the above calculations.

Increasing the Ca^{2+} buffer strength by increasing EGTA from 2 to 15 mM with a proportional increase in total Ca^{2+} (free Ca^{2+} is unaltered) does not significantly alter the secretory response.[14] Thus, Ca^{2+}/EGTA

[10] J. Bjerrum, G. Schwarzenbach, and L. G. Sillen, "Stability Constant of Metal–Ion Complexes, with Solubility Products of Inorganic Substances. Part I: Organic Ligands." Chem. Soc., London, 1957.

[11] D. Chang, P. S. Hsieh, and D. C. Dawson, $Comput.\ Biol.\ Med.$ **18**, 351 (1988).

[12] H. Portzehl, P. C. Caldwell, and J. C. Reugg, $Biochim.\ Biophys.\ Acta$ **79**, 581 (1964).

[13] A. E. Martell and R. M. Smith, "Critical Stability Constants. Vol. 1: Amino Acids." Plenum, New York, 1974.

[14] M. A. Bittner, R. W. Holz, and R. R. Neubig, $J.\ Biol.\ Chem.$ **261**, 10182 (1986).

buffers rapidly equilibrate with the interior of digitonin-permeabilized cells.

Digitonin interacts with membrane cholesterol.[15] At the low concentrations used in secretion experiments, digitonin selectively permeabilizes the plasma membrane and has little direct effect on chromaffin granule integrity[16] or on Ca^{2+} stores.[17] Chromaffin granules in digitonin-permeabilized cells do not leak catecholamine, have the same osmotic stability as in intact cells, and maintain H^+ electrochemical gradients that drive catecholamine influx into the granule.[16] The stability of chromaffin granule membranes in digitonin-permeabilized cells is paradoxical, because the granule membranes contain approximately the same amount of cholesterol as the plasma membrane,[18] and chromaffin granules in homogenates release most of their catecholamine when incubated with 10 μM digitonin.[16] Experiments demonstrate a much greater sensitivity of chromaffin cells to digitonin-containing solutions perfused onto cells compared to unstirred digitonin-containing solutions bathing cells (J. A. Jankowski and R. M. Wightman, University of North Carolina, personal communication, 1992). The concentration of digitonin in contact with monolayer cells in static solutions may be less than in bulk solution because of a concentration gradient of digitonin between the interface with the monolayer cells and the unstirred bulk solution.

Digitonin-permeabilized cells are freely permeant not only to low molecular weight salts and Ca^{2+} buffers but also to carbohydrates at least as large as the tetrasaccharide stachyose.[16] This characteristic allowed the investigation of osmotic effects on chromaffin granules during exocytosis without cell shrinkage under conditions in which the intracellular milieu was controlled. The relative ineffectiveness of solutions of high osmolality to inhibit secretion indicated that osmotic swelling of the secretory vesicle is not a critical event preceding exocytosis.[19,20] This conclusion is consistent with direct observations of mast cell secretory vesicles that do not swell before exocytosis.[21,22]

Digitonin-treated cells are permeable to protein, although not as freely permeable as to low molecular weight solutes. Lactate dehydrogenase, a

[15] H. Gogelein and A. Huby, *Biochim. Biophys. Acta* **773**, 32 (1984).
[16] R. W. Holz and R. A. Senter, *J. Neurochem.* **45**, 1548 (1985).
[17] S. J. Stoehr, J. E. Smolen, R. W. Holz, and B. W. Agranoff, *J. Neurochem.* **46**, 637 (1986).
[18] S. P. Wilson and N. Kirshner, *J. Neurochem.* **27**, 1289 (1976).
[19] R. W. Holz and R. A. Senter, *J. Neurochem.* **46**, 1835 (1986).
[20] R. W. Holz, *Annu. Rev. Physiol.* **48**, 175 (1986).
[21] J. Zimmerberg, M. Curran, F. S. Cohen, and M. Brodwick, *Proc. Natl. Acad. Sci. USA* **84**, 1585 (1987).
[22] L. J. Breckenridge and W. Almers, *Proc. Natl. Acad. Sci. USA* **84**, 1945 (1987).

134-kDa cytosolic marker, exits from chromaffin cells permeabilized with 20 μM digitonin with a half-time of 5–15 min.[3,4] Protein kinase C[23] and other soluble proteins[3,24] also exit. Associated with the loss of cytosolic protein is a loss of the secretory response with a 50% loss of secretion within 5–15 min.[3,25] In digitonin-permeabilized PC12 cells the loss of the secretory response is also correlated with the loss of cytosolic protein.[9]

Typical Experiment with Chromaffin Cells

1. Dispersed chromaffin cells are prepared from bovine adrenal medullas by collagenase digestion,[8] purified by differential plating[26] to yield preparations that are 90–95% pure, and maintained for 3 days in modified Eagle's medium (MEM) with 10% (v/v) fetal calf serum, 10 μM cytosine arabinoside to inhibit proliferation of contaminating cells and antibiotics [100 units/ml penicillin, 100 μg/ml streptomycin, 25 μg/ml gentamicin, and 1.3 μg/ml Fungizone (Squibb, Princeton, NJ)]. We have found that because such pure cultures do not attach well to plastic, the culture surface should be coated with sterile calf skin collagen (5 μg/cm^2) (Vitrogen 100; Celtrix Laboratories, Palo Alto, CA) prior to plating. Sterile calf skin collagen solution, 50 μl/cm^2 of a 32-μg/ml solution (in 0.01 N HCl) is applied to each well. Plates without lids dry overnight in a sterile culture hood. Cell density is 250,000–500,000 cells/cm^2. Secretion experiments can be performed with 96-well culture dishes with 150,000 cells/well (470,000 cells/cm^2). After 3 days, the medium is replaced with fresh medium without Fungizone (to reverse toxic effects of the antibiotic). Experiments are performed 3–8 days after cell preparation.

2. Catecholamine stores can be labeled with [^{3}H]norepinephrine by incubating cells in MEM containing 0.5 mM ascorbate and 2 μCi/ml [^{3}H]norepinephrine for 1–4 hr. When preparing the solution [H]norepinephrine should be added to the ascorbate-containing medium to prevent breakdown of the catecholamine. Cells are then washed three times with physiological salt solution [PSS: 145 mM NaCl, 5.6 mM KCl, 2.2 mM CaCl$_2$, 0.5 mM MgCl$_2$, 15 mM HEPES (pH 7.4), 5.6 mM glucose, and 0.5 mM ascorbate]. Secretion can also be measured by release of endogenous catecholamine.[3]

3. Cells are permeabilized by incubation with KGEPM (see above)

[23] D. R. TerBush and R. W. Holz, *J. Biol. Chem.* **261**, 17099 (1986).

[24] K. L. Kelner, K. Morita, J. S. Rossen, and H. B. Pollard, *Proc. Natl. Acad. Sci. USA* **83**, 2998 (1986).

[25] T. Sarafian, D. Aunis, and M. F. Bader, *J. Biol. Chem.* **262**, 16671 (1987).

[26] J. C. Waymire, W. F. Bennett, R. Boehme, L. Hanteins, K. Gilmer-Waymire, and J. Haycock, *J. Neurosci. Methods* **7**, 329 (1983).

with 2 mM MgATP and 20 μM digitonin (Fluka, Ronkonkoma, NY) at 25°. Digitonin is prepared as a 2 mM stock solution in deionized water. The solution is heated to 95° to dissolve the digitonin before it is diluted into KGEPM. After a 3- to 10-min incubation with cells, the solution is replaced with KGEPM with 2 mM MgATP. Calcium ion is buffered from 0 to 30 μM. Secretion is determined by measuring the radioactivity released into the medium and the radioactivity remaining in cells [lysed with 100 μl 1% (v/v) Triton X-100] and is expressed as the percentage of the total radioactivity in the well. Cells become sufficiently permeable to respond to micromolar Ca^{2+} and release lactate dehydrogenase after 2 min of incubation with digitonin-containing solution. Cells are irreversibly permeabilized and continue to release lactate dehydrogenase after digitonin-containing solution is removed. Incubation volumes are 40–100 μl in wells of a 96-well plate.

4. The secretory response decreases with time after permeabilization. Typically, 15–25% of the catecholamine is released in 15 min after a 4-min permeabilization and drops to 4–8% after a 10-min permeabilization. Components to be tested can be added to the incubation with digitonin if they do not interfere with permeabilization. Cytosolic fractions (see below) often interfere with permeabilization and must be added after incubation with digitonin.

5. Many variations of the protocol can be used. Cells permeabilized in the presence of 10 μM Ca^{2+} secrete more vigorously than cells permeabilized in the absence of Ca^{2+} and subsequently exposed to Ca^{2+}. Secretion can be as much as 50% of the total catecholamine. Secretion occurs after a 15–30 sec delay when Ca^{2+} is present in the permeabilization solution. The delay reflects the time necessary for permeabilization. It is absent when Ca^{2+} is introduced after the cells are permeabilized. Numerous sequential incubations can also be performed after permeabilization. In some experiments, cells are permeabilized in Ca^{2+}-free KGEPM, then incubated for 5–10 min with a substance of interest, and finally stimulated to secrete by incubation with Ca^{2+}-containing solution. Care must be taken to properly control for the decay in the secretory response.

Experiments with PC12 cells are conducted in much the same way as with bovine chromaffin cells with the same solutions and with similar protocols.[9] PC12 cells are labeled with [³H]norepinephrine and washed twice over 3 hr before beginning the experiment. PC12 cells are sensitive to manipulations in the small wells of 96-well plates. Typically, the 16-mm diameter wells of 24-well plates are used with 0.25- to 1 ml-solution volumes. Because a small number of PC12 cells become dislodged from the monolayer during experimental manipulations, incubation solutions are briefly centrifuged after an experiment to pellet dislodged cells (1000 g for 5 min at 4°); radioactivity in an aliquot of the supernatant is determined.

Secretion from Digitonin-Permeabilized Cells

Calcium and ATP

The extent of secretion[3] from digitonin-permeabilized cells is half-maximal between 1 and 2 μM Ca^{2+} and maximal by 10 μM. These characteristics are virtually identical to those originally obtained in electropermeabilized cells.[1,2] Early rates of secretion saturate at greater than 10 μM Ca^{2+}.[26a]

The biochemical basis for the ability of micromolar Ca^{2+} to stimulate secretion is unknown. The use of digitonin-permeabilized cells allows the investigation of specific peptide reagents that interact with Ca^{2+}-dependent enzymes but do not enter intact cells. The protein kinase C pseudosubstrate inhibitor PKC(19–31), under conditions in which it inhibits *in situ* protein kinase C activity (measured by phosphorylation of an exogenous peptide substrate), has no effect on Ca^{2+}-dependent secretion in the absence of exogenous protein kinase C activators.[27] Thus, although the activation of protein kinase C can enhance secretion,[27-30] it is not necessary for secretion. A calmodulin-binding peptide derived from Ca^{2+}/calmodulin kinase II, under conditions in which it inhibits *in situ* Ca^{2+}-dependent phosphorylation, also has no effect on Ca^{2+}-dependent secretion.[27] There is additional evidence from experiments in intact and permeabilized cells that calmodulin is not the primary target of Ca^{2+}.[27,31,32] However, this conclusion is controversial.[33,34]

It was originally shown in electropermeabilized cells that MgATP is required for optimal secretion[1,2] and subsequently confirmed in digitonin-permeabilized chromaffin cells.[3,4,35] ATP hydrolysis is required.[2] The published studies indicate that half-maximal stimulation of secretion occurs at

[26a] M. A. Bittner and R. W. Holz, *J. Biol. Chem.* **267**, in press (1992).
[27] D. R. TerBush and R. W. Holz, *J. Biol. Chem.* **265**, 21179 (1990).
[28] K. W. Brocklehurst, K. Morita, and H. B. Pollard, *Biochem. J.* **228**, 35 (1985).
[29] S. L. Pocotte, R. A. Frye, R. A. Senter, D. R. TerBush, S. A. Lee, and R. W. Holz, *Proc. Natl. Acad. Sci. USA* **82**, 930 (1985).
[30] S. A. Lee and R. W. Holz, *J. Biol. Chem.* **261**, 17089 (1986).
[31] D. R. TerBush and R. W. Holz, *J. Neurochem.* **58**, 680 (1992).
[32] H. J. G. Matthies, H. C. Palfrey, and R. J. Miller, *FEBS Lett.* **229**, 238 (1988).
[33] R. L. Kenigsberg and J. M. Trifaro, *Neuroscience* **14**, 335 (1985).
[34] D. E. Clapham and E. Neher, *J. Physiol. (London)* **353**, 541 (1984).
[35] K. Morita, S. Ishii, H. Uda, and M. Oka, *J. Neurochem.* **50**, 644 (1988).

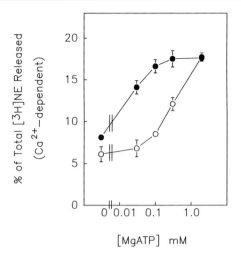

[MgATP] mM

FIG. 1. An ATP-regenerating system shifts the MgATP dose-response relation for secretion to lower MgATP concentrations. [^{3}H]Norepinephrine ([^{3}H]NE)-labeled cells were permeabilized for 6 min in Ca^{2+}-free KGEPM (pH 6.60) with 5 mg/ml BSA, 20 μM digitonin without ATP. Cells were then incubated $\pm 10\ \mu M$ Ca^{2+} with various concentrations of MgATP with (filled circles) or without (open circles) an ATP-regenerating system. The regenerating system consisted of 5 mM phosphocreatine and 25 U/ml creatine phosphokinase (Sigma, St. Louis, MO). Secretion was determined after 20 min. $n = 4$ wells/group. Secretion in the absence of Ca^{2+} was 3–4%. The presence of the regenerating system shifted the apparent EC$_{50}$ from 300 to 30 μM MgATP.

0.3–0.5 mM and maximal secretion at 1–2 mM.[2,36] However, experiments with an ATP-regenerating system utilizing phosphocreatine and phosphocreatine kinase indicate that because of ATP hydrolysis the submillimolar concentrations of ATP have been overestimated. In the presence of the regenerating system, half-maximal stimulation occurs at 30 μM and maximal stimulation at approximately 100 μM (Fig. 1). The regeneration system was not necessary to obtain maximal secretion at 2 mM MgATP.

Intact cells are primed by intracellular ATP so that immediately on permeabilization, there is a component of secretion that is independent of added MgATP.[36] MgATP partially maintains the primed state after permeabilization by acting before Ca^{2+} in the secretory pathway.[36] Secretion that occurs in the absence of medium MgATP (primed secretion) is rapid

[36] R. W. Holz, M. A. Bittner, S. C. Peppers, R. A. Senter, and D. A. Eberhard, J. Biol. Chem. 264, 5412 (1989).

but labile and usually disappears within 8 min of permeabilization. Secretion stimulated by the continuous presence of MgATP is slow but stable and results in a constant rate of secretion for approximately 12 min after permeabilization. The MgATP-dependent reaction is unknown. The MgATP dependency of secretion may reflect the phosphorylation of critical proteins[30,37-39] or the need to maintain the polyphosphoinositides for a function that is independent of the role of these lipids as substrates for the generation of inositol trisphosphate (IP_3) and diacylglycerol.[40]

Guanine Nucleotides

The presence of guanine nucleotides in the medium is not necessary for Ca^{2+}-dependent secretion from electropermeabilized or digitonin-permeabilized chromaffin[41-43] or PC12 cells.[9,44] GTPγS inhibits Ca^{2+}-dependent secretion from electropermeabilized bovine chromaffin cells when added together with Ca^{2+} [41] and from digitonin-permeabilized bovine chromaffin cells when a low concentration of GTPγS is incubated with permeabilized cells prior to stimulation with Ca^{2+}.[45] These effects are consistent with the effects of GTPγS to inhibit transfer and fusion of vesicles in earlier steps in the biosynthetic pathway of secretory proteins.[46-49] It is possible that one or more of the numerous GTP-binding proteins between 20 and 30 kDa on secretory vesicles including chromaffin granules[45,50,51] is responsible for the effects. However, guanine nucleotides have additional effects depending on species, method of plasma membrane permeabilization, and experimental design. For example, nonhydrolyzable guaine nucleotides stimulate Ca^{2+}-dependent secretion in electropermeabilized chicken chromaffin cells[41] and in bovine chromaffin cells permeabilized by *Staphylococcus aureus* α-toxin.[43] GTPγS also stimulates Ca^{2+}-

[37] C. M. Amy and N. Kirshner, *J. Neurochem.* **36**, 847 (1981).
[38] A. Cote, J. P. Doucet, and J. M. Trifaro, *Neuroscience* **19**, 629 (1986).
[39] P. D. Wagner and N.-D. Vu, *J. Biol. Chem.* **265**, 10352 (1990).
[40] D. A. Eberhard, C. L. Cooper, M. G. Low, and R. W. Holz, *Biochem. J.* **268**, 15 (1990).
[41] D. E. Knight and P. F. Baker, *FEBS Lett.* **189**, 345 (1985).
[42] A. Morgan and R. D. Burgoyne, *Biochem. J.* **269**, 521 (1990).
[43] M.-F. Bader, J.-M. Sontag, D. Thierse, and D. Aunis, *J. Biol. Chem.* **264**, 16426 (1989).
[44] A. G. Carroll, A. R. Rhoads, and P. D. Wagner, *J. Neurochem.* **55**, 930 (1990).
[45] R. W. Holz, J. Senyshyn, and M. A. Bittner, *Ann. N.Y. Acad. Sci.* **635**, 382 (1991).
[46] C. J. Beckers and W. E. Balch, *J. Cell Biol.* **108**, 1245 (1989).
[47] P. Melancon, B. S. Glick, V. Malhotra, P. J. Weidman, T. Serafini, M. L. Gleason, L. Orci, and J. E. Rothman, *Cell* **51**, 1053 (1987).
[48] J. E. Rothman and L. Orci, *FASEB J.* **4**, 1460 (1990).
[49] H. R. Bourne, *Cell* **53**, 669 (1988).
[50] R. D. Burgoyne and A. Morgan, *FEBS Lett.* **245**, 122 (1989).
[51] T. L. Burgess and R. B. Kelly, *Annu. Rev. Cell Biol.* **3**, 243 (1987).

independent secretion from digitonin-permeabilized bovine chromaffin cells[14,42] and PC12 cells.[44] Guanine nucleotides in combination with Ca^{2+} strongly stimulate secretion in mast cells.[52-54] The multitude of effects of nonhydrolyzable guanine nucleotides on regulated secretion is not surprising because guanine nucleotides can activate a host of enzymes either directly or indirectly (e.g., adenylate cyclase, phospholipase C, phospholipase A_2, protein kinase C) that could influence the secretory response. Thus, it is unclear which, if any, of the effects of guanine nucleotides on secretion requires GTP-binding proteins on secretory vesicles. Furthermore, it is possible that the GTP-binding proteins associated with secretory vesicles are not directly involved in secretion but with other functions of the vesicles related to biosynthesis of the membrane[55] or recycling after exocytosis.

Effects of Proteins on Secretion

One of the distinguishing features of digitonin- and streptolysin O-permeabilized cells is the ability of proteins to enter and leave permeabilized cells. This characteristic has permitted the study of exogenous proteins, including trypsin,[56] antibodies to fodrin,[57] bacterial phosphoinositol (PI)-specific phospholipase C,[40] and clostridial neurotoxins,[58-62] on the regulated secretory pathway.

Because loss of the secretory response from digitonin- (and streptolysin O)-permeabilized cells is associated with the loss of cytosolic proteins, considerable effort[3,25,35] has been made to identify protein components of cytosolic fractions from adrenal chromaffin cells and brain that protect against or reverse rundown. This approach assumes that loss of protein rather than the method of permeabilization is responsible for the decay of the secretory response. Although crude cytosolic fractions do enhance secretion,[25] it has been difficult to purify factors that enhance secretion from digitonin-permeabilized chromaffin cells. Part of the problem may be

[52] S. K. Fisher, R. W. Holz, and B. W. Agranoff, *J. Neurochem.* **37**, 491 (1981).
[53] Y. Churcher and B. D. Gomperts, *Cell Regul.* **1**, 337 (1990).
[54] W. R. Koopman and R. C. Jackson, *Biochem. J.* **265**, 365 (1990).
[55] S. A. Tooze, U. Weiss, and W. B. Huttner, *Nature (London)* **347**, 207 (1990).
[56] R. W. Holz and R. A. Senter, *Cell. Mol. Neurobiol.* **8**, 115 (1988).
[57] D. Perrin, O. K. Langley, and D. Aunis, *Nature (London)* **326**, 498 (1987).
[58] M. A. Bittner and R. W. Holz, *J. Neurochem.* **51**, 451 (1988).
[59] M. A. Bittner, W. H. Habig, and R. W. Holz, *J. Neurochem.* **53**, 966 (1989).
[60] M. A. Bittner, B. R. DasGupta, and R. W. Holz, *J. Biol. Chem.* **264**, *10354 (1989).*
[61] G. Ahnert-Hilger, U. Weller, M. E. Dauzenroth, E. Habermann, and M. Gratzl, *FEBS Lett.* **242**, 245 (1989).
[62] G. Ahnert-Hilger, M. F. Bader, S. Bhakdi, and M. Gratzl, *J. Neurochem.* **52**, 1751 (1989).

variability in the cell preparations. It is our experience that different preparations of cells lose their secretory response at different rates after permeabilization. For example, some preparations lose 50% of their secretory response within 5 min of permeabilization while others lose less than 50% of their secretory response after 15 min. Even among cell preparations that run down quickly, some respond strongly to incubation with a cytosolic fraction (1–5 mg protein/ml) while others respond only weakly to the same cytosolic preparation. There is a range of enhancement of 50 to 300%.

Appropriate controls are essential in protein reconstitution experiments. Bovine serum albumin (BSA) often is used. However, we have investigated a number of purified proteins and have found that they can either increase or decrease the secretory response if present before stimulation with Ca^{2+} (Table I). Therefore, protection against rundown by a crude

TABLE I

EFFECTS OF VARIOUS PROTEINS ON SECRETION FROM DIGITONIN-
PERMEABILIZED CELLS[a]

Proteins	Ca^{2+}-dependent secretion[b] (% of total radioactivity)
None	4.7 ± 0.7
Bovine serum albumin	4.3 ± 0.3
Histone	2.5 ± 0.4
Chicken ovalbumin	6.1 ± 0.9
Casein	7.0 ± 0.4
Phosvitin	9.5 ± 0.9[c]
Dialyzed chromaffin cell cytosol	9.0 ± 0.5[c]

[a] [³H]Norepinephrine-labeled chromaffin cells were permeabilized with KGEPM with 2 mM MgATP and 20 μM digitonin for 6 min. Cells were then incubated for 18 min in KGEPM with 2 mM MgATP with or without 5 mg/ml of the indicated proteins. Cells were then incubated in the continuing absence or presence of protein in KGEPM with 2 mM MgATP $\pm$ 10 μM Ca^{2+}. The percentage of radioactivity released into the medium was determined after 18 min. The following protease inhibitors were present in the last two incubations: phenylmethylsulfonyl fluoride, 1 mM; leupeptin, 1 μg/ml; pepstatin, 50 ng/ml; antipain, 1 μg/ml; chymostatin, 0.5 μg/ml.
[b] Data are expressed as mean $\pm$ standard error of the mean. $n = 3$ wells/group.
[c] $p < 0.01$ vs no proteins.

cytosolic fraction may not reflect the specific action of a critical component.

Part of the difficulty in reconstituting secretion with cytosolic factors may be that digitonin-permeabilized cells retain proteins important in the final steps of the regulated pathway. This may not be the case with GH3 cells[63] or PC12 cells (T. F. J. Martin, personal communication, 1991) permeabilized with a ball homogenizer ("cell cracker"). Suspended cells are forced through a cylinder containing a precision bearing with very small clearance with the cylinder wall. When the mechanically disrupted cells are washed several times Ca^{2+}-dependent secretion is almost completely dependent on added cytosolic proteins. Most importantly, a 40-fold purification of a 250 to 350-kDa protein responsible for this effect was achieved.[63]

There is evidence that different phases of the secretory response in permeabilized cells have different rate-limiting steps.[36] A factor in the cytosol that is required for secretion will have substantial effects only if the step at which it acts is limiting the rate of secretion.

The annexins (also called lipocortins or calcimedins) bind to membranes in a Ca^{2+}-dependent manner[64-67] and have been implicated in secretion. Incubation of digitonin-permeabilized chromaffin cells with the annexin calpactin I [or its 36-kDa heavy chain (p36)] protects against the loss of secretory response.[68-70] It is possible that p36 participates in the interaction of the chromaffin granule with the plasma membrane.[71]

Comparison of Different Permeabilization Methods

Table II is a summary of the characteristics of chromaffin cells permeabilized by different methods. Several points can be made.

1. Cells that are leaky to proteins rapidly lose their secretory response. Thus, cells permeabilized by digitonin, streptolysin O, or a ball homogen-

---

[63] T. F. J. Martin and J. H. Walent, *J. Biol. Chem.* **264**, *10299 (1989)*.

[64] R. D. Burgoyne and M. J. Geisow, *Cell Calcium* **10**, 1 (1989).

[65] C. E. Creutz, D. S. Drust, H. C. Hamman, M. Junker, N. G. Kambouris, J. R. Klein, M. R. Nelson, and S. L. Snyder, *in* "Stimulus–Response Coupling: The Role of Intracellular Calcium" (V. Smith *et al.,* eds.), p. 279. CRS Press, Boca Raton, Florida, 1990.

[66] C. E. Creutz, C. J. Pazoles, and H. B. Pollard, *J. Biol. Chem.* **253**, 2858 (1978).

[67] D. S. Drust and C. E. Creutz, *Nature (London)* **331**, 88 (1988).

[68] S. M. Ali, M. J. Geisow, and R. D. Burgoyne, *Nature (London)* **340**, 313 (1989).

[69] S. M. Ali and R. D. Burgoyne, *Cell. Signall.* **2**, 265 (1990).

[70] T. Sarafian, L.-A. Pradel, J.-P. Henry, D. Aunis, and M.-F. Bader, *J. Cell Biol.* **114**, *1135 (1991)*.

[71] T. Nakata, K. Sobue, and N. Hirokawa, *J. Cell Biol.* **110**, 13 (1990).

TABLE II
COMPARISON OF DIFFERENT PERMEABILIZATION METHODS[a]

Permeabilization method (cell type)	Freely permeable to low molecular weight species	Permeable to proteins	Ca^{2+} concentration at half-maximal secretion	Significant decay of secretory response with time	Ref.
Electropermeabilization (chromaffin cells)	Yes	No	$1-2\,\mu$	No	2
Digitonin (chromaffin cells)	Yes	Yes	$1-2\,\mu$	Yes	3,4
Staphylococcus aureus α-toxin (chromaffin cells)	?[b]	No	30^c	No	6,43
Streptolysin O (chromaffin cells)	Yes	Yes	$1-2\,\mu$	Yes	7
Ball homogenizer (cell cracker) (GH3 cells)	Yes	Yes	$0.1-1\,\mu$	Yes	73

[a] Experiments performed with bovine adrenal chromaffin cells except for permeabilization with the ball homogenizer, which was performed with prolactin-secreting GH3 cells.

[b] See text for discussion of Ca^{2+} buffering in S. aureus α-toxin-permeabilized cells.

[c] Staphylococcus aureus α-toxin-treated PC12 cells show two ranges of responsiveness to Ca^{2+}: one is $0.1-1\,\mu M$ and the other is $10-100\,\mu M$.[74]

izer lose their secretory response much more rapidly than cells permeabilized by high-voltage discharges (electropermeabilization) or S. aureus α-toxin. This comparison has stimulated the search for cytosolic proteins necessary for the final steps of regulated secretion.

2. Both digitonin[15] and streptolysin O[72] interact with membrane cholesterol and permeabilize cells with similar characteristics[7,26a]

3. The Ca^{2+} dependency for secretion is similar for all the methods (approximately 1 μM Ca^{2+} for half-maximal secretion) except for S. aureus α-toxin, for which at least a 10-fold greater Ca^{2+} concentration is required. One possible explanation for the lower sensitivity to Ca^{2+} is that the Ca^{2+} buffers do not rapidly equilibrate with the cell interior in S. aureus α-toxin-permeabilized chromaffin cells, which would result in a lower Ca^{2+} concentration within the cell than in the extracellular medium. Indeed, we have found that increasing the buffer strength threefold by increasing the

[72] S. Bhakdi and J. Tranum-Jensen, Rev. Physiol. Biochem. Pharmacol. **107**, 147 (1987).

concentration of Ca^{2+} chelators with a proportional increase in the total Ca^{2+} increased the secretory response to 100 μM Ca^{2+} two- to threefold over a wide range of α-toxin concentrations. However, it has been reported that increasing the medium buffer strength does not change the Ca^{2+} dependency of secretion from α-toxin-permeabilized chromaffin cells.[43] Investigators who use *S. aureus* α-toxin should determine the effects of different Ca^{2+} buffer strengths on the secretory response in their own preparation.

Acknowledgments

R.W.H. and M.A.B. were supported by grants from the Public Health Service (RO1 DK27959) and the National Science Foundation (BNS-9008685), respectively.

[73] T. F. J. Martin and J. A. Kowalchyk, *J. Biol. Chem.* **264**, 20917 (1989).
[74] G. Ahnert-Hilger, M. Brautigam, and M. Gratzl, *Biochemistry* **26**, 7842 (1987).

[18] Regulated Exocytotic Secretion from Permeabilized Cells

By BASTIEN D. GOMPERTS and PETER E. R. TATHAM

Background

Regulated exocytosis is controlled by a sequence of tightly coupled, biochemical steps and membrane–membrane interactions that occur within the cell interior. Investigations using whole cells or tissues, and using agonists, ionophores, and various pharmacological agents, have in many cases revealed a requirement for Ca^{2+} and an intact cellular metabolism (implying a requirement for intracellular ATP). However, because the sites at which the membrane fusions occur are intracellular, it has been necessary to gain experimental access to the cytosol to investigate the processes that lead up to the exocytotic event. This has been achieved by various methods of cell permeabilization. Here we report techniques for studying the exocytotic process in rat peritoneal mast cells. In our own work we have also investigated other myeloid cells and these methods, and more importantly the basic considerations, are widely applicable.

Permeabilized Cells

Cell permeabilization can take many forms and at the outset it should be pointed out that each method has its own distinctive features. Broadly, there are two possible outcomes: plasma membrane damage may be slight and the cells may retain much of their integrity so that only low molecular mass solutes are exchanged by dialysis, offering a measure of control of $[Ca^{2+}]_i$ (by providing Ca^{2+} buffers) and the opportunity to apply agents such as guanine nucleotides and their analogs; alternatively, membrane damage may be substantial, allowing rapid control of $[Ca^{2+}]_i$ and nucleotide concentrations and also permitting the loss of cytosol proteins to provide a simplified system. Whatever the method, the aim is to expose the exocytotic site to normally impermeant solutes while ensuring that the cells retain sufficient structural integrity to undergo a normal secretory process.[1]

An example in the first category is treatment of susceptible cells with extracellular ATP.[2,3] The permeabilizing ligand is the free acid form of ATP, i.e., ATP^{4-},[4,5] and for responsive cells concentrations in the micromolar range are typically sufficient to permit entry of Ca^{2+} buffers, fluorescent probes, nucleotides, etc., without significant loss of soluble proteins. This technique is limited in its applicability because it is restricted to those cells that possess appropriate receptors (possibly the gap junction protein, connexin 43[6]). ATP^{4-} permeabilization has been applied to the study of secretion in mast cells[7] and phagocytosis in macrophages.[8] A particular feature of the use of ATP^{4-} as a permeabilizing agent lies in the possibility of resealing the induced lesions. This can be achieved within seconds by addition of an excess of Mg^{2+} to the system.[9,10] In this way one can load and trap exogenous solutes into the cytosol of otherwise fully intact cells and it was by this strategy that the first indications of G protein involvement (G_E) in the exocytotic mechanism were obtained.[9]

To pose direct questions about the later events in the exocytotic pathway, it is necessary to reduce the complexity of the cells still further and to

[1] B. D. Gomperts and J. M. Fernandez, *Trends Biochem. Sci.* **10**, 414 (1985).
[2] B. D. Gomperts, in "Developments in Cell Biology" (R. T. Dean and P. Stahl, eds.), Vol. 1, p. 18. Butterworth, London, 1985.
[3] L. A. Heppel, G. A. Weisman, and I. Friedberg, *J. Membr. Biol.* **86**, 189 (1985).
[4] S. Cockcroft and B. D. Gomperts, *Nature (London)* **279**, 541 (1979).
[5] S. Cockcroft and B. D. Gomperts, *J. Physiol. (London)* **296**, 229 (1979).
[6] E. C. Beyer and T. H. Steinberg, *J. Biol. Chem.* **266**, 7971 (1991).
[7] J. P. Bennett, S. Cockcroft, and B. D. Gomperts, *J. Physiol. (London)* **317**, 335 (1981).
[8] F. Di Virgilio, B. C. Meyer, S. Greenberg, and S. C. Silverstein, *J. Cell Biol.* **106**, 657 (1988).
[9] B. D. Gomperts, *Nature (London)* **306**, 64 (1983).
[10] P. E. R. Tatham, N. J. Cusack, and B. D. Gomperts, *Eur. J. Pharmacol.* **147**, 13 (1988).

have the ability to manipulate the protein composition of the cell interior, yet still maintain the gross structural features of the cells. This may be achieved by applying the bacterial cytolysin streptolysin O, which forms lesions large enough to allow rapid leakage of lactate dehydrogenase (LDH).[11] It is therefore possible to deplete the cells of their cytosol proteins and, conversely, it is also possible to introduce exogenous proteins (such as antibodies) into the interior of secreting cells.[12]

While we describe the application of only two methods in detail in this chapter, we would stress the importance of applying more than one technique of cell permeabilization, so that the effects of retaining and also of releasing macromolecules from the cytosol can be discerned. Note that for any experiment involving cells in a permeabilized condition (resealed cells can be regarded as being fully intact) the use of chelating buffers to regulate the concentration of free Ca^{2+} becomes mandatory.[13] Mainly these are based on ethylene glycol-bis(β-aminoethyl ether)-N,N,N',N'-tetraacetic acid (EGTA), which discriminates strongly between Ca^{2+} and Mg^{2+}, although when Mg^{2+} is excluded from the system or when it is desirable to work at elevated pH (e.g., when using ATP^{4-} as a permeabilizing ligand) then we have used hydroxyethylethylenediaminetetraacetic acid (HEDTA; see below).

Procedures

Preparation of Mast Cells

Mast cells may be obtained from rats by peritoneal lavage using ~ 30 ml of 150 mM NaCl containing 1 mg/ml bovine serum albumin. The washings are centrifuged (approximately 250 g for 5 min) and after resuspension the mast cells are isolated by centrifuging this suspension through a cushion of Percoll (Pharmacia LKB, Milton Keynes, UK). Typically 7 ml of suspension are layered over 2 ml of density medium [1.114 g/ml, e.g., 87.8% (v/v) of original density, 1.124 g/ml Percoll solution in buffered saline] and centrifuged (250 g) for 5 min at room temperature. The top layers are then removed by aspiration and the pellet, which should consist of mast cells at over 95% purity, is washed twice in a buffered salt solution comprising 137 mM NaCl, 2.7 mM KCl, 1 mg/ml bovine serum albumin, and 20 mM piperazine-N,N'-bis(2-ethanesulfonic acid) (PIPES) or N-2-hydroxyethylpiperazine-N'-2-ethanesulfonic acid (HEPES), depending on

---

[11] T. W. Howell and B. D. Gomperts, *Biochim. Biophys. Acta* **927**, 177 (1987).
[12] D. Perrin, O. K. Langley, and D. Aunis, *Nature (London)* **326**, 498 (1987).
[13] P. E. R. Tatham and B. D. Gomperts, *in* "Peptide Hormones—A Practical Approach" (K. Siddle and J. C. Hutton, eds.), Vol. 2, p. 257. IRL Press, Oxford, 1990.

the final pH required. The washings from 1 large Sprague-Dawley rat should provide at least 10^6 cells, quite sufficient for 100 secretion determinations.

Permeabilization and Mast Cell Loading Using ATP^{4-}

Permeabilization by the free acid ATP^{4-} is most conveniently carried out at pH 7.8 (using HEPES buffer) and in the absence of Ca^{2+} and Mg^{2+} (which form inactive complexes with ATP). To ensure low concentrations of divalent cations, the buffered saline should be supplemented with a low concentration (25 μM) of ethylenediaminetetraacetic acid (EDTA). If it is desirable to conserve the amount of material for loading, the working volumes can be kept very small. Having resuspended the cells in this medium, ATP is added (as its Na or Tris salt) to a final concentration of 3 to 5 μM and the cells incubated at 37° or room temperature for 5 min. Under these conditions plasma membrane lesions form rapidly and substances with molar masses of at least 1 kDa (e.g., Fura 2, Indo 1, and other indicator probes as their free anions; for further details see Ref. 13). can be taken up by the cells. The process is terminated by removal of ATP^{4-}, which may be achieved by adding excess Mg^{2+} to the suspension (1 mM $MgCl_2$). Full recovery can be achieved by allowing a further period of incubation (10 to 15 min). The cells should then be washed three times in the normal buffered saline to remove all traces of the extracellular material (although if they have been loaded with guanine nucleotides, the extracellular $[Ca^{2+}]$ should be kept low to avoid premature activation[7]).

Permeabilization of Mast Cells Using Streptolysin O

Here we describe in detail the method we use to assess the dependence on Ca^{2+} and guanine nucleotide for secretion of β-N-acetylglucosaminidase (hexosaminidase) from mast cells permeabilized by streptolysin O. We would stress that it should be easy to adapt these methods to a wide range of other secretory products, cell types, and tissues.

Dependence of Exocytosis on Ca^{2+} and $GTP\gamma S$ Concentration

To make a simultaneous investigation of the dependence of secretion on the concentrations of the two principal effectors, Ca^{2+} and guanine nucleotide, a matrix of incubations is set up. The rows and columns of this matrix correspond to increasing $[Ca^{2+}]$ and $[GTP\gamma S]$, respectively (in duplicate pairs). It is most convenient to carry out such incubations using standard 96-well microtiter plates. Additional plates may be used to enlarge the format of the matrix. An experiment is illustrated in Fig. 1 as an example. Here Ca^{2+}, $GTP\gamma S$, and streptolysin O are applied simulta-

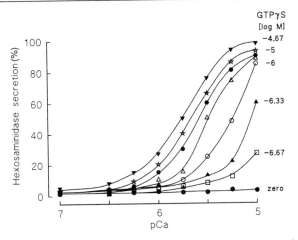

FIG. 1. Model experiment to determine Ca^{2+} and GTPγS dependence of secretion from streptolysin O-permeabilized rat mast cells. Cells pretreated with metabolic inhibitors for 5 min are added to tubes containing streptolysin O (0.4 IU/ml), 3 mM CaEGTA buffers (pCa 7 to 5), and GTPγS at $10^{-4.67}$, 10^{-5}, $10^{-5.33}$, $10^{-5.67}$, 10^{-6}, $10^{-6.33}$, $10^{-6.67}$ M, and zero, as indicated. All concentrations are final; further details are given in the text.

neously to mast cells that have been pretreated with metabolic inhibitors. ATP is not provided. In the procedure described below, each plate well contains equal volumes of solutions of Ca^{2+} buffer, GTPγS, streptolysin O, and of cell suspension (30 μl of each is a convenient quantity). Each should therefore be prepared at four times its final concentration in saline buffered at pH 6.8 (PIPES) as follows.

Ca^{2+} *Buffers.* A set of CaEGTA buffers (see below) should be available as nominally 100 mM stock solutions ([EGTA]), covering the range pCa 7 to pCa 5; from these make a set of 12 mM working stocks and add 30 μl to paired columns (i.e., 1 and 2, 3 and 4, etc.) on the microtiter plates. Select buffers to maintain pCa 7, pCa 6.5, pCa 6.25, pCa 6, pCa 5.75, pCa 5.5, pCa 5.25, and pCa 5. The final [EGTA] will be 3 mM.

GTPγS Solution. As with the calcium, it is convenient to use a logarithmically graded set of concentrations of the guanine nucleotide. Prepare seven test tubes containing 1 ml of buffer and in an eighth tube prepare 1 ml of 86.2 μM GTPγS (i.e., $4 \times 10^{-4.67}$ M). Transfer 0.866 ml of this to the seventh tube and mix (to give 4×10^{-5} M); now transfer 0.866 ml of this to the sixth tube and repeat the process so that, finally, the first tube contains the lowest concentration ($4 \times 10^{-6.67}$ M). (This series has three steps per decade; transferring 0.462 ml into 1 ml at each stage will give two steps per decade.) Pipette 30 μl of the appropriate GTPγS solution into each well of a row of the microtiter plate. Note that Boehringer Mannheim (BCL, Ltd., Lewes, UK) supplies the thio analogs as 100 mM solutions of high purity. In this form they are stable for many months if kept frozen.

Streptolysin O Solution. Dissolve streptolysin O (Wellcome Diagnostics, Beckenham, UK) by adding 2 ml of water to give a stock of 20 IU/ml. (Although it becomes cloudy, this solution may be stored for several days at 4°.) Make a working stock of 1.6 IU/ml and pipette 30 μl into each of the wells on the microtiter plate.

Cell Suspension. Dilute the cell suspension in 10 ml of buffer (glucose free) containing 2-deoxyglucose (6 mM) and antimycin A (10 μM) and incubate at 37° for 5 min. At the end of this time the cells will be refractory to stimulation by receptor-directed agonists (IgE-directed ligands or compound 48/80) or to calcium ionophores.

To commence the experiment, place the microtiter plate on a suitable support in a water bath (37°) and, using a repeating pipette, add 30 μl of cell suspension to each well. To calibrate the secretion, it is important to reserve cells for lysis by Triton X-100 (0.2%, v/v), which releases the total hexosaminidase. After incubation for 10 min, 150 μl of ice-cold buffered NaCl solution (120 mM NaCl, 10 mM EGTA, 25 mM HEPES buffer, pH 8) are added to each well to quench any residual release. The cells are then sedimented by centrifugation (1200 rpm for 5 min using a standard refrigerated benchtop centrifuge with microtiter plate adaptors) and supernatant samples (50 μl) removed for assay. (*Note:* In the experiment described it is not strictly necessary to stop the ongoing reactions with quench solution, but the additional volume of fluid in each well improves the accuracy of supernatant removal.)

Measurement of Secreted Hexosaminidase. This may be accomplished by enzyme assay using appropriate chromogenic or fluorogenic substrates and either an absorption or fluorescence microtiter plate reader. Absorption assays (e.g., measurement of p-nitrophenolate) require longer incubations with substrate to achieve the same sensitivity as fluorescence assays (e.g., measurement of 4-methylumbelliferone). It is also possible to carry out the enzyme reaction in test tubes and to record fluorescence or absorption on a tube-by-tube basis using a conventional spectrophotometer or fluorimeter. We use black plastic 96-well plates and a fluorescence plate reader to measure 4-methylumbelliferone. Both methods are described below.

Use a multichannel pipette to transfer 50 μl of supernatant, containing secreted hexosaminidase, directly into the corresponding wells of the receiving plate. To calibrate the system, include one row (or column) of reagent blanks, and a row (or column) of appropriate amounts of lysed cells. Add 50 µl of 1 mM 4-methylumbelliferyl N-acetyl-β-D-glucosaminide (fluorogenic substrate) or p-nitrophenyl-N-acetyl-β-D-glucosaminide (chromogenic substrate), dissolved in 0.2 M citrate buffer, pH 4.5, to each well and incubate at 37° for about 2 or 15 hr, respectively. Quench the reactions by addition of 150 µl of 1 M Tris and measure the developed colors at the appropriate wavelengths (360–450 nm for fluorescence, and 400 nm for absorption). Most microtiter plate readers can be coupled to computers so that the data may be read, processed, and presented graphically within minutes by using a suitable spreadsheet program.

The simple experiment outlined above is amenable to almost limitless variation and forms a good basis for work with other cell types. As examples, we have tested the effects of providing ATP and other nucleoside triphosphates,[14] substitution of glutamate for Cl⁻ in the formulation of buffer solutions,[15] the use of [³H]inositol- or [³H]arachidonate-labeled cells to measure phospholipase C or phospholipase A_2 activity in parallel with exocytosis,[16,17] the effects of applying Ca^{2+} and GTPγS to permeabilized cells after a timed delay,[18] and many other variations. We have also adapted the method to measure secretion from rat basophilic leukemia cells, cultured in the microwells overnight: here the effectors are added to the intact adherent cells and addition of the permeabilizing agent provides the trigger.

Time Course of Secretion from Permeabilized Mast Cells

The time course of secretion from streptolysin O-permeabilized mast cells is affected by the concentrations of Ca^{2+} and of GTPγS as well as by other factors. Here we describe in detail the method we use to assess the effect of varying the concentration of Ca^{2+} on the progress of secretion from permeabilized mast cells. An example is shown in Fig. 2. In this experiment six reactions proceed in parallel. The water bath temperature is set to 30° to slow the secretory process, so that by sampling at 3-sec intervals a reasonable time resolution is achieved.

At the start the water bath holds six reaction tubes, each of which contains streptolysin O (0.4 IU/ml final) in 0.5 ml of buffer. They also contain 10 µM GTPγS and 0.2 mM CaEGTA (pCa 8). Adjacent is an ice

[14] T. H. W. Lillie, T. D. Whalley, and B. D. Gomperts, *Biochim. Biophys. Acta* **1094**, 355 (1991).

[15] Y. Churcher and B. D. Gomperts, *Cell Regul.* **1**, 337 (1990).

[16] S. Cockcroft, T. W. Howell, and B. D. Gomperts, *J. Cell Biol.* **105**, 2745 (1987).

[17] Y. Churcher, D. Allan, and B. D. Gomperts, *Biochem. J.* **266**, 157 (1990).

[18] Y. Churcher, I. M. Kramer, and B. D. Gomperts, *Cell Regul.* **1**, 523 (1990).

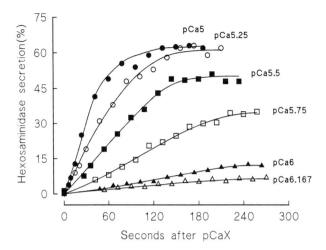

FIG. 2. Ca^{2+} dependence of the time course of secretion from streptolysin O-permeabilized rat mast cells equilibrated with GTPγS. Cells pretreated with metabolic inhibitors are added to a mixture of streptolysin O (0.4 IU/ml) and GTPγS. After an equilibration period (1 to 2 min), the reaction is started by the addition of Ca^{2+} buffer (pCa as indicated). Samples are then withdrawn at timed intervals and quenched. (Details are given in the text.) Note the time resolution of the sampling at the early time points.

bath that holds a rack containing 6 lines of 14 tubes, each containing 0.5 ml of phosphate-buffered saline (PBS) to quench the exocytotic reaction.

Before commencing, the cells ($\sim 0.5 \times 10^6$) are metabolically inhibited by incubating them for 5 min in 3 ml of (glucose-free) buffer containing 6 mM 2-deoxyglucose, 5 μM antimycin A, and 0.2 mM CaEGTA, pCa 7. They are then permeabilized by transferring 0.5-ml aliquots at 3-sec intervals to the six reaction tubes, starting with the one that will receive the lowest concentration of Ca^{2+} (i.e., the tube in which it is anticipated the reaction will proceed the slowest).

After 30 sec a "zero-time" sample (70 μl) is withdrawn from each tube and quenched. At 1 min, the exocytotic reaction is triggered by addition of 0.1 ml of 30 mM Ca^{2+} buffer (pCa 6.167 to pCa 5, final [EGTA] is 3 mM; see Fig. 2). This is again done in sequence at 3-sec intervals such that the fastest responding cells (i.e., those treated with the highest concentrations of Ca^{2+}) are triggered last and then sampled and quenched first in a reverse sequence, starting 3 sec later. Thereafter the six tubes are sequentially sampled and quenched as described at 3-sec intervals (i.e., a cycle of 18 sec). Because these operations require the undivided attention of the experimenter an audible prompt is necessary to time the sampling. This is most conveniently provided by a clockwork metronome (setting *andante*, 80 beats/min, 4/4 time). Finally, the quenched cells are sedimented and the supernatant sampled for determination of secreted hexosaminidase. This can be accomplished using microtiter plates as described above.

Introduction of Exogenous Proteins into SL-O-Permeabilized Cells

One of the advantages of streptolysin O-permeabilized cells is that they offer the possibility of introducing exogenous proteins into the cell interior, with the opportunity of modifying the exocytotic process by agents such as specific antibodies. Little has so far been achieved in this direction with mast cells, but certain features of the permeabilized cells have important implications for such experiments. To achieve effective permeation and binding to internal structures, the cells should be left in an open condition for an extended period of time. However, experiments have shown that the responsiveness of mast cells to stimulation by Ca^{2+} and GTPS declines after several minutes of permeabilization.[19] Inclusion of ATPMg (1 mM) in the permeabilizing medium has the effect of prolonging cell responsiveness for a period of about 5 min, and a useful response can still be obtained up to about 20 min. To extend this time we have found that it is possible to keep the permeabilized cells on ice (for an indefinite period) and then to activate them by applying the stimulus (Ca^{2+} plus GTPγS) and simultaneously rewarming to 37°. Note that while streptolysin O binds to the cells in the cold, permeabilization does not occur, so this must be achieved first at 37° before cooling. It is then necessary to ascertain that the permeability lesions remain patent at ice temperature. The simplest test, which measures membrane permeability to low molecular weight solutes, is to monitor the uptake and binding of (normally impermeant) ethidium bromide to DNA, using a fluorimeter with a cooled cuvette. To detect protein loading, a method involving detection of uptake and binding of an antibody to a fixed intracellular structure should be devised.

Preparation of Calcium Buffers

Calcium buffers are solutions of an appropriate chelating agent together with Ca^{2+}. The relative proportions of each determine the concentration of free Ca^{2+}. The most familiar chelators for Ca^{2+} are EDTA and EGTA. EDTA, however, has poor selectivity for Ca^{2+} over Mg^{2+} and its affinity for Ca^{2+} (k_{app} 6.77) at pH 6.8 is too high to allow satisfactory regulation in the range pCa 7–pCa 5. Both of these chelators suffer from the drawback of having significant affinity for hydrogen ions at physiological pH, and this

---

[19] T. W. Howell, I. Kramer, and B. D. Gomperts, *Cell. Signall.* **1,** 157 (1989).

affects their ability to buffer Ca^{2+}. For example, at pH 7 the CaEGTA buffering system will give control in the range pCa 7.5 to pCa 5.5 and the buffering capacity is poor at $[Ca^{2+}] >$ pCa 5.75. If pH is maintained at 6.8, however, CaEGTA (k_{app} 6.19) gives effective control over the range of Ca^{2+} concentrations most appropriate for investigations of intracellular processes (i.e., between pCa 7 and pCa 5). For experiments in which it is important to eliminate Mg^{2+}, hydroxyethylethylenediaminetriacetic acid (HEDTA) may replace EGTA, because this chelator discriminates poorly between Ca^{2+} and Mg^{2+} and has a lower affinity for Ca^{2+} than EDTA. HEDTA buffers are also useful in ATP-permeabilization experiments, as they provide control over the range pCa 7 to 5 at pH 7.8 (k_{app} 6.14), which allows full dissociation of ATP^{4-}. Because the affinity constants of these chelators are sensitive to the hydrogen ion concentration control of pH is very important, and an accurately calibrated pH meter, reading to at least two decimal places, should be employed.

In the method described here the buffer stock solutions are prepared from solutions of CaEGTA and EGTA of identical molar concentrations and with the same pH. These are combined in varying proportions and the relative amounts of each can be determined using one of the many available computer programs. We use LIGANDY, which is an IBM PC-compatible modification of the program CHELATE, provided by Dr. Sherwin Lee (University of Pennsylvania, Philadelphia). It uses the algorithm of Perrin and Sayce[20] to compute $[Ca^{2+}]_{total}$ and $[Mg^{2+}]_{total}$ from given values of $[Ca^{2+}]_{free}$ and $[Mg^{2+}]_{free}$ and vice versa.

Commercially supplied EGTA is customarily only 95% pure and for maximum accuracy it is desirable to titrate it first against standard $CaCl_2$ solution,[21] in order to prepare the standard EGTA and CaEGTA solutions. The Ca^{2+} buffer stock solutions obtained by mixing the standard solutions are stored frozen. A detailed recipe is as follows.

1. Prepare 500 ml of 0.25 M $CaCl_2$ by diluting commercially supplied 1 M solution (analytical grade).

2. Prepare 500 ml of a solution containing approximately 0.2 M EGTA and 40 mM PIPES. To dissolve the EGTA mix the acid form with approximately four times the molar equivalent of NaOH. Adjust the pH to neutrality.

3. Clean all volumetric glassware (burette and pipettes) in chromic acid or a detergent such as Decon90 (Decon Laboratories, Hove, UK) and rinse

[20] D. Perrin and I. G. Sayce, *Talanta* **14**, 833 (1967).
[21] D. J. Miller and G. L. Smith, *Am. J. Physiol.* **246**, C160 (1984).

extensively with water. Before filling, rinse the glassware with the appropriate solution.

4. Titrate 0.25 M CaCl$_2$ (in the burette) with 50 ml of the buffered chelator after adding 5 ml of 10 M NaOH so as to keep the pH above 10. The purpose of this is to maintain the chelator fully dissociated throughout the titration, during which protons are displaced by Ca^{2+}, and it ensures a sharp end point after addition of approximately 40 ml of CaCl$_2$. A suitable indicator is provided by 2.5 ml of 1 M potassium oxalate, titrating until cloudiness persists for 1 min.

5. Repeat this titration twice.

6. Now prepare the CaEGTA stock solution using the same pipette and burette to mix 50 ml of chelator solution (but omit both the alkali and the indicator) with the exact amount of CaCl$_2$ indicated by the titrations. The total volume should be approximately 90 ml.

7. Prepare the EGTA stock solution by addition of 40 ml water to 50 ml EGTA. *Note:* Although not strictly necessary, it is sometimes convenient at this stage to add Mg^{2+} as MgCl$_2$ to each of these stock solutions. For example, for a final buffer concentration of 3 mM with 2 mM Mg^{2+}, MgCl$_2$ should be present at two-thirds the final chelator concentration.

TABLE I
RECIPE FOR CA^{2+} BUFFER SOLUTIONS ^{a}

| | [Ca]$_{total}$ | Volume (ml) | |
pCa	[EGTA]$_{total}$	CaEGTA	EGTA
8	0.014	0.112	7.888
7	0.124	0.996	7.004
6.75	0.202	1.614	6.386
6.5	0.310	2.481	5.519
6.25	0.444	3.555	4.445
6.0	0.587	4.698	3.302
5.75	0.717	5.736	2.264
5.5	0.819	6.552	1.448
5.25	0.891	7.125	0.875
5	0.938	7.501	0.499

^{a} These values are calculated for a final [EGTA]$_{total}$ = 3 mM, [Mg^{2+}] = 2 mM and pH 6.8. Eight milliliters of each buffer stock solution are prepared by mixing the CaEGTA and EGTA solutions described in the text, in the proportions indicated in the last two columns. For use, the buffer stocks (which contain Mg^{2+} at two-thirds of the total [EGTA]) are diluted to 3 mM.

This can be achieved by adding approximately 6.67 ml of 1 M MgCl$_2$ (commercially supplied). However, because EGTA binds Mg^{2+} to a small extent, a more accurate 2:3 molar ratio of Mg^{2+} to EGTA over the complete series is provided when the ion is added to the CaEGTA stock at 2.01:3 and to the EGTA stock at 2.22:3. The pH is then adjusted to 6.80 and the solutions brought to exactly 100 ml in volumetric flasks.

8. At this stage we have prepared two solutions, CaEGTA and EGTA, both at pH 6.80. The concentration of each is the same (nominally 100 mM) and they can be stored for many months if they are frozen in tightly capped plastic containers. Preparing the buffers from these stocks is achieved simply by mixing them in proportions determined by the computer program, entering the concentration of free Ca^{2+} required. An example is shown in Table I. Note that there will be a small displacement of Ca^{2+} by added Mg^{2+} and this could be compensated for, but the effect is negligible except at the lowest levels of Ca^{2+}.

Acknowledgments

This work was supported by grants from the Wellcome Trust. We thank the Royal Society and the Gower Street Secretory Mechanisms Group for funds to purchase microtiter plate readers.

[19] Transfer of Bulk Markers from Endoplasmic Reticulum to Plasma Membrane

By Felix Wieland

Which steps connecting the sequential compartments of the endoplasmic reticulum (ER)–Golgi–plasma membrane system require special signals to determine the fate of a protein? In one view, now widely accepted and termed the "bulk flow" model, no special signal is needed to allow constitutive forward movement toward the cell surface; rather, retention signals are needed to enable resident proteins to stay in place en route, and diversion signals are needed at the trans-Golgi network for destinations other than the cell surface (for reviews, see Refs. 1–3). Alternative views include the idea that forward movement of newly synthesized proteins is selective, requiring a special signal.[4]

[1] R. B. Kelly, *Science* **230**, 25 (1985).
[2] S. R. Pfeffer and J. E. Rothman, *Annu. Rev. Biochem.* **56**, 829 (1987).
[3] J. E. Rothman, *Cell* **50**, 521 (1987).
[4] H. F. Lodish, *J. Biol. Chem.* **263**, 2107 (1988).

In favor of the latter hypothesis is the observation that secretory and other proteins can exit the ER at widely different rates.[5-10] Major lines of evidence in favor of the bulk flow hypothesis are that prokaryotic proteins (lacking ER-to-Golgi transport signals) are secreted when redirected to the lumen of the ER,[11] that transplantable retention signals have now been identified for both ER membrane and luminal proteins,[12-15] and that despite 10 years of effort no definitive signal for export from the ER was found. A crucial prediction of the bulk flow hypothesis is that transport from the organelles ER and Golgi to the cell surface should occur by default. Consistent with this are the findings of retention signals in ER-resident proteins (e.g., see Ref. 15) and the observations that mutations or deletions in certain Golgi resident proteins result in mislocalization to the cell surface.[16,17] What is needed to test the bulk flow model most directly in a cell of interest is a measurement of the rate of externalization of inert compounds introduced into the lumina of the ER and Golgi cisternae in comparison to the rate of export of proteins from the same sites. Any compound present in these organelles will be secreted at some rate. The key question is kinetic competency: whether the rate of bulk flow is as fast as transit of the fastest proteins.

Described here in a more general way is a concept about how to introduce small molecules as probes for the secretory pathway; also presented are luminal markers to measure the rate of transfer from the ER to the plasma membrane and from the Golgi to the plasma membrane, respectively.

General Considerations

An artificial luminal marker to measure the rate of vesicular transport requires the following characteristics: (1) it must be generated exclusively in the lumen of the organelle of interest; (2) it must be generated synchro-

[5] T. Fitting and D. Kabat, J. Biol. Chem. **257**, 14011 (1982).
[6] H. F. Lodish, N. Kong, M. Snider, and G. J. A. M. Strous, Nature (London) **304**, 80 (1983).
[7] B. E. Ledford and D. F. Davis, J. Biol. Chem. **258**, 3304 (1983).
[8] K.-T. Yeo, J. B. Parent, T. K. Yeo, and K. Olden, J. Biol. Chem. **260**, 7896 (1985).
[9] G. Scheele and A. Tartakoff, J. Biol. Chem. **260**, 926 (1985).
[10] D. B. Williams, S. J. Swiedler, and G. W. Hart, J. Cell Biol. **101**, 725 (1985).
[11] M. Wiedmann, A. Huth, and T. A. Rapoport, Nature (London) **309**, 637 (1984).
[12] S. Paabo, B. M. Bhat, W. S. M. Wold, and P. A. Perterson, Cell **50**, 311 (1987).
[13] M. S. Poruchinsky, C. Tyndall, G. W. Both, F. Sato, A. R. Ballamy, and P. A. Atkinson, J. Cell Biol. **101**, 2199 (1985).
[14] M. S. Poruchinsky and P. A. Atkinson, J. Cell Biol. **107**, 1697 (1988).
[15] S. Munroe and H. R. B. Pelham, Cell **48**, 899 (1987).
[16] R. S. Fuller, A. J. Brake, and J. Thorner, Science **246**, 482 (1989).
[17] K. J. Colley, E. U. Lee, B. Adler, J. K. Browne, and J. C. Paulson, J. Biol. Chem. **264**, 17619 (1989).

nously in this lumen in millions of cells to allow biochemical analysis; (3) it must be trapped in the membrane system of interest so as to exclude leaking out of the cells by diffusion; (4) its generation must occur at a constant rate in the course of the experiment, i.e., its precursor must be available in excess; and (5) it must not contain any structures that could be recognized by a tentative transport receptor. Figure 1A shows this concept. A substrate (or substrate analog) that is able to diffuse through biological membranes (S_d) is added to the medium of cultured cells. It will penetrate all membranes and therefore equilibrate with the lumina of all organelles. However, the enzyme for this substrate is restricted to the organelle of interest, and therefore only in this lumen will the substrate be reactive.

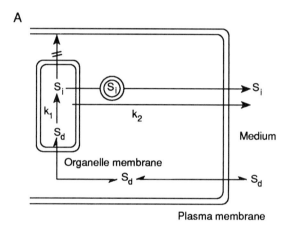

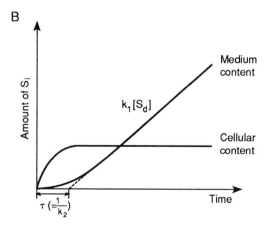

FIG. 1. (A) Concept for the generation of a luminal marker to measure the rate of vesicular flow. (B) Evaluation of the transport rates from an experiment as explained in (A). (From Ref. 18.)

Here the amphilic substrate S_d is converted to yield a product (S_i) with hydrophilic characteristics (such as ionic charges or saccharides) that inhibit diffusion through biological membranes of this substance $(S_i$, indiffusible substance). k_1 is the rate constant of formation of S_i; k_2 is the rate constant of vesicular transport of S_i from the organelle to the plasma membrane. After expression at the plasma membrane, S_i cannot remain in the outer leaflet because of its hydrophilic character and therefore is released into the medium. Evaluation of a kinetic experiment is described in Fig. 1B. Substrate S_d is added in excess and its diffusion is much faster than the subsequent reactions, so that in the cell [S] = constant. Transport follows first-order kinetics, and at steady state S_i is linearly accumulated into the medium. Extrapolation of the resulting straight line to the X axis yields a lag time that represents the mean residence time of S_i in the cell, $\tau = 1/k_2$. With first-order kinetics

$$t_{1/2} = \frac{\ln 2}{k}$$

$$t_{1/2} = \tau \ln 2$$

Therefore τ is easily correlated with the half-time of transport. Practically, after addition of the labeled substrate S_d aliquots of a cell suspension are removed at various times and the content of S_i in the cells and corresponding medium is determined. Extrapolation to the time axis must occur from a straight line of accumulation of S_i into the medium, and as a control, the cellular level of S_i during the time of linear accumulation into the medium must be constant.

A luminal marker for the ER and one for the proximal Golgi will be discussed in the following section.

N-Glycosylated Tripeptides as Markers for Lumen of Endoplasmic Reticulum[18]

The ER is the site of N-glycosylation,[19] and it has been shown *in vitro* that small peptides with the consensus structure X-Asn-Y-Thr/Ser- are sufficient to serve as acceptors for the unique oligosaccharide that is provided as a dolichol pyrophosphate-linked precursor.[20] We have synthesized the tripeptide Asn-Tyr-Thr and blocked its ionic charges by octanoylation of its α-amino group and by amidation of its carboxy terminus. For

---

[18] F. T. Wieland, M. L. Gleason, T. Serafini, and J. E. Rothman, *Cell* **50**, 289 (1987).

[19] R. Kornfeld and S. Kornfeld, *Annu. Rev. Biochem.* **54**, 631 (1985).

[20] J. K. Welphy, P. Shenbagamurthi, W. J. Lennarz, and F. Naider, *J. Biol. Chem.* **258**, 11856 (1983).

convenient quantitation the tripeptide was labeled with ^{125}I in its tyrosine residue. This compound readily diffuses into cells and is glycosylated in the ER, and the resulting glycotripeptides do not diffuse through membranes and therefore represent markers for the lumen of this organelle. Due to the minimized structure of the consensus peptide as well as the unique structure of the oligosaccharide shared by all N-glycosylated proteins in the ER, the glycopeptides are not expected to contain any signal structure that could be recognized by a putative transport receptor.

Preparation and Iodination of Tripeptide

Solid-phase synthesis is performed according to Ref. 21. Benzhydrylamine-threonine-resin is used. Boc-protected amino acids as well as the resin are from Peninsula Laboratories (Belmont, CA). The peptide is cleaved from the resin by solvolysis with anhydrous hydrogen fluoride. The resultant peptide is purified by washing the resin with ether and subsequent extraction with 50% acetic acid. The dried extract is further purified by high-performance liquid chromatography (HPLC). The tripeptide H_2N-Asn-Tyr-Thr-NH_2 elutes from an RP-18 column at 21% (v/v) acetonitrile, 0.1% (v/v) trifluoroacetic acid (TFA) in a linear gradient of 5–65% acetonitrile in 0.1% TFA at 1 ml/min. The purified peptide is acylated with octanoic acid[22] using octanoic *p*-nitrophenyl ester. Purification is achieved by HPLC on RP-18 under conditions as described for the purification of the free tripeptide. The octanoylated peptide elutes at about 35% acetonitrile.

For radioiodination, up to 50 nmol of peptide in 50 μl acetonitrile is added to 100 μl of 0.5 M NaP$_i$ (pH 7.5). Between 0.5 and 10 mCi Na^{125}I (carrier free; ICN Pharmaceuticals, Irvine, CA) is added. To this solution, 100 μl of chloramine-T (Sigma, St. Louis, MO) (2 mg/ml) in 0.05 M NaP$_i$ (pH 7.5) is added. After 1–2 min at room temperature, the reaction is stopped by the addition of 400 μl of a solution of sodium bisulfite (2.4 mg/ml) in 0.05 M NaP$_i$ (pH 7.5). An additional 600 μl of water is added, and the solution is loaded onto a Sep-Pack C$_{18}$ cartridge (Waters, Milford, MA) and washed first with 20 ml of 0.1% TFA and then with 20 ml of 5% acetonitrile in 0.1% TFA. The radiolabeled peptide is eluted with 60% acetonitrile in 0.1% TFA. Yields of iodination are between 25 and 50%. Specific radioactivities range from 1×10^7 to 4×10^8 cpm/nmol. The purity of the iodinated peptide is confirmed by HPLC (see above) and by thin-layer chromatography on silica gel plates [Si250-PA (Baker, Phillips-

---

[21] J. M. Stewart and J. D. Young, Pierce Chem. Co., Rockford, Illinois, 1984.

[22] M. Bodanski, "The Peptides: Analysis, Synthesis, Biology" (E. Gross and J. Meienhofer, eds.), Vol. 1, p. 106. Academic Press, New York, 1979.

burg, NJ) in butanol/acetic acid/ water (5:2:2, v/v/v)]. After drying in a Speed-Vac (Savant, Hicksville, NY) concentrator centrifuge, the radiolabeled peptides are dissolved in dimethyl sulfoxide (DMSO) at about 5×10^6 to 10×10^8 cpm/μl.

Standard Assay for Quantitation of Glycosylated Tripeptide

Aliquots of 200 μl are removed from cell suspensions at $0.5-1 \times 10^7$ cells/ml, chilled on ice, centrifuged at 6000 g for 1 min, and the supernatant media are separated from the cell pellets. Cell pellets are extracted with 200 μl of buffer A [10 mM Tris-HCl, pH 7.4, 0.15 M NaCl, 1 mM CaCl$_2$, 1 mM MnCl$_2$, 0.5% (v/v) Triton X-100], and the supernatant after centrifugation in a microfuge is saved for further analysis. The media are made 0.5% (v/v) in Triton X-100, and 1 mM CaCl$_2$ and 1 mM MnCl$_2$ are added from stock solutions.

Equivalent aliquots of Triton extracts of cells and media (typically 100- to 200-μl volume) are passed through small columns (200- to 400-μl bed volume) of concanavalin A-Sepharose (Pharmacia, Piscataway, NJ) in buffer A. The columns are washed with five successive 1-ml portions of buffer A and then eluted with three 500-μl portions of 0.5 M α-methylmannoside in buffer A. [125]I radioactivity in the combined α-methylmannoside eluates is determined. ^{125}I-Labeled glycopeptides intended for further analysis (e.g., thin-layer chromatography) are prepared similarly, but after the washes with buffer A, five washes are performed with 1 ml each of buffer A without Triton X-100. Elution with α-methylmannoside is without Triton X-100 as well.

Generation of Luminal Marker for Golgi Apparatus

The Golgi apparatus is known to be the site of sphingolipid biosynthesis.[23,24] Sphingolipids are generated by conversion of ceramide either with phosphorylcholine (from phosphatidylcholine) to yield sphingomyelin or with glucose from UDPglucose to yield glucocerebroside. This glycolipid is further converted to a variety of higher glycosphingolipids in different cell types.

We have developed an amphiphilic ceramide truncated in both the sphingosine and the fatty acyl residue to a chain length of only eight carbon atoms. This truncated ceramide (tCA) is able to diffuse through membranes and, due to its short hydrophobic chains, will not intercalate in a bilayer. Therefore the substance equilibrates into all organelles of a living

---

[23] R. E. Pagano, *Trends Biochem. Sci.* **13**, 202 (1988).
[24] G. Van Meer, *Annu. Rev. Cell Biol.* **5**, 247 (1989).

cell. In the Golgi, it is converted into a truncated sphingomyelin (tSPH) and a truncated glucocerebroside (tGlcCer).[25,26] Truncated SPH turned out not to diffuse through membranes and therefore can serve as a marker for the transport of lumenal content of the Golgi apparatus. In semiintact cells tSPH transport was shown to be temperature dependent and inhibited by the unhydrolyzable GTP analog, GTPγS.[27] For convenient analysis we have synthesized a radiolabeled tCA with ^{3}H at carbon atoms 2 and 3 of its octanoyl residue.

Synthesis of D-erythro-trans-Sphingosine C_8 [(2S, 3R, 4E)-2-Amino-4-octene-1,3-diol]

Synthesis of the truncated sphingosine is essentially as described for long-chain erythrosphingosines,[28] with the following modifications: the Wittig reaction is performed using butyltriphenylphosphonium bromide, and thin-layer chromatography (TLC) of the reaction product is performed on silica gel with petroleum ether/acetic acid ethyl ester, 9:1 (R_f 0.16). The same solvent is used for analysis of the products after introduction of the azido group (R_f 0.68). Removal of the benzylidene-protecting group is performed in dry methanol without addition of dichloromethane. The products are analyzed by TLC on silica gel in dichloromethane/methanol, 95:5 (R_f 0.31). Reduction of the azido group is performed in pyridine/water, 1:1 (rather than 2:1). Analysis is carried out by TLC on silica gel in chloroform/methanol, 1:1 (R_f 0.19).

For storage, the resulting aminocompound is transformed to its hydrochloride by acidification with methanolic HCl.

Synthesis of [^{3}H]tCA C_8C_8

To 100 mCi 2,3-[^{3}H]octanoic acid (2 μmol) in 10 ml of dichloromethane 48 mg of D-*erythro-trans*-sphingosine C_8 hydrochloride (250 μmol) in 1 ml of methanol, 48 μl triethylamine (500 μmol), and 226 mg N-ethoxycarbonyl-2-ethoxy-1,2-dihydroquinoline (1 mmol) are added. This

[25] A. Karrenbauer, D. Jeckel, W. Just, R. Birk, R. R. Schmidt, J. E. Rothman, and F. T. Wieland, *Cell* **63**, 259 (1990).
[26] D. Jeckel, A. Karrenbauer, R. Birk, R. R. Schmidt, and F. T. Wieland, *FEBS Lett.* **261**, 155 (1990).
[27] J. B. Helms, A. Karrenbauer, K. W. A. Wirtz, J. E. Rothman, and F. T. Wieland, *J. Biol. Chem.* **265**, 20027 (1990).
[28] P. Zimmermann and R. R. Schmidt, *Liebigs Ann. Chem.* p. 663 (1988).

mixture is kept in a water bath and stirred at 37°. The reaction is controlled by TLC on silica gel [chloroform/methanol/formic acid, 80:20:0.1 (v/v/v)]. After termination of the reaction (~60 hr), the solvents are removed with a stream of nitrogen, and the yellow residue is resuspended in 0.1% TFA in water (~10 ml) and applied to three combined Sep-Pak C_{18} (Waters) cartridges, preconditioned by treatment with 80% acetonitrile and washing with 0.1% TFA in water. After washing three times with 10-ml portions of 0.1% TFA and four times with 10-ml portions of 20% acetonitrile, 0.1% TFA, the product is eluted with 20 ml of 60% acetonitrile, 0.1% TFA. One-milliliter fractions are collected. Fractions 3–16 are dried down in a Speed-vac (Savant) concentrator and the residues redissolved in 120 μl ethanol and combined. This solution is applied to an HPLC reversed-phase column (RP-18, Lichrospher, 100RP-18, 5 μm, 125 × 4 mm) and a gradient of 20 to 45% acetonitrile, 0,1% TFA in water is applied in 50 min at a flow rate of 1.0 ml/min. After 30 min, 0.25-ml fractions are collected. N-[2,3-[^{3}H$_2$]Octanoyl-D-erythro-trans-sphingosine C_8 ([^{3}H]CA) elutes in fractions 42–49, which are combined, dried, and resuspended in 10 ml ethanol. The yield is 34 mCi (34% of the ^{3}H radioactivity used) and the specific radioactivity is ~50 μCi/nmol.

Standard Assay for Quantitation of [^{3}H]tSPH and [^{3}H]tGlcCer

Typically, 30–300 μCi of [^{3}H]tCA together with 50–100 nmol unlabeled tCA is added per 1 ml of stirred Chinese hamster ovary (CHO) cells in α-modified Eagle's medium (α-MEM) at a density of 0.5 to 1 × 10^7 cells/ml. Aliquots (20 to 200 μl) are taken at various times, centrifuged, and the supernatant media are separated from the cell pellets. The pellets are extracted by the addition of the original volume of 50% methanol, centrifuged in a microfuge, and the supernatant is removed (cell extract). Five microliters of cell extracts and of medium is spotted onto the loading zone of a Whatman (Clifton, NJ) silica gel plate (LK6DF, 20 × 20 cm). Chromatography is performed in tanks preequilibrated with butanone/acetone/water/formic acid, 30:3:5:0.1 (v/v). For fluorography the chromatograms are prepared according to Ref. 29. For determination of radioactivity, the chromatograms are either analyzed on a Berthold two-dimensional radioactivity scanner or radioactive spots are scraped using the corresponding fluorogram as a template, and the scraped material is analyzed by liquid scintillation counting.

[29] K. Randerath, *Anal. Biochem.* **34**, 188 (1970).

Comparison in Chinese Hamster Ovary Cells of Transport Rates from Endoplasmic Reticulum and Proximal Golgi to Plasma Membrane

Cell Culture

Chinese hamster ovary wild-type cells are grown in suspension cultures in α medium (Biochrom KG) as described. The spinner bottles are kept in an incubator (5% CO_2) at 37°. For [^{3}H]tCA incubations, cells are harvested at a density of between 5×10^5 and 8×10^5 cells/ml, washed with α-MEM medium, 20 mM N-2-hydroxyethylpiperazine-N'-2-ethane sulfonic acid (HEPES), pH 7.4, without added fetal calf serum and antibiotics, and resuspended in the same medium to give a final cell density of 0.5 $\times 10^7$ to 1×10^7 cells/ml.

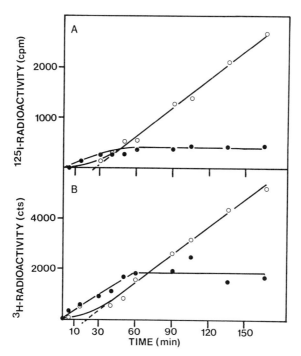

FIG. 2. Transport rates of (A) the ER marker octanoyl glycopeptides (OTP) and (B) the proximal Golgi marker truncated sphingomyelin (tSPH) in CHO cells.[18] For experimental details see text.

Incubation of Cells with [³H]tCA

In a typical experiment, 10^7 CHO wild-type cells per milliliter α-MEM (20 mM HEPES buffer, pH 7.4) is gently stirred at 37 or 30° in a 5% CO_2 atmosphere. [³H]tCA or ^{125}I-labeled octanoyl tripeptide (OTP) is added at time point 0 min, and thereafter, 100-μl aliquots are withdrawn at the time points indicated and quickly centrifuged in the cold (30 sec, 6000 g, 4°). The supernatants (media) are separated, and the pellets extracted with 100 μl 50% (v/v) methanol. After centrifugation (2 min, 10,000 g, 4°) the supernatants are removed (cell extracts).

^{125}I-Labeled glycotripeptides and tSPH are determined as described above. The result is shown in Fig. 2. At 30°, the proximal Golgi marker tSPH is transported to the cell surface with a lag time of 20 min, whereas the lag time of transport of the ER markers is about 28 min. This corresponds to half-times of transport of about 14 and 21 min, respectively. Thus, the half-time of transport from the ER to the proximal Golgi can be estimated to be 7 min at 30°.

The bulk transport rates measured with the two markers presented here indicate that no special signal–receptor system is involved in the secretion of proteins from the ER in the Golgi apparatus, but rather the proteins travel with the bulk of the material, and proteins destined for individual stations of the pathway must be sorted out from the unsignaled flow.[30]

[30] H. R. B. Pelham, *Annu. Rev. Cell Biol.* **5**, 1 (1989).

[20] Reconstitution of Endocytosis and Recycling Using Perforated Madin–Darby Canine Kidney Cells

By Benjamin Podbiliewicz and Ira Mellman

Endocytosis in animal cells is characterized by the internalization of extracellular fluid and macromolecular ligands bound to cell surface receptors. In general, endocytosis involves the formation of small (0.1–0.2 μm) transport vesicles that form from the invagination and pinching off of clathrin-coated pits of the plasma membrane. While some toxins and lectins appear to be internalized in uncoated vesicles, in many cells, it is apparent that the formation of coated vesicles represents the quantitatively most important route of entry into the cell.

After the formation, coated vesicles uncoat and fuse with acidic early endosomes (pH 6.0–6.3) within minutes. Often, but not always, ligands dissociate from their receptors, which recycle together with fluid and membrane back to the plasma membrane. Some receptors are cross-linked by polyvalent ligands and remain associated in a complex that does not recycle. In many cells endocytic vesicles are translocated from the periphery along microtubules to the perinuclear region. Late endosomes are more acidic (pH < 5.5) and contain hydrolytic enzymes. Arrival of endocytosed material to late endosomes takes 15–60 min and is not as synchronous as the recycling pathway to the plasmalemma (5–15 min). Late endosomes transfer their contents to a terminal endocytic organelle (lysosomes) where most of the degradation occurs.

Eukaryotic cells regulate the composition of the plasmalemma by continuously inserting and removing selected proteins and lipids on a process referred to as membrane recycling.[1,2] This cycle involves the delivery of membrane by an exocytic fusion event of a vesicular carrier with the plasmalemma and the import or removal of membrane by an endocytic event or the invagination of small domains of the plasmalemma followed by budding of vesicles into the cytoplasm. This constitutive process must be regulated to maintain the identity of different membrane domains and organelles along the endocytic and biosynthetic pathways that meet in the plasmalemma. One approach to study membrane recycling has been to follow the fate of receptors during the cycle. Transferrin receptor (TfnR) is a good model system where the ligand, transferrin (Tfn), recycles bound to its receptor and is discharged only after getting back to the plasma membrane.

Transferrin receptor binds two iron-loaded Tfns with high affinity at neutral pH. Independently of the binding of ligand,[3] TfnR localizes to clathrin flat lattices[4] and coated pits on the surface of the plasma membrane.[5] After invagination and coated vesicle formation TfnR and Tfn–TfnR complexes remain in a neutral compartment (pH 7) for 1–2 min[6] before delivery to an acidic early endosome where iron dissociates from Tfn that remains bound to its receptor.[7-11] The mechanisms of iron release

[1] R. M. Steinman, S. E. Brodie, and Z. A. Cohn, *J. Cell Biol.* **68**, 665 (1976).
[2] R. M. Steinman, I. S. Mellman, W. A. Muller, and Z. A. Cohn, *J. Cell Biol.* **96**, 1 (1983).
[3] C. Watts, *J. Cell Biol.* **100**, 633 (1985).
[4] K. Miller, M. Shipman, I. S. Trowbridge, and C. R. Hopkins, *Cell* **65**, 621 (1991).
[5] B. J. Iacopetta, S. Rothenberger, and L. C. Kuhn, *Cell* **54**, 485 (1988).
[6] D. M. Sipe and R. F. Murphy, *Proc. Natl. Acad. Sci. USA* **84**, 7119 (1987).
[7] M. Karin and B. Mintz, *J. Biol. Chem.* **256**, 3245 (1981).
[8] J. Van Renswoude, K. R. Bridges, J. B. Harford, and R. D. Klausner, *Proc. Natl. Acad. Sci. USA* **79**, 6186 (1982).
[9] R. D. Klausner, J. van Renswoude, G. Ashwell, C. Kempf, A. N. Schechter, A. Deang, and K. R. Bridges, *J. Biol. Chem.* **258**, 4715 (1983).

and translocation across the endosomal membrane to the cytosol are not completely understood. Acidification, reduction, and an Fe^{2+} transporter system have been implicated.[12] After ~ 5 min at $37°$ Tfn reaches a minimum pH 6 followed by alkalinization and final fusion of recycling vesicles with the plasmalemma.[6] Apotransferrin rapidly exchanges with iron-loaded Tfn that is present in excess (mg/ml) in the serum.

Several in vitro assays have been used to reconstitute different steps in the endocytic cycle. For example, when permeabilized A431 cells were incubated at $32°$ in the presence of ATP and cytosol an increase in the number of coated pits containing horseradish peroxidase (HRP)–Tfn was observed.[13] In a different assay planar membranes were stripped of clathrin with a high pH wash and clathrin reassembled into cages that looked similar to coated pits. This process occurred in the absence of ATP, at $4°$, and does not require addition of cytosol.[14,15] These results suggest that ATP is not required for de novo clathrin assembly into coated pits. Recycling of clathrin from coated vesicles back to the plasma membrane is catalyzed in vitro by Hsc70, an uncoating ATPase.[16-18] ATP and cytosolic factors are also needed for in vitro fusion of endosomes,[19-23] fusion between plasma membrane-derived vesicles and early endosomes,[24] and for fusion involving lysosomes.[25-28] Sorting of receptors in early endosomes and budding of

---

[10] A. Ciechanover, A. L. Schwartz, A. Dautry-Varsat, and H. F. Lodish, J. Biol. Chem. **258**, 9681 (1983).

[11] D. Yamashiro, B. Tycko, S. Fluss, and F. R. Maxfield, Cell **37**, 789 (1984).

[12] M.-T. Nunez, V. Gaete, J. A. Watkins, and J. Glass, J. Biol. Chem. **265**, 6688 (1990).

[13] E. Smythe, M. Pypaert, J. Lucocq, and G. Warren, J. Cell Biol. **108**, 843 (1989).

[14] M. S. Moore, D. T. Mahaffey, F. M. Brodsky, and R. G. W. Anderson, Science **236**, 558 (1987).

[15] D. T. Mahaffey, M. S. Moore, F. M. Brodsky, and R. G. W. Anderson, J. Cell Biol. **108**, 1615 (1989).

[16] S. L. Schmid, W. A. Braell, and J. E. Rothman, Nature (London) **311**, 228 (1984).

[17] W. A. Braell, D. M. Schlossman, S. L. Schmid, and J. E. Rothman, J. Cell Biol. **99**, 734 (1984).

[18] J. E. Rothman and S. L. Schmid, Cell **46**, 5 (1986).

[19] W. A. Braell, Proc. Natl. Acad. Sci USA **84**, 1137 (1987).

[20] J. Davey, S. M. Hurtley, and G. Warren, Cell **43**, 643 (1985).

[21] R. Diaz, L. S. Mayorga, P. J. Weidman, J. E. Rothman, and P. Stahl, Nature (London) **339**, 398 (1989).

[22] J. E. Gruenberg and K. E. Howell, EMBO J. **5**, 3091 (1986).

[23] P. G. Woodman and G. Warren, Eur. J. Biochem. **173**, 101 (1988).

[24] L. S. Mayorga, R. Diaz, and P. D. Stahl, J. Biol. Chem. **263**, 17213 (1988).

[25] L. Altstiel and D. Branton, Cell **32**, 921 (1983).

[26] A. Raz and R. Goldman, Nature (London) **247**, 206 (1974).

[27] P. J. Oates and O. Touster, J. Cell Biol. **68**, 319 (1976).

[28] B. M. Mullock, W. J. Branch, M. van Schaik, L. K. Gilbert, and J. P. Luzio, J. Cell Biol. **108**, 2093 (1989).

vesicles are ATP dependent.[29] Recycling of preinternalized Tfn and its receptor also requires ATP and cytosolic factors *in vivo* and *in vitro*.[30]

It appears that the complete endocytic cycle is more dependent on ATP than early stages in internalization. Some steps in the pathway are probably efficient in the utilization of ATP or do not require metabolic energy. The enzymology of endocytosis and recycling is far from being solved but it is clear that many cellular factors and ATP-requiring steps are involved in this biochemical cycle.

Studies of cell surface binding, internalization, and recycling of transferrin have provided important information regarding the phenomenology of receptor-mediated endocytosis.[10,31] The assay described in this chapter is designed to advance our understanding of endocytosis a step further. Using permeabilized cells it is possible to directly investigate the requirements and components involved in receptor-mediated endocytosis.[30,32]

One of the advantages of using permeabilized cells to study endocytosis is that the cellular organization is maintained. A direct comparison between the morphological and biochemical characteristics of the endocytic machinery in intact and disrupted cells is possible. Therefore it is important to describe the assays for binding, internalization, and recycling of Tfn in intact cells. It is necessary to master all the parameters of the assays and their controls *in vivo,* before attempting to permeabilize cells and measure a biochemical assay *in vitro* that should mimic the intact cells.

This chapter describes the methods to study cell surface binding, internalization, and recycling of transferrin in intact polarized epithelial cells and then presents the optimized procedure to permeabilize Madin–Darby canine kidney (MDCK) cell monolayers using nitrocellulose filters. The assays to reconstitute receptor-mediated endocytosis and recycling to permeabilized cells should help to understand the enzymology of endocytosis.

Cells and Cell Culture

To ensure reproducible results when working with disrupted MDCK cells, it is critical to maintain low-passage cell stocks according to a standard cell culture regimen. Accordingly, MDCK cells (strain II) are maintained on plastic flasks (75 cm²) in Dulbecco's modified Eagle's medium (DMEM) supplemented with 10% (v/v) heat-inactivated fetal calf serum (FCS), 20 mM HEPES, pH 7, 100 U/ml penicillin, and 100 mg/ml streptomycin at 37° in 5% CO_2. The cells are split 1 : 5 every week and used for

[29] M. W. Resnick and W. A. Braell, *J. Biol. Chem.* **265**, 690 (1990).
[30] B. Podbilewicz and I. Mellman, *EMBO J.* **9**, 3477 (1990).
[31] C. R. Hopkins and I. S. Trowbridge, *J. Cell Biol.* **97**, 508 (1983).
[32] G. B. Warren, P. Woodman, M. Pypaert, and E. Smyth, *TIBS* **13**, 462 (1988).

< 15 passages. For experiments, confluent monolayers of cells are washed twice with Dulbecco's phosphate-buffered saline (PBS) and then dissociated with 0.05% trypsin–0.02% (w/v) ethylenediaminetetraacetic acid (EDTA) in PBS. Cells (1.5×10^6) in 1.5 ml of fresh growth medium are plated on 24-mm polycarbonate filter units, and 1.4×10^7 cells in 14 ml are plated on 100-mm filter units (0.4-μm pore size, tissue culture treated; Transwells 3412 and 3419, Costar, Cambridge, MA). The basal compartments of the 24- and 100-mm dishes contain 2.5 and 14 ml of growth medium, respectively. After plating, the cells are left in the hood for about 20 min to allow uniform cell attachment to the filter, before transferring to the incubator. Confluent monolayers are used on the third day in culture.

Notes and Precautions. The way the cells are plated and the length of the incubations are critical to get uniform permeabilizations. When the cells are used after 4 or more days in culture the number of transferrin receptors decreases. When the cells are used before reaching 3 days in culture the permeabilization is variable because the cells are too fragile.

Preparation of Transferrin

Human and canine apotransferrin (> 98% iron free; Sigma, St. Louis, MO) are converted to fully saturated, chelate-free iron(III)-transferrin (holotransferrin) using nitrilotriacetic acid.[33] Holotransferrin is desalted by chromatography on Sephadex G-25 (Pharmacia, Piscataway, NJ) in 0.1 M sodium perchlorate followed by dialysis (or a second column) in 20 mM HEPES, pH 7, and concentrated using Centricon microconcentrators (M_r 10,000 cutoff; Amicon, Danvers, MA). Holotransferrin concentrations are obtained by determining absorption at 457 nm (human holo-Tfn $E^{1\%} = 0.56$; canine holo-Tfn $E^{1\%} = 0.72$). Aliquots are frozen and stored at $-80°$. Transferrin is iodinated ($1-5 \times 10^6$ cpm/μg) using Iodogen (Pierce, Rockford, IL) as described.[34] To obtain more than 95% (v/v) trichloroacetic acid (TCA)-precipitable [^{125}I]Tfn, the iodination mixture is run over two consecutive 0.3 ml packed Dowex (Sigma) columns.

Binding and Elution of ^{125}I-Labeled Transferrin

Filter-grown MDCK cells are washed twice with PBS containing Ca^{2+} and Mg^{2+} (PBS$^+$) and incubated for 45 min at 37° in DMEM supplemented with 0.5% (w/v) bovine serum albumin (BSA) and 20 mM

[33] G. W. Bates and M. R. Schlabach, *J. Biol. Chem.* **248**, 3228 (1973).
[34] I. S. Mellman, H. Plutner, R. M. Steinman, J. C. Unkeless, and Z. A. Cohn, *J. Cell Biol.* **96**, 887 (1983).

HEPES, pH 7 (binding medium) to deplete bovine Tfn. The cells are then transferred to an ice box containing a metal plate covered with wet paper towels. Binding medium containing 1–5 μg/ml canine ^{125}I-labeled holo-Tfn is added to the apical or basal compartments for 1.5 hr. More than 90% of the canine Tfn receptors (cTfnR) are in the basolateral domain of MDCK.[30] Therefore for most binding, internalization, and recycling assays, 80 or 600 μl (for 24- and 100-mm filters, respectively) is placed on Parafilm and the cell monolayers rest on top. The apical compartments contain 0.5 and 3.5 ml, respectively. The tightness of each cell monolayer is determined by counting the media of both compartments in a γ counter. Less than 1% of the label diffuses to the opposite compartment under these conditions. Both sides of the filter are washed with seven 2-ml changes of ice-cold PBS[+] over 35 min, the filters cut with a razor blade from the Transwell unit, and counted. Identical results are obtained if the cells are harvested by scraping, indicating that no radioactivity is associated with the filter. More than 85% of the radiolabel bound at 4° can be eluted from cells by alternating PBS[+] and low pH washes (25 mM acetic acid, 150 mM NaCl, pH ∼3) over 35 min for 5 min each. Nonspecific binding and internalization is routinely determined by subtracting the amounts of radiolabel bound to parallel filters incubated in the presence of a 100-fold excess of unlabeled Tfn; nonspecific binding should be <10% of the specific binding. Experiments are performed in duplicate and generally differ by <10%. For Scatchard analysis (Fig. 1), cells are incubated in a constant amount of [125I]Tfn (0.4 μg) and increasing concentrations of unlabeled Tfn (0.4 to 80 μg/ml; 5 to 1000 nM).

Internalization and Recycling of Prebound Transferrin

After incubation with [125I]Tfn on ice, both sides of the filters are washed 7 times with 2 ml of cold PBS[+]. The filters are transferred to six-well dishes that contain 1 ml of binding medium/well on the basal compartment; 1 ml is then added to the apical compartment. The cells are preincubated for 15 min on ice (with or without permeabilization; see below), then transferred to a water bath at 37° and rapidly cooled to 4° at various times by adding ice-cold PBS[+] and immediately transferring the cells back to an ice bath. The apical and basal media are collected and released [125I]Tfn measured. Internalized [125I]Tfn is determined from the amount resistant to release from the cell surface acid washing, as described above. A parallel set of filters is washed with ice-cold PBS[+] and used to determine total cell-associated radioactivity. Surface-bound radioactivity is calculated by subtracting the acid-resistant radioactivity from the total cell associated. Surface-bound radioactivity can be determined directly by

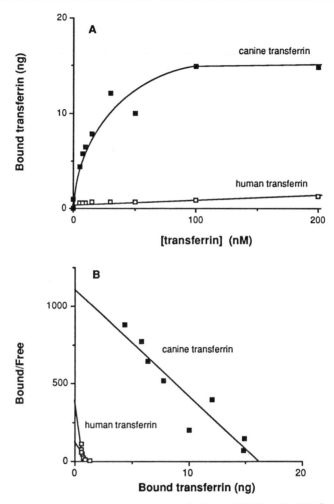

FIG. 1. Binding of canine and human transferrin to MDCK cells. (A) Concentration dependence of canine vs human Tfn. Confluent filter-grown cells were washed to remove serum-derived Tfn and incubated in 5 nM canine or human [^{125}I]Tfn in the presence of increasing concentrations of unlabeled ligand (in the basolateral compartment). After 90 min on ice, filters were washed with cold PBS$^+$ and cell-associated radioactivity determined using a γ counter. Each point represents the mean specific binding to two filters (<10% variation); nonspecific binding was determined in the presence of 1000-fold unlabeled Tfn. The specific activities were 3.3×10^6 and 1.2×10^6 cpm/μg for human and canine [^{125}I]Tfn, respectively. (B) Scatchard analysis of canine vs human Tfn binding. From Scatchard analysis of the data in (A), we obtained a $k_d = 1.46 \times 10^{-8}$ M and 3.48×10^4 cTfn-binding sites per cell. Each filter contains $3.55 \times 10^6 \pm 0.58$ cells ($n = 6$). (■), Canine transferrin; (□), human transferrin.

measuring the amount of [^{125}I]Tfn released by the acid wash. Specific binding, release, and internalization on duplicate filters should differ by < 10%. Figure 2 shows an example of internalization of prebound cTfn in intact and perforated cells.

Recycling of Preinternalized Transferrin

Filter-grown cells are incubated in binding medium containing 5 μg/ml [^{125}I]cTfn in the basal medium for 1 hr at 37°. Both sides of the filters are washed seven times with 2 ml of ice-cold PBS+ to remove unbound Tfn and incubated in binding medium containing 10 μg/ml unlabeled cTfn for 2.5 min at 37° to internalize remaining surface-bound [^{125}I]cTfn. After this treatment, virtually 100% of the cell-associated [^{125}I]cTfn is resistant to removal by low pH. The cells (with or without permeabilization; see below) are preincubated for 15 min on ice in binding medium containing 10 μg/ml unlabeled cTfn to allow internalized [^{125}I]cTfn to recycle and be released into the medium. To start the reaction they are transferred to a water bath at 37°. After various times, the reaction is stopped with 1 ml ice-cold PBS+ in the apical and basal compartments, and the filters transferred to ice. Cell-associated [^{125}I]cTfn and radioactivity in the apical and the basal media are determined. Nonspecific counts (< 10% of total) are determined using parallel filters allowing the internalization of [^{125}I]cTfn in the presence of excess unlabeled Tfn (500 μg/ml). For convenience, filters can be cut into 2–4 or 14–40 equal pieces (for 24- and 100-mm filters, respectively) after Tfn uptake (and/or disruption) and placed in separate vials or 12-well dishes before the second 37° incubation. Although > 95% of the internalized Tfn is released from intact cells into the basolateral compartment, when the filters are cut after internalization the polarity of released Tfn is not maintained. Figure 3 shows an example of a kinetic experiment of cTfn recycling *in vivo* and *in vitro*.

Permeabilization of Apical Membrane

The following sections provide a detailed guide for the preparation of nitrocellulose-permeabilized cells as well as for the preparation of the various reagents required for reconstituting transferrin recycling following disruption.

Principle

The use of nitrocellulose filters to perforate cells in culture is based on the ability of nitrocellulose to bind to plasmalemma phospholipids and proteins. This binding is controlled to obtain uniform and reproducible

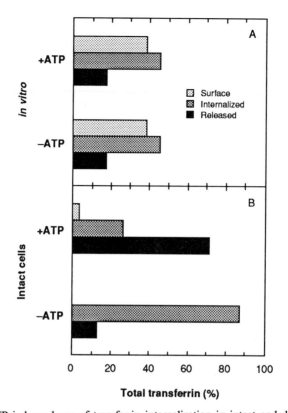

FIG. 2. ATP independence of transferrin internalization in intact and disrupted cells. Confluent monolayers of filter-grown MDCK cells were allowed to bind [^{125}I]cTFN (5 μg/ml) in DMEM/BSA/HEPES (+ATP) or in glucose-free medium containing 20 mM 2-deoxyglucose (dGlc) and 10 mM NaCN (−ATP) added to the basolateral compartment for 1 hr at 4° and then washed to remove unbound radioactivity. (A) Internalization in disrupted cells. The cells were apically disrupted using nitrocellulose (as described in the section Permeabilization of Apical Membrane), and incubated for 15 min at 4° in ICT buffer containing an ATP-regenerating system (+ATP) or an ATP-depleting system (glucose–hexokinase; −ATP). Filters were then rapidly warmed to 37° for 15 min. After this time, the reaction was stopped with 1 ml cold PBS$^+$, the incubation medium was removed to determine the amount of released [^{125}I]Tfn, and the filters transferred to ice and washed with cold PBS$^+$. Surface-bound and internalized Tfn were determined by accessibility or resistance (respectively) to elution by washing the filters with low pH buffers. Although the ATP concentration in the depleted cells was <0.5% of control (at the end of the 4° incubation), there was no effect of depletion on the internalization of [^{125}I]Tfn into an acid-resistant compartment. (B) Internalization in intact cells. After binding [^{125}I]cTfn, washed intact cells were incubated for 15 min at 4° in DMEM/BSA/HEPES (+ATP) or in glucose-free medium containing 20 mM dGlc and 10 mM NaCN (−ATP). The filters were then warmed to 37° for 15 min and surface-bound, internalized, and released [^{125}I]Tfn determined as above. Treatment with energy poisons reduced ATP concentrations to ~10% of control after the 4° incubation and to <0.5% of control after 15 min at 37°. In control cells (+ATP), almost 100% of [^{125}I]Tfn was internalized and/or released into the medium (due to recycling of previously internalized ligand). In ATP-depleted cells (ATP), ~85% of the cell-associated [^{125}I]Tfn was internalized, but only 13% was released into the medium.

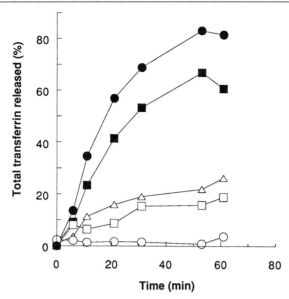

FIG. 3. Transferrin recycling in disrupted cells is dependent on ATP and cytosol. Intact cells after 3 days in culture (one 100-mm and two 24-mm filters) were allowed to internalize 5 μg/ml [125I]cTfn from the basolateral compartment for 1 hr at 37°. One of the 24-mm filters was incubated in the presence of 1 mg/ml of unlabeled hTfn to subtract nonspecific internalization. The filters were washed and then incubated an additional 2.5 min at 37° in medium containing 10 μg/ml unlabeled Tfn to deplete remaining surface-bound [125I]cTfn. After washing with cold PBS, the cells in the 100-mm filter and the one with excess cold Tfn were disrupted using nitrocellulose (see the section Permeabilization of Apical Membrane). The 24-mm filters were cut in halves and the 100-mm filter in 14 equal squares. The disrupted cells were washed at 4° immersed in 50 ml of 150 mM KCl/50 mM HEPES/KOH (pH 7) for 30 min to further remove remaining cytosol. The filters were incubated for 15 min at 0–4° in scintillation vials containing 0.5 ml of assay buffer containing 10 μg/ml Tfn under the indicated conditions: +ATP = 1 mM ATP, 8 mM creatine phosphate, and 32 IU/ml creatine phosphokinase; +cytosol = 2.5 mg/ml dialyzed CHO cytosol; +BSA = 2.5 mg/ml; −ATP = 50 U/ml apyrase; intact cells = DMEM/0.5% (v/v) BSA/20 mM HEPES, pH 7. The release of [125I]cTfn into the medium was quantified from 50-μl aliquots collected at different times and corrected to the total remaining volume. The total amount of [125I]cTfn internalized/ piece of filter was ~1.3 × 10^4 cpm. Nonspecific uptake, determined in the presence of 200-fold excess unlabeled Tfn, was 1.3 × 10^3 cpm/half filter. The specific activity was 3.9 × 10^6 cpm/μg. All points represent the mean of duplicate filters and varied by < 10%. (●) Intact cells; (■) + ATP + cytosol; (△) + ATP + BSA; (□) − ATP + BSA; (○) + ATP + cytosol (0°).

perforation and was first used to study membrane transport phenomenon *in vitro* by Simons and Virta.[35] Basically, nitrocellulose filters are adsorbed to the apical domain of MDCK cells and then are removed together with fragments of the plasma membrane. Under optimal conditions, the cells remain attached to the substrate and the basolateral domain remains intact. This allows access to the interior of the cell together with the maintenance of the cellular organization and polarity. The method described below produces apically permeabilized monolayers that are biologically active with respect to basolateral endocytosis and recycling of Tfn. The cells remain attached to the polycarbonate substratum without loss of activity for up to 8 hr on ice and after incubating at 37° for up to 2 hr. The holes remain open during this time; they probably fail to be resealed because the lesions introduced are too large.

To establish a consistent and reproducible permeabilization procedure to be used for a biochemical assay, it is necessary to control a number of important variables: (1) Reproducible cells and cell culture to obtain maximum attachment to the polycarbonate substrate in which they are grown; (2) assay conditions must be rigidly standardized to allow meaningful comparisons of recycling and endocytosis activities between intact and permeabilized cells; (3) the strength and uniformity of nitrocellulose filter binding to the cells is dependent on their humidity, the amount of contact between the two surfaces, and temperature.

Reagents

Intracellular transport buffer (ICT) to permeabilize and for incubations *in vitro:* 78 mM KCl, 4 mM MgCl$_2$, 8.37 mM CaCl$_2$, 10 mM EGTA/KOH, 1 mM dithiothreitol (DTT) (added fresh from a 1 M stock solution), 50 mM HEPES/KOH, pH 7.[36] The final free Ca^{2+} is 2.67 μM and the final free Mg^{2+} is 2.95 mM. Higher free Ca^{2+} inhibits the assay

Nitrocellulose circles [24-mm diameter, 0.45-μm pore size (BA85) and 75-mm diameter (BA85/20)] with a grid to help maintain a uniform humidity (Schleicher & Schuell, Inc., Keene, NH)

Whatman (Clifton, NJ) 3MM filter paper to blot the wet nitrocellulose filters

Whatman microfiber filters (25- and 75-mm diameter) to remove excess buffer

Rubber stoppers: 24 mm attached to a 20-ml syringe and a ~75-mm

[35] K. Simons and H. Virta, *EMBO J.* **6,** 2241 (1987).
[36] B. Burke and L. Gerace, *Cell* **44,** 639 (1986).

diameter (#15) to maximize contact between nitrocellulose filters and
the cells

KH buffer to wash away cytosol in permeabilized cells: 150 mM KCl,
50 mM HEPES/KOH, pH 7

A 150-mm plastic petri dish with a painted grid to use as a cutting board,
razor blades, and fine forceps to cut and handle all the filters

Procedure

The perforation of MDCK cells[35] was optimized for cells grown on
polycarbonate filters.[30] All steps are performed at 0–4° on an ice box
containing a metal plate covered with wet paper towels, in a well-ventilated
fume hood containing an iodine trap. Monolayers on 24-mm filter units
are washed once with 2 ml ice-cold PBS[+] on the apical and the basal
compartments and once with 1 ml of ICT buffer. The apical medium is
then aspirated and replaced by a 24-mm diameter nitrocellulose filter;
before use, nitrocellulose filters are presoaked 1 min in a petri dish containing ~ 50 ml deionized water, than 1 min in ~ 50 ml ICT buffer, and
blotted 1 min between four Whatman 3MM filter papers under a 1-liter
bottle containing PBS. A 25-mm glass microfiber filter is placed on top of
the nitrocellulose filter to remove the excess ICT buffer. The basal buffer is
removed and uniform gentle pressure is applied for 10 sec using a 24-mm
rubber stopper attached to a 20-ml syringe (cell disrupter; patent pending).
After 1 min, 1 ml of ICT buffer is added to the apical and the basal sides.
After an additional 1 min, the microfiber filter is removed, the polycarbonate filter is excised from the Transwell unit with a razor blade, and the
polycarbonate filter with the nitrocellulose filter on top is placed on a
plastic petri dish. Finally, the nitrocellulose filter is slowly ripped off using
a forceps; the polycarbonate filter containing permeabilized cells remains
adherent to the petri dish and can be cut in two to four pieces. Permeabilized cells can be kept in ICT buffer on ice for up to 8 hr without loss of
activity.

The permeabilization of MDCK monolayers grown on 75-mm filter
units is useful for biochemical studies and is different in a few steps due to
their large size. The filters are washed once with 5 ml ice-cold PBS[+] on the
apical and the basal compartments and once with 5 ml of ICT buffer. The
apical medium is then aspirated and replaced by a 75-mm diameter nitrocellulose filter with the printed grid facing up; before use, nitrocellulose
filters were presoaked and blotted as above. The filter is cut with a razor
blade from the unit and is placed on top of a 100-mm petri dish lid with the
nitrocellulose filter facing up. A 75-mm glass microfiber filter is placed on
top the nitrocellulose filter to remove the excess ICT buffer. A rubber

stopper (#15) is placed on top of the glass microfiber filter for 10 sec to ensure maximum contact between the nitrocellulose and the apical membranes. After 1 min, 5 ml of ICT buffer is added. After an additional 1 min, the microfiber filter is removed, and the polycarbonate filter with the nitrocellulose filter on top is placed on a 150-mm plastic petri dish. Finally, the nitrocellulose filter is slowly ripped off using a forceps, the polycarbonate filter containing permeabilized cells remains adherent to the petri dish and can be cut in 14–40 pieces using fine forceps and a razor blade. Cutting into similarly sized pieces is assisted by using a dish on which a grid was drawn with an indelible marker.

In the case of transferrin release, the cytosol dependence of the assay is increased if the disrupted cells are then subjected to a mild salt wash at this stage. This is accomplished by incubating the disrupted cells for 30 min at 0–4° immersed in 5 or 50 ml of KH buffer (for the 24- and 75-mm filters, respectively) and then transferred back to fresh ICT buffer.

The efficiency of permeabilization is routinely assessed by trypan blue (0.04%, v/v) exclusion; also, the nitrocellulose filters are incubated with trypan blue to assess if any cells came off attached to the nitrocellulose. The increased ability of disrupted cells to accumulate 10 μg/ml propidium iodide in their nuclei, and the accessibility of cytoplasmic myosin and clathrin to antibodies added to unfixed cells, are determined by immunofluorescence. To quantify disruptions, the fraction of propidium iodide-stained nuclei ($n = \sim 600$) surrounded by fluorescent staining due to anti-myosin are counted. Monolayers should be $>85\%$ permeable by these criteria. About 80% of the total lacate dehydrogenase (LDH) activity should be found in the medium after permeabilization at 4°. The LDH is determined as described.[37]

Incubation of Intact and Permeabilized Cells

After removal from the Transwell units, 24-mm filters containing disrupted cells with prebound or preinternalized Tfn are placed cell side up in six-well dishes on 80-μl droplets of ICT buffer containing the indicated additions; 160 μl of the identical buffer is added to the apical side. When the filters are cut in smaller pieces, the volume of assay mixture can be reduced accordingly. For example, small filter pieces of ~ 0.5 cm^2 are placed in 12-well dishes (Costar) on 10-μl droplets with 20 μl apically. The filters are preincubated on ice for 15 min in ICT supplemented with 0.5–8 mg/ml desalted cytosol or 3–8 mg/ml ovalbumin or BSA. For incubations in the presence of ATP, an ATP-regenerating system is prepared fresh from

[37] M. K. Bennett, A. Wandinger-Ness, and K. Simons, *EMBO J.* 7, 4075 (1988).

stocks ($100-200\times$) stored at $-80°$; the final concentrations in the reaction are 1 mM ATP (Boehringer Mannheim, Indianapolis, IN), pH 7, 8 mM creatine phosphate (CP), and 32 IU/ml creatine phosphokinase (CPK). To deplete ATP, either 25 U/ml hexokinase pelleted from an $(NH_4)_2SO_4$ suspension and resuspended in a final concentration of 5 mM glucose, or 50 U/ml apyrase (Sigma), is added. Both intact and permeabilized cells are preincubated in the reaction buffer for 15 min on ice before starting the recycling or internalization assays by transferring the plates to a 37° water bath. The apical and basal total volumes used for the permeabilized cells were the same as for the intact cells.

Cytosol Preparation

Madin–Darby canine kidney cells are plated on 150-mm diameter tissue culture dishes. When confluent, each dish is washed twice with 20 ml ice-cold PBS⁻ and twice with ICT buffer in the cold room. The dishes are drained for 30 sec and scraped into the remaining buffer. The scraped cells are resuspended with a 10-ml plastic pipette and then passed four times through a 22-gauge syringe needle.

Chinese hamster ovary (CHO) cells grown in suspension are harvested and washed by centrifugation twice with 20 ml ice-cold PBS⁻ and twice with ICT buffer at 1000 g for 5 min at 4°. Packed cells are resuspended in 5 vol of ICT buffer and homogenized at 4° by passing the suspension five times through a stainless steel ball-bearing homogenizer with ~23-μm clearance.[38] Yeast cytosol is prepared essentially as described.[39]

Aliquots are observed under the light microscope with trypan blue and there should be no broken nuclei and more than 90% breakage. After centrifugation at 1000 g for 5 min at 4° the postnuclear supernatants are collected and centrifuged for 1 hr at 100,000 g at 4°. The supernatants are harvested, desalted in Sephadex G-25 columns, or dialyzed overnight against two changes of ICT buffer. The desalted cytosols are frozen in aliquots in liquid nitrogen and stored at $-80°$. Protein is determined as described[40] and should be between 5 and 10 mg/ml for different preparations of MDCK cytosol, and ~5 mg/ml for the CHO cytosol.

[38] W. E. Balch, W. G. Dunphy, W. A. Braell, and J. E. Rothman, *Cell* **39**, 405 (1984).
[39] H. Ruohola, A. K. Kabcenell, and S. Ferro-Novick, *J. Cell Biol.* **107**, 1465 (1988).
[40] M. M. Bradford, *Anal. Biochem.* **12**, 248 (1976).

[21] Morphological Studies of Formation of Coated Pits and Coated Vesicles in Broken Cells

By MARC PYPAERT and GRAHAM WARREN

Introduction

Clathrin-coated pits and vesicles are involved in the receptor-mediated endocytosis of various ligands in animal and plant cells.[1] Studies *in vivo* have led to the identification of the different organelles involved and the sequence of events that take place during endocytic uptake. Receptors, with or without bound ligand, cluster in coated pits. The curvature of the plasma membrane in the coated pit is then progressively increased, a process termed *invagination.* This is followed by detachment of the coated pit from the plasma membrane, a process referred to as *scission,* to form a free coated vesicle in the cytoplasm. The coated vesicle rapidly loses its coat, and then goes on to fuse with an endosome, whereas the coat subunits are reused for further rounds of internalization.

Coated pits are not accessible to direct experimental manipulation in intact cells. Previous attempts to overcome this problem included the introduction of anti-clathrin antibodies into living cells[2] or the use of inhibitors that affect the normal functioning of coated pits.[3,4] However, these approaches have provided little new information on invagination and scission. Our own approach has been to try and reproduce these events in a cell-free system. Using broken A431 cells, we were able to show that the formation of coated pits, invagination, and scission to form coated vesicles occurred in broken cells.[5] Furthermore, we found that the formation of new coated pits and scission *in vitro* required both added ATP and cytosol, whereas invagination required neither.[5] A combination of a biochemical assay and a quantitative morphological analysis was used during this study. The aim of this chapter is to describe in more detail the morphological methods that were used to measure the formation of coated pits and vesicles from the plasma membrane of broken A431 cells.

[1] J. S. Rodman, R. W. Mercer, and P. D. Stahl, *Curr. Opinions Cell Biol.* **2,** 664 (1990).
[2] S. J. Doxsey, F. M. Brodsky, G. S. Blank, and A. Helenius, *Cell* **50,** 453 (1987).
[3] J. M. Larkin, W. C. Donzell, and R. G. W. Anderson, *J. Cell Physiol.* **124,** 372 (1985).
[4] K. Sandvig, S. Olsnes, O. W. Petersen, and B. Van Deurs, *J. Cell Biol.* **105,** 679 (1987).
[5] E. Smythe, M. Pypaert, J. Lucocq, and G. Warren, *J. Cell Biol.* **108,** 843 (1989).

Experimental Procedure

Incubation of Broken Cells

A431 cells were labeled with a conjugate of transferrin and horseradish peroxidase (transferrin–HRP) for 2 hr at 4°, washed six times in KSHM buffer [100 mM KCl, 85 mM sucrose, 20 mM N-2-hydroxyethylpiperazine-N'-2-ethanesulfonic acid (HEPES), 1 mM magnesium acetate, adjusted to pH 7.4 with NaOH], and then broken by scraping from the dish using a rubber policeman, as previously described.[5]

Aliquots (50 μl) of broken cells were incubated for various times at 0 or 37° in the presence of 5 μl of an ATP-regenerating system. The ATP-regenerating system comprised 1 vol of 40 mM MgATP (Boehringer Mannheim UK, Ltd., Lewes, England), adjusted to pH 7.0 with NaOH, 1 vol of 200 mM creatine phosphate (Boehringer Mannheim), and 2 vol of 0.2 mg/ml creatine phosphokinase (Boehringer Mannheim) prepared from a 10-fold concentrated solution in 50% (v/v) glycerol. After incubation, 100 μl of KSHM was added and each tube was kept on ice for 90 min. The tubes were then centrifuged at 1000 g_{av} for 10 min at 4° and the pellets fixed and processed for electron microscopy.

Processing for Electron Microscopy

Pellets were fixed in 1 ml of fixative comprising 1 vol of KSHM and 1 vol of 1% (w/v) glutaraldehyde (Fluka BioChemika, Glossop, England) in 0.2 M cacodylate buffer, pH 7.2. Fixation was for 1 hr at room temperature. Fixed cells were washed three times in phosphate-buffered saline (PBS), pH 7.4, and then resuspended in 1 ml of 1 mg/ml 3,3'-diaminobenzidine (tetrahydrochloride salt; Sigma Chemical Co., Poole, England) and 0.3% (v/v) hydrogen peroxide in PBS, pH 7.4. After incubation for 30 min at room temperature, cells were washed three times in PBS and then pelleted in a J2-21 centrifuge (Beckman Instruments, Inc., Palo Alto, CA) at 7100 g_{av} for 10 min at 4° so that flat pellets were obtained. The pellets were postfixed for 30 min at room temperature in 1% (w/v) osmium tetroxide (Taab Laboratories Equipment, Ltd., Aldermaston, England) in 0.1 M cacodylate buffer, pH 7.2, then washed once in distilled water.

Dehydration and embedding were as follows:

Ethanol (70%), 2 × 5 min
Ethanol (90%), 2 × 5 min
Ethanol (100%), 3 × 10 min
Propylene oxide, 1 × 10 min

Propylene oxide/Epon 812 (Taab), 2 hr on rotor
Epon 812, 2 hr on rotor
Epon 812, 2 hr on rotor
Polymerization for 8 hr at 60°, or 16 hr at 55°

Serial Sectioning

One stack of serial sections (> 100) was obtained from a random location inside each pellet. The plane of section was perpendicular to the bottom of the pellet to ensure that each section contained cell profiles from both the top and the bottom of the pellet. Serial sections were cut using a Reichert–Jung ultramicrotome. The average section thickness was 30 nm, as estimated by the fold method.[6] Sections were mounted onto carbon/ Formvar-coated slot grids essentially using the method of Galey and Nilsson.[7] They were stained for 15 min with a 3% (w/v) aqueous solution of uranyl acetate and for 5 min with a solution of lead citrate according to the method of Reynolds.[8] They were finally viewed at 60 kV in a Jeol 1200 EX electron microscope.

Morphological Quantitation

Requirements

To quantify the extent of formation of coated pits and coated vesicles from the plasma membrane of broken A431 cells, several conditions must be met.

1. Newly formed coated vesicles must be differentiated from preexisting ones. The use of transferrin–HRP or other markers to label cells before breaking makes this possible, because only coated pits are labeled before incubation of broken cells. Labeled coated vesicles can appear only by the invagination and scission of labeled coated pits during the incubation.

2. Serial sections must be used, because an earlier morphological study has shown that more than 50% of apparent coated vesicles seen at the electron microscope are in fact coated pits with necks outside the plane of section.[9]

3. A sampling method must be used that gives all structures an identi-

---

[6] J. V. Small, *in Proc. Eur. Congr. Electron Microsc., 4th* (D. S. Bocciarelli, ed.), p. 609. Tipografia Poliglotta, Rome, 1968.

[7] F. R. Galey and S. E. G. Nilsson, *J. Ultrastruct. Res.* **14,** 405 (1966).

[8] E. S. Reynolds, *J. Cell Biol.* **17,** 208 (1963).

[9] O. W. Petersen and B. Van Deurs, *J. Cell Biol.* **96,** 277 (1983).

cal chance of being sampled. This is necessary because the probability of any given structure being sampled by a two-dimensional sampling frame placed on a single section is proportional to its size. Whereas some shallow coated pits can be larger than 500 nm in diameter,[10] coated vesicles and deeply invaginated coated pits just about to pinch off have a diameter in the region of 100 nm. One would therefore expect sampling with a two-dimensional sampling frame (e.g., a rectangular frame on a single section) to overestimate the number of shallow coated pits and underestimate the number of coated vesicles.

4. The most commonly used method to compare numbers of coated pits found under different conditions is to express them as numbers per unit length of plasma membrane (e.g., Ref. 10). However, the length of plasma membrane could be difficult to estimate in broken cells. Furthermore, there is the possibility that some of the plasma membrane would be lost during the incubation, causing an overestimation of the numbers of coated pits and vesicles after incubation. To avoid these problems, the numbers of coated pits and vesicles per cell should be estimated, ideally using a stereological method that does not require the sectioning of entire cells.

Some of the conditions listed above (1 and 2) were met in our study by using transferrin–HRP to label coated pits and by the use of serial sections to identify free coated vesicles. To meet the other two conditions, a new stereological method, the disector, was used. This method is described below.

The Disector

The disector is a three-dimensional counting method with which unbiased estimates of particle numbers in a specimen can be obtained.[11,12] The principle of the disector can be summarized as follows. Any two sections from a specimen are chosen. Particles are counted only if they intersect with a two-dimensional sampling frame on the first section and do not intersect with the second section, termed the look-up section (see Fig. 1A). The number (ΣQ^-) of particles that have disappeared in the look-up gives an estimate of the total number of particles (N) inside the reference space (the pellet in this case), using the following equation:

$$N = \sum Q^- V_{ref}/V_{dis} \tag{1}$$

[10] J. M. Larkin, W. C. Donzell, and R. G. W. Anderson, *J. Cell Biol.* **103**, 2619 (1986).
[11] D. C. Sterio, *J. Microsc.* **134**, 127 (1984).
[12] H. J. G. Gundersen, *J. Microsc.* **143**, 3 (1986).

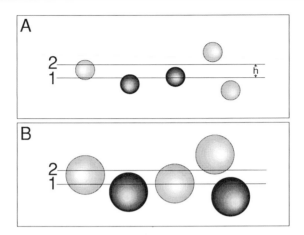

FIG. 1. Principle of sampling with the disector. (A) Particles are sampled if they intersect the first section (section 1) and do not intersect the look-up section (section 2) separated from the first section by a distance (h) smaller than the smallest particle size. Two particles are sampled in this case (dark spheres). (B) The diameter of the particles shown in (A) was doubled. Although the number of particles intersecting sections 1 or 2 is increased in (B), the number sampled by the disector (dark spheres) does not change.

where V_{ref} is the volume of the reference space (pellet), and V_{dis} is the volume of the disector. The two sections used for the disector do not need necessarily to be consecutive, the conditions being that the separation between the first and the last sections (h) is known, and that this distance is not greater than the smallest size of the particles to be counted. Ideally, h would be about one-third of the mean diameter of the particles.[12] If a stack of serial sections is used, then h can be much greater than this, but the section thickness must be smaller than the smallest particle diameter. In such a case, every section of the stack must be examined, and all particles that intersect the first section or appear in consecutive sections are counted only if they do not intersect the look-up section.

The greatest advantage of this method is that every particle has the same chance of being sampled, irrespective of its size or shape. This is illustrated in Fig. 1B. Whereas increasing the size of particles increases their chance of being included in a section and therefore counted by a two-dimensional stereological method, the number of particles sampled by the disector remains the same.

In our study, estimates of numbers of coated pits and vesicles per cell were required. However, because the cells were broken such estimates were difficult to obtain, so that numbers per nucleus were estimated instead. Because A431 cells have only one nucleus, numbers per nucleus should be equivalent to numbers per cell. By dividing the estimate of the number of coated pits and vesicles (n) in the reference space (the pellet, see above) by the same estimate for nuclei (N), it is possible to calculate the numbers of coated structures per nucleus using the following equation:

$$n/N = \sum q^- V_{dis} / \sum Q^- v_{dis} \tag{2}$$

where Σq^- is the number of coated structures counted in the disector; ΣQ^- is the number of nuclei; v_{dis} is the volume of the disector used to count coated structures; and V_{dis} is the volume of the disector used to count nuclei. V_{ref} has disappeared from Eq. (2) because both n and N were estimated from the same pellet. V_{dis} and v_{dis} can be estimated using the following equations:

$$V_{dis} = AKT \tag{3}$$

and

$$v_{dis} = akt \tag{4}$$

where A is the surface area of the sampling frame, K is the number of sections between the first plane and the look-up, T is the section thickness in disectors used to count nuclei, and $a,k,$ and t are the same parameters in disectors used to count coated structures (see Fig. 2). If V_{dis} and v_{dis} in Eq. (2) are replaced by Eqs. (3) and (4), the number of coated structures per nucleus can be obtained using Eq. (5):

$$n/N = \sum q^- AK / \sum Q^- ak \tag{5}$$

T and t have been canceled in Eq. (5) because the same stack of serial sections was used for both estimates. This method therefore allows the counting of structures without the need to measure the volume of the pellet or the section thickness. The reference in this case is nuclei, but other structures could be chosen such as organelles (e.g., mitochondria), cells, or groups of cells.

Sampling Procedure

Disectors to Count Nuclei. Two sections with an average separation of 76 sections (corresponding to 2.3 μm, which is smaller than the smallest nuclei) were sampled from each pellet. The first and the last section were used in turn as the look-up. This yields twice as much information from the same series of sections, without counting the same particles twice. Micrographs of these two sections were taken at $\times 200$ and printed with a three- to fourfold magnification. A rectangular sampling frame was placed over the micrographs, and nuclei from broken cells were counted. The surface area of the sampling frame was in the range $2-4 \times 10^4 \; \mu m^2$.

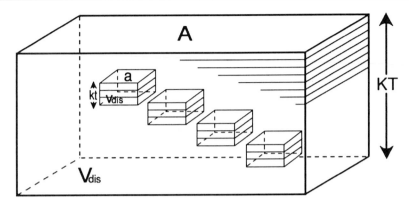

FIG. 2. Sampling protocol to estimate numbers of coated pits and vesicles per nucleus. Two sections on either side of a stack of serial sections were used to count nuclei, using a two-dimensional sampling frame placed on the first section. The volume of this disector (V_{dis}) $= AKT$, where A is the surface area of the sampling frame, K is the number of sections in the stack, and T is the section thickness. To count coated pits and coated vesicles, four smaller disectors were placed at random within the stack, each starting on a different section to account for the variability in section thickness. Each of these small disectors comprised three consecutive sections. The volume of the smaller disectors (v_{dis}) $= akt$, where a is the surface area of the sampling frame, k is the number of sections, and t is the section thickness. In the two equations above, $T = t$ because both the large and the small disectors were made from the same stack of serial sections.

Disectors to Count Coated Pits, Coated Vesicles, and Uncoated Structures. The stack of sections used to count nuclei was then used to count coated and uncoated structures (see Fig. 2). Disectors were placed across the sections in a systematic fashion with a random start, and were placed on different sets of sections each time to take account of the variability in section thickness. Each disector was composed of three consecutive serial sections, the first and the third being used in turn as the look-up. Micrographs were taken of all three sections at a magnification of ×3000 and printed at threefold magnification. A rectangular sampling frame was placed over the sections, and coated pits and coated vesicles were counted if they appeared in the first or second section, but not in the look-up. The surface area of the sampling frame was 316 μm^2. Often counting was carried out at the electron microscope rather than on micrographs. This allowed more precise identification of coated vesicles, because narrow necks could have been difficult to observe on low-magnification micrographs.

Measurement of Degree of Invagination of Coated Pits. A representative sample of coated pits were photographed at ×25,000 and measured as follows. The points where the coat began and ended were determined, and

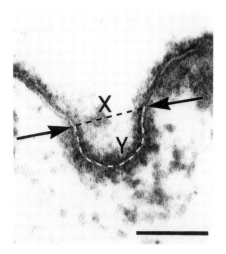

FIG. 3. Measurement of coated pits. For each coated pit sampled, the points where the coat starts and ends were identified (arrows), and the direct distance (X) and the length of the plasma membrane (Y) between them measured. Bar: 100 nm.

the length of plasma membrane (Y) and the diameter of the neck (X) were measured between these two points using a digitizer (Fig. 3). X and Y were often measured on different sections through the same pit, so that the maximal value was obtained for both these measurements. The ratio of these two values was used to estimate the extent of invagination. The ratio (X/Y) decreases as invagination proceeds so for purposes of convenience the results were expressed as $1 - (X/Y)$, so that flat coated pits have a value of 0 and vesicles just about to pinch off a value of 1.[13]

The coated pits were placed in one of two categories: shallow or deeply invaginated.[5] The profiles of shallow coated pits had neck diameters greater than one-fifth of the coat length $[0 \leq 1 - (X/Y) < 0.8]$. Deeply invaginated coated pits had neck diameters equal to or less than one-fifth of the coat length $[0.8 \leq 1 - (X/Y) < 1]$. Once the percentages of coated pits in these two categories were determined for each condition tested, they were used to estimate the total numbers of shallow and deeply invaginated coated pits per nucleus.

Estimation of Numbers of Coated Pits and Vesicles per Nucleus

A431 cells were labeled with transferrin–HRP at 4°, broken, and then incubated for 15 min at 0 or 37° in the presence of an ATP-regenerating

[13] M. Pypaert, J. M. Lucocq, and G. Warren, *Eur. J. Cell Biol.* **45,** 23 (1987).

TABLE I
BROKEN A431 CELLS INCUBATED FOR 15 min AT 0°[a]

Cell component	$A \, (\mu m^2)$	K	ΣQ^-
Nuclei	3.56×10^4	72	35

	$a \, (\mu m^2)$	k	Σq^-
Coated pits	316	2	10.3 ± 4.6
Coated vesicles	316	2	0
Uncoated structures	316	2	0

[a] Results are from one (nuclei) or four (coated pits and vesicles and uncoated structures) disectors. When four disectors were used, results are shown as mean ± standard deviation. Note that although three sections were used for the disectors to count coated and uncoated structures, $k = 2$ as the particles present in the third section (the look-up) are not counted. The same principle applies to the nuclei disector.

system. They were fixed and processed for electron microscopy as described above. Horseradish peroxidase-labeled coated pits and vesicles were counted. Uncoated HRP-labeled vesicles also appeared during incubation at 37°. These probably represent coated vesicles that have lost their coat, and elements of the endosomal compartment, the next stage along the endocytic pathway.[5] They too were counted.

Example: 15 min, 0°. The number of coated pits per nucleus was calculated from the values shown in Table I using Eq. (5). No HRP-labeled coated vesicles or uncoating structures were found in these cells (see also Ref. 5). The results were 1194 ± 533 coated pits per nucleus,[14] of which 88.8% were shallow and 11.2% deeply invaginated.

Example: 15 min, 37°. The number of coated pits and vesicles and uncoated structures per nucleus were calculated from the values shown in Table II using Eq. (5). The results were 967 ± 312 coated pits per nucleus,[14] of which 15.0% were shallow and 85.0% deeply invaginated; 267 ± 193 coated vesicles per nucleus[14]; and 193 ± 74 uncoated structures per nucleus.[14]

[14] Results are expressed as mean of four estimates ± standard deviation.

TABLE II
BROKEN A431 CELLS INCUBATED FOR 15 min AT $0°^a$

Cell component	A (μm^2)	K	ΣQ^-
Nuclei	3.76×10^4	70	28
	a (μm^2)	k	Σq^-
Coated pits	316	2	6.5 ± 2.1
Coated vesicles	316	2	1.8 ± 1.3
Uncoated structures	316	2	1.3 ± 0.5

[a] Results are from one (nuclei) or four (coated pits and vesicles and uncoated structures) disectors. When four disectors were used, results are shown as mean $\pm$ standard deviation.

Conclusions

Several conclusions can be derived from the results shown above (see also Ref. 5). The number of labeled coated pits decreased slightly during the incubation at 37°. This was matched by the appearance of labeled coated vesicles and uncoated structures. Interestingly, the total number of coated structures (coated pits plus coated vesicles) remained constant during the incubation. This previously led us to suggest that only one round of internalization occurred in broken A431 cells.[5] From the results above, it can be calculated that about 21% of coated pits were converted into coated vesicles during the incubation at 37°. However, this is a minimum estimate because some labeled uncoated structures were also formed during the incubation, probably by uncoating of coated vesicles.

Although > 900 coated pits per nucleus remain after 5 min at 37°, 822 (= 85%) of them are deeply invaginated, compared to only 11% when cells are kept at 0°. This indicates that during the incubation invagination proceeded to a greater extent that scission. We have previously shown that invagination in this system is not dependent on added cytosol or ATP, in contrast to formation of coated pits or scission.[5]

The numbers of coated pits or vesicles per cell have never been estimated before in cultured mammalian cells. However, the numbers found here are consistent with the earlier observation that ~ 1600 coated vesicles are formed per minute in BHK-21 cells.[15] If the half-life of coated pits is

[15] M. Marsh and A. Helenius, *J. Mol. Biol.* **142**, 439 (1980).

<1 min, BHK-21 cells would have 1000 to 1500 coated pits per cell, a number similar to what we found in A431 cells.

As mentioned earlier, sampling with the disector is not affected by the size or shape of the particles to be counted. This was especially relevant during these experiments as a change of shape of coated pits occurred during the incubation at 37°, with >80% of coated pits becoming deeply invaginated after 15 min (see also Ref. 5). Necks of deeply invaginated coated pits are much narrower than those of shallower coated pits. Because sampling of coated pits using a two-dimensional sampling frame (on a single section) requires the necks of coated pits to intersect the plane of section, one might expect such a sampling to be biased toward shallow coated pits. To test this, the number of coated pits sampled using a two-dimensional sampling frame placed on the first section in each stack of three sections was divided by the number of coated pits sampled in the same stack using the disector. The ratio of these two values was 2.9 for cells kept at 0° for 15 min and 1.1 for cells incubated at 37° for the same period of time. This indicates that on average the necks of coated pits were visible in three consecutive serial sections in cells kept at 0°, and in only one section in cells incubated at 37°. Sampling using a two-dimensional sampling frame would therefore have led to a threefold reduction in the number of coated pits found on broken cells following incubation at 37°, instead of observing a relatively constant number of coated structures (coated pits plus coated vesicles) during incubation as was the case when using the disector. This in turn could have led us to conclude either that more than half of the coated pits had been lost from the plasma membrane of broken cells during the incubation, or that more than 70% of them had undergone scission and that most of the newly formed coated vesicles had escaped through holes in the plasma membrane or transferred their ligand to endosomes. This would have been quite different from the conclusions we actually reached from this work,[5] and clearly illustrates the importance of choosing the right stereological method when using electron microscopy as a tool to solve problems in cell biology.

Acknowledgment

The authors wish to thank Tom Misteli for valuable comments on the manuscript.

[22] Receptor-Mediated Endocytosis in Semiintact Cells

By ELIZABETH SMYTHE, THOMAS E. REDELMEIER, and
SANDRA L. SCHMID

Introduction

Many biologically important macromolecules are internalized into cells via the complex, multistep process of receptor-mediated endocytosis. Bound ligands and their receptors are first clustered into specialized coated pit regions of the cell surface. Although poorly understood, this sorting step presumably involves specific interactions between the cytoplasmic tails of receptors and the protein constituents of the coat. The major coat proteins are clathrin heavy and light chain trimers (consisting of three 180-kDa and ~35-kDa polypeptides) and heterotetrameric adaptor or assembly protein complexes (consisting of two ~100-kDa adaptin molecules, 50-kDa and 16-kDa subunits).[1,2] Cell surface receptors internalized via coated pits fall into two classes. One class, including, transferrin and low-density lipoprotein (LDL) receptors, are constitutively localized in coated pits and efficiently internalized at rates independent of bound ligand. In contrast, unoccupied receptors such as epidermal growth factor (EGF) and insulin receptors are internalized at low basal rates. For this class of receptors, ligand binding triggers efficient association with coated pits and subsequent internalization. Thus initial stages of receptor-mediated endocytosis involve assembly of the coat proteins at the cell surface and selective inclusion of receptors into newly formed coated pits. These events not only require the addition of coat constituents but are accompanied by morphological changes in the coat, which progresses from an initially planar structure to a highly curved, invaginated pit. Finally, a membrane fission event occurs and the coat is sealed around the newly budded coated vesicle.

Molecular dissection of these biochemically diverse events will be facilitated by the use of cell-free assays that allow distinct stages to be assayed independently. Here are described assays (diagrammed in Fig. 1) used to examine receptor-mediated endocytosis in A431 cells. These assays allow independent measurement of coated pit assembly, invagination, and coated vesicle budding.

[1] F. M. Brodksy, *Science* **242**, 1396 (1988).
[2] B. M. F. Pearse and M. S. Robinson, *Annu. Rev. Cell Biol.* **6**, 151 (1990).

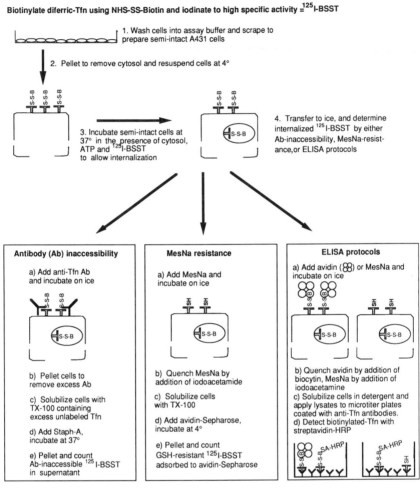

FIG. 1. Diagram of methods used for detection of Tfn receptor-mediated endocytosis into semiintact A431 cells.

Experimental Procedures

Preparation of Biotinylated Ligands

Biotinylated Transferrin. Diferric transferrin (Tfn; Boehringer Mannheim Biochemicals, Indianapolis, IN) is dialyzed into 50 mM NaP$_i$, pH 7.5, 100 mM NaCl and stored at 10 mg/ml ($E_{465}^{1\%} = 0.57$) in aliquots at $-70°$. A 32-mg/ml stock solution of NHS-SS-biotin (Pierce Chemicals,

Rockford, IL) is prepared in dimethyl sulfoxide (DMSO) just before use. For biotinylation, 2 μl of NHS-SS biotin (10.5×10^{-2} μmol, a sevenfold molar excess over Tfn) is added to 120 μl of transferrin (1.2 mg, 1.5×10^{-2} μmol) and incubated for 60 min at room temperature. Unconjugated NHS-SS-biotin is removed by gel filtration on a Sephadex G-25 spin–desalt column equilibrated in phosphate-buffered saline (PBS). The biotinylated transferrin (BSST) is iodinated in a 12×75 cm glass tube coated with 10 μg Iodogen (as described by Pierce) using 1 mCi Na^{125}I (Amersham, Arlington Heights, IL). Free ^{125}I is removed by chromatography over 0.5 ml Dowex AG-1X8 resin (Sigma, St. Louis, MO), preequilibrated with PBS containing 0.2% bovine serum albumin (BSA) and poured in a Pasteur pipette plugged with glass wool. The Dowex AG-1X8 resin is prepared by washing on a sintered glass funnel sequentially with 0.5 N NaOH, water to neutrality, 1 M HCl, water to neutrality. The resin is stored dry until use. To minimize loss of biotin residues due to disulfide exchange or spontaneous reduction, the [^{125}I]BSST is aliquoted and stored at $-70°$. Specific activity of [^{125}I]BSST is $1-3 \times 10^6$ cpm/μg. We estimate that this procedure results in the addition of about five biotins per transferrin molecule. Higher ratios of biotin to Tfn resulted in higher backgrounds. Eighty to 90% of the total [^{125}I]BSST could be specifically adsorbed to avidin-Sepharose. Biotinylation of transferrin did not affect its specific binding to cell surface transferrin receptors.

Other Ligands. Similar biotinylation and iodination procedures have been successfully applied to a variety of other ligands, including monoclonal antibodies (Ab) specific to both the transferrin receptor and the EGF receptor ectodomains. These [^{125}I]BSS–Ab reagents have been used as ligands to follow receptor internalization using both the Ab inaccessibility and the 2-mercaptoethanesulfonic acid, sodium salt (MESNa) resistance assays described below.[3] For each ligand, the ratio of biotin to protein should be optimized. Biotinylated ligands should be analyzed with respect to the following parameters. For all assays biotinylation should not interfere with ligand binding to receptor. For both the Ab-inaccessibility assay, and enzyme-linked immunosorbent assays (ELISAs), biotinylation should not interfere with binding to anti-ligand antibodies. For both MESNa and ELISA-based assays, there should be sufficient biotinylation of ligands to ensure efficient adsorption to avidin-Sepharose or detection with horseradish peroxidase (HRP)-streptavidin. At the same time, it is important that the residual background of biotinylated ligand after MESNa reduction be minimized. For these reasons, a minimum ratio of biotin residues to ligand should, in general, be used.

[3] S. L. Schmid and E. Smythe, *J. Cell Biol.* **114,** 869 (1991).

Preparation of Cytosol Fractions from K652 Cells and Bovine Brain

Cytosol from a variety of sources has been used to support Tfn internalization into semiintact A431 cells. These include A431 cells, K562 human erythroleukemic cells, bovine brain, rat liver, yeast, and insect tissue culture cells. Cytosol fractions from K562 cells and bovine brain are prepared as follows. K562 cells are washed in KSHM [100 mM potassium acetate, 85 mM sucrose, 1 mM magnesium acetate, and 20 mM N-2-hydroxyethylpiperazine-N'-2-ethanesulfonic acid (HEPES)–NaOH, pH 7.4] and homogenized in ~2 vol KSHM using a ball-bearing homogenizer.[4] Cytosol is collected from the postnuclear supernatant (1500 g_{av} for 5 min at 4°) after centrifugation in an SW41 rotor (Beckman Instruments, Inc., Palo Alto, CA) at ~200,000 g_{av} for 1 hr at 4°. Bovine brains (obtained from a local slaughterhouse), are frozen on dry ice immediately after slaughter and stored at −70° until use. For cytosol preparations, ~50 g of brain is thawed over ~30–60 min and washed in KSHM. Brain is homogenized in ~2 vol KSHM using either a Dounce homogenizer (Wheaton, Millville, NJ) or a Potter homogenizer with motor-driven Teflon pestle. The homogenate is centrifuged at 10,000 g for 15 min and the supernatant is subjected to a second centrifugation at 200,000 g_{av} for 1 hr at 4°. Cytosolic fractions from both K562 cells and bovine brain are rapidly frozen in liquid nitrogen and stored in aliquots at −70° for use. Cytosolic fractions prepared at high concentrations (~10–15 mg/ml) are most active.

Preparation of Semiintact A431 Cells

Procedure. A431 cells, obtained from G. Warren (ICRF, London) are seeded at 6.5 × 10^6 cells per 150 × 25 mm culture dish (Falcon, Becton Dickinson, Oxnard, CA) 20–24 hr before experiments. Best results are obtained when the cells are uniformly spread, forming an 80–90% confluent monolayer. Cells are preincubated for 30–60 min in serum-free medium [Dulbecco's modified Eagle's medium (DMEM) containing 20 mM HEPES, pH 7.4 and 0.2% (v/v) bovine serum albumin (BSA)] to remove bound transferrin. Dishes are washed four times at 4° with KSHM and briefly drained before scraping toward the bottom lip of each dish with a flexible, natural rubber policeman (Cat. No. 7835-Q65; Thomas Scientific, Swedesboro, NJ). Scraped cells are harvested using a 1-ml Pipetman and diluted to 15 ml in KSHM. After incubation for ~5 min on ice to permit loss of cytosolic contents, semiintact cells are pelleted at 800 g for 4 min at 4°. The semiintact cells are then gently resuspended at 4° using a 1-ml Pipetman into 250–300 μl (final volume) KSHM containing 20

---

[4] W. E. Balch and J. E. Rothman, *Arch. Biochem. Biophys.* **240**, 413 (1985).

μg/ml [^{125}I]BSST and 5 mg/ml BSA. Scraped cells are >90–95% semiintact as assessed by trypan blue permeability. Each 15-cm dish provides sufficient cells for 25–30 assays.

General Comments. Consistency in cell plating is important for reducing day-to-day variability in assay efficiencies. Semiintact cell preparations showing efficient endocytosis have been prepared from several cell lines, including SV589 fibroblasts and mouse L cells. Preswelling cells in hypoosmotic buffer before scraping, as is routinely done for the preparation of semiintact Chinese hamster ovary (CHO) cells,[5,6] was found to inhibit endocytosis. Therefore, for endocytosis studies semiintact cells should be prepared without the use of osmotic shock by scraping more tenaciously adherent cell lines. Chinese hamster ovary, SV589 fibroblasts, and other less adherent cells can be plated on poly(L-lysine) before scraping, as described.[5,6]

Cell morphology should be routinely assessed by phase-contrast light microscopy. Clearly intact, almost continuous plasma membranes should define the boundaries of individual cells. Because many organelles appear to be held in place around the nucleus by cytoskeletal connections, it is possible to have cell preparations with intact nuclei surrounded by cytoplasmic organelles that nonetheless lack significant portions of plasma membrane. The degree of semiintactness is most easily and routinely determined by trypan blue staining. However, it is more accurately assessed by determining the latency of cytoplasmic enzymes such as lactate dehydrogenase, as described.[7]

Incubation Conditions

Assays are performed in 1.5-ml microfuge tubes (Cat. No. 72,690; Sarstedt), which are kept in aluminum blocks on ice. Assays are performed in KSHM and, in general, all components are mixed before addition of semiintact A431 cells. Thus the time between resuspension of semiintact A431 cells in [^{125}I]BSST and BSA and their use in assays is minimized. Complete assay mixtures (final volume, 40 μl) typically contain 10–20 μl cytosol, 4–5 μg/ml [^{125}I]BSST, and either an ATP-regenerating system or an ATP-depleting system. The ATP-regenerating system (consisting of 0.8 mM ATP, 5 mM creatine phosphate, and 0.2 IU creatine phosphokinase) is prepared as a 20× stock solution, aliquoted, and stored at −70°. The ATP-depleting system (consisting of 5 mM glucose and 40 U/ml

[5] S. Pind, H. Davidson, R. Schwaninger, C. J. M. Beckers, H. Plutner, S. L. Schmid, and W. E. Balch, this series, in press.

[6] R. Schwaninger, H. Plutner, H. Davidson, S. Pind, and W. E. Balch, this volume [12].

[7] E. Smythe, M. Pypaert, J. Lucocq, and G. Warren, *J. Cell Biol.* **108**, 843 (1989).

hexokinase) is prepared daily as a 20× stock. Ten microliters of an ammonium sulfate suspension of yeast hexokinase (~4000 U/ml, Cat. No. H-5875; Sigma) is pelleted in a microfuge. The supernatant is removed and the pellet is resuspended in 10 μl of 500 mM glucose (stored at −20°) and 40 μl KSHM. To each microfuge tube containing the remaining assay reagents in 30 μl is added 10 μl of the semiintact cell suspension (~2 × 10⁵ semiintact cells). The contents are gently mixed by "flicking" each tube, which is then transferred to 37° for incubation. Cells that settle out during incubation are periodically resuspended by shaking the rack that holds them. Vigorous vortexing of cells is avoided. Following incubation at 37°, the tubes are returned to ice and the extent of [125I]BSST internalization is quantitated by the assays described below.

Assays for Detection of Ligand Internalization

Antibody-Inaccessibility Assay

Procedure. In this assay cell surface [125I]BSST is immuneprecipitated by anti-transferrin antibodies.[7] Surface-bound [125I]BSST–anti-transferrin immunocomplexes are adsorbed to *staphylococcus aureus* cells and pelleted. Internalized, Ab-inaccessible [125I]BSST is quantitated by counting the supernatants. Specifically, following internalization at 37° and return to ice, cells are pelleted for 20 sec in a refrigerated microfuge and resuspended in 100 μl PBS containing 3 μl of sheep anti-transferrin serum (obtained from the Scottish Antibody Production Unit, Carluke, Scotland). Antibodies are allowed to bind during a 90-min incubation at 4° with mixing. One milliliter PBS is then added to each tube and unbound antibody is removed by pelleting the cells for 30 sec in a refrigerated microfuge. The cells are lysed by resuspension in 100 μl PBS containing 1% (v/v) Triton X-100, 1 mM MgCl$_2$, 0.2% (v/v) BSA, and 5 μg/ml unlabeled transferrin (to scavenge any free transferrin binding sites). Thirty microliters of a 10% (v/v) suspension of *S. aureus* cells (Pansorbin; Calbiochem San Diego, CA) washed in lysis buffer is then added and the lysates are incubated for 30–60 min at 37°. [125I]BSST–Ab immunocomplexes adsorbed to the *S. aureus* cells are pelleted for 5 min at 6000 rpm in a microfuge. The supernatants are removed and counted on a γ counter.

Bulky antibodies appear to be excluded from deeply invaginated pits that retain continuity with the plasma membrane.[3,7] Therefore, as discussed below, this assay appears to measure both internalization into *bona fide* sealed endocytic vesicles and sequestration of ligand from antibody into deeply invaginated, but not yet budded, coated pits.[3,8]

[8] S. L. Schmid and L. L. Carter, *J. Cell Biol.* **111**, 2307 (1990).

Expression of Results. Total cell-associated [^{125}I]BSST is determined by adding the counts in the *S. Aureus* pellets with those in the supernatants of samples incubated at 37°. Counts in the supernatant of samples kept on ice are considered background and are subtracted from the results. Backgrounds typically range from 2 to 5% of cell associated counts after binding at 4°. Assay results are expressed as the percentage of total cell-associated [^{125}I]BSST that is inaccessible to Ab precipitation [i.e., counts per minute (cpm) in *S. aureus* supernatant/cpm in *S. aureus* pellet + cpm in *S. aureus* supernatant × 100%]. Typical efficiencies using this assay to determine [^{125}I]BSST internalization into semiintact cells range from 40 to 60% over 4° backgrounds. Internalization following incubation in the absence of ATP (with or without cytosol) is typically 8–12% over 4° backgrounds. When ATP-dependent internalization is to be assessed, this value can be subtracted as background. Internalization in the absence of cytosol is typically 10–20% over 4° backgrounds. [^{125}I]BSST internalization under each of these conditions (i.e., complete reaction, ATP independent, cytosol independent) is strongly temperature dependent. There is no detectable internalization below 10°.[3] All the factors that affect day-to-day variabilities in assay efficiencies have not been identified. However, it is most likely that these can be ascribed to variations in semiintact cell preparations.

2-Mercaptoethanesulfonic Acid-Resistance Assay

Procedure. In this assay, biotin residues on cell surface [^{125}I]BSST are cleaved by reduction with the membrane-impermeant reducing agent 2-mercaptoethanesulfonic acid (MESNa, Cat. No. M1511; Sigma). Internalized, MESNa-resistant [^{125}I]BSST is quantitated following adsorption to avidin-Sepharose. Following incubation at 37°, cells are pelleted for 20 sec in a refrigerated microfuge. The supernatants are aspirated and the pellets resuspended in 50 μl of 10 m*M* MESNa in 50 m*M* Tris, pH 8.6, 100 m*M* NaCl, 1 m*M* ethylenediaminetetraacetic acid (EDTA), 0.2% (v/v) BSA. Cells are incubated for 30 min at 4° with gentle mixing. A second bolus of MESNa is added (12.5 μl of a 50 m*M* stock, freshly prepared before addition) and the cells incubated 30 min. Finally, a third bolus of MESNa (16 μl of a 50 m*M* stock) is added and the cells incubated for an additional 30 min at 4°. Excess MESNa is quenched by addition of 25 μl of 500 m*M* iodoacetamide (IAA; Sigma). After a 10-min incubation, the cells are solubilized by addition of 100 μl 2% (w/v) Triton X-100 in PBS containing 0.5% (v/v) BSA and 50 μl of a 50% (v/v) suspension of avidin-Sepharose (Pierce). Following a 45- to 60-min incubation with shaking, the internalized, unreduced [^{125}I]BSST adsorbed to the avidin-Sepharose beads is collected by centrifugation (30 sec in a microfuge). The beads are washed three times with PBS/1% (w/v) Triton X-100/0.5% (w/v) BSA and counted on a Beckman γ counter.

The MESNa appears to have access to deeply invaginated coated pit structures that exclude antibody. Thus, as discussed below, this assay appears to measure only coated pit budding and internalization into sealed endocytic vesicles.[3,8]

Expression of Results. To determine total cell-associated [^{125}I]BSST, samples incubated at 37° are pelleted to remove unbound [^{125}I]BSST and resuspended in 100 μl PBS containing 50 mM iodoacetamide. These unreduced samples remain on ice while the other samples are reduced. They are then lysed and processed along with the other samples to determine total cell-associated [^{125}I]BSST adsorbed to avidin-Sepharose. Data, in general, are expressed as the percentage of cell-associated [^{125}I]BSST that is resistant to reduction by MESNa (i.e., cpm in avidin-Sepharose pellet after MESNa reduction/cpm in avidin-Sepharose pellet in nonreduced sample $\times$ 100%). Backgrounds, which correspond to MESNa-resistant [^{125}I]BSST bound to cells at 4°, vary between 5 and 10% of total surface-bound [^{125}I]BSST and are, in general, subtracted from the results. Typical efficiencies using this assay to determine [^{125}I]BSST internalization into semiintact cells range from 10 to 25% over 4° backgrounds. Internalization following incubation in the absence of ATP (with or without cytosol) is typically 0–3% over 4° background and is considered insignificant. Internalization in the absence of cytosol is typically 10–15% over 4° backgrounds. As for Ab inaccessibility, both the cytosol-dependent and cytosol-independent internalization of [^{125}I]BSST is temperature dependent. There is no detectable internalization below 10–15°.[3] Internalization as assessed by acquisition of MESNa resistance is less efficient than acquisition of Ab inaccessibility (see Interpretation of Results, below). The efficiency of acquisition of MESNa resistance is also more variable in our hands than acquisition of Ab inaccessibility and appears more sensitive to day-to-day variations in semiintact cell preparations and to the efficiency of reduction by MESNa. This latter problem has been reduced by the ELISA-based assay described below.

ELISA-Based Assay for Internalization of BSST into Semiintact A431 Cells. We have developed an alternative approach to measuring receptor-mediated internalization into semiintact cells. As above, internalized ligands are distinguished from cell surface ligands by their inaccessibility to either a bulky probe (in this case, avidin) or the small membrane-impermeant reducing agent, MESNa. Internalized biotinylated ligands are then detected by an ELISA assay using streptavidin-HRP. This approach offers several advantages over the assays previously developed: it requires less anti-ligand antibody, it circumvents the use of radioactivity, it displays greater sensitivity so that less cells are used and it is more readily applicable to a greater variety of ligands. In addition, backgrounds due to incomplete reduction by MESNa are reduced because the number of biotins remaining bound to them are measured directly, as opposed to quantitating the number of the molecules that still have at least one biotin moiety.

Preparation of ELISA Plates. Rabbit anti-human Tfn (IgG fraction; Boehringer Mannheim) is coated onto ELISA plates (Maxisorp, Cat. No. 4-69949; Nunc, Thousand Oaks, CA) in 200 μl of 50 mM NaHCO$_3$ (pH 9.6) at a protein concentration of 2 μg/ml. The plates are incubated for 3 hr at 37°, washed twice with PBS, and incubated for 1 hr at 37° with blocking buffer [0.2% (w/v) BSA, 1% (w/v) Triton X-100, 0.1% (w/v) sodium dodecylsulfate (SDS), 1 mM EDTA, 50 mM NaCl, 10 mM Tris-HCl, pH 7.4]. Plates may be stored for up to 1 week at 4° in blocking buffer. ELISA plates prepared in this manner are linear between 0.4 and 4 ng BSST per well after detection using HRP-streptavidin. This plate-coating protocol should be optimized, especially with respect to antibody concentration, for each new antibody used. Conditions are sought that minimize backgrounds and maximize the binding capacity and linear range of the plates.

Procedure. Internalization assays are essentially conducted as described in the previous section except that the scraped cells are typically resuspended at a concentration of 5–8 μg/ml unlabeled BSST and 50–75 assays can be performed from a 15 cm dish of cells. After internalization and pelleting, the cells are resuspended in 100 μl of 100 μg/ml avidin (Canadian Lysozyme, Inc., Vancouver, Canada), 0.2% (w/v) BSA, KSHM and shaken for 60 min at 4°. The excess avidin is quenched by the addition of 10 μl of 1 mg/ml biocytin (Cat. No. B 4261; Sigma) followed by a further 15-min incubation. The samples are solubilized by the addition of 300 μl of blocking buffer and 200 μl of the lysate is plated on the ELISA plates. Plates are incubated overnight at 4° or for a minimum of 3 hr at 37°. For MESNa resistance, the cells are treated with MESNa, followed by IAA as described above. The cells are lysed in 300 μl blocking buffer and 200 μl of the lysate is plated onto ELISA plates. In either case it is important that the amount of total ligand applied to the ELISA plates is not beyond the linear range of the plates because this may cause an overestimation of the efficiency of the reaction. This should be checked by plating serial dilutions of samples representing total cell-associated BSST (i.e., those not treated with either avidin or MESNa). The total amount of ligand applied to the plates should approach the total plate capacity so as to maximize signal to noise.

The amount of biotinylated ligand bound to the ELISA plates is quantitated using HRP-streptavidin (Cat. No. 1089 153; Boehringer Mannheim). After ligand binding, the plates are washed three times in PBS, incubated in blocking buffer for 5 min, and washed a further three times in PBS. Streptavidin-HRP (200 μl of a 1/5000 dilution in blocking buffer) is added to each well and allowed to bind for 1 hr at room temperature. The plates are subsequently washed (as before) and bound HRP is quantitated by incubating at room temperature with 200 μl of 10 mg/25 ml o-phenylenediamine (Cat. No. P1526; Sigma), 10 μl/25 ml 30% (v/v) H_2O_2 prepared fresh in 51 mM Na_2PO_4, 27 mM citric acid, pH 5.0. The reaction is terminated (typically after 2–4 min) by the addition of 50 μl of 2 M H_2SO_4. The absorbance at 490 nm is determined using an ELISA plate reader.

Expression of Results. As above, the results are expressed as the percentage of BSST inaccessible to MESNa or avidin relative to the total BSST that binds to the cells during a typical incubation. The absorbances are corrected for the amount of BSST that is inaccessible after binding at 4°. In general, BSST internalization as assayed by the two ELISA-based procedures shows similar characteristics to that previously described for the antibody-inaccessibility assay and the MESNa assay. Internalization is temperature, cytosol, and ATP dependent. The efficiency of the internalization is greater when measured for avidin inaccessibility (50%) as compared to the MESNa resistance (25%). It is likely that the internalization event as measured by avidin inaccessibility is quantitatively similar to that reported for antibody inaccessibility. Similarly, MESNa resistance as determined by either assay is likely to measure the same events.

Applicability to Other Ligands

To provide further insight into the process of receptor-mediated endocytosis it would be advantageous to study the internalization of ligands other than [^{125}I]BSST into semiintact cells. In particular, it would be of interest to measure internalization of receptors, such as the EGF receptor, which are not constitutively localized in coated pits. This would facilitate the identification of putative components specific for the internalization of a given ligand or class of receptors and in addition it may lend insight into the overall process of ligand-induced internalization.

Each of the assays described above is sensitive and should allow the detection of internalization of a variety of ligands. We have successfully measured EGF internalization as well as internalization of antibodies directed toward the ectodomain of both the Tfn and EGF receptors. The following considerations are important in establishing these assays using other ligands. The Ab-inaccessibility assay requires that the antibody used can efficiently (>90–95%) bind to cell surface ligand (i.e., receptor-bound ligand). For example, antibody bound inefficiently to the small ligand, EGF, while bound to its receptor leading to unacceptably high backgrounds. This problem is not encountered when the ELISA assay procedure is used, because the cells are solubilized before plating. In addition, for the Ab-inaccessibility assay, it is important that excess unlabeled ligand present during cell solubilization does not displace bound [^{125}I]labeled ligand. Antibodies used for the ELISA assay protocol should be either high-titer IgG fractions or monoclonal antibodies. Because the binding capacity for each well is limited, higher titer antibodies will ensure higher binding capacity and greater linear range of the plates. Finally, for both the MESNa and ELISA assays, biotinylation procedures for each new ligand should be optimized as discussed above. In general, the ELISA-based protocols are more amenable to adaptation to other ligands.

Interpretation of Results

The data in Fig. 2 demonstrate a typical time course and the ATP dependence of [^{125}I]BSST internalization as determined by acquisition of Ab inaccessibility and MESNa resistance. Acquisition of MESNa resistance is less efficient than acquisition of Ab inaccessibility. Similar differences are observed using the ELISA-based assay when comparing avidin vs MESNa treatments. This result suggests that Ab inaccessibility measures additional events not scored by acquisition of MESNa resistance. Morpho-

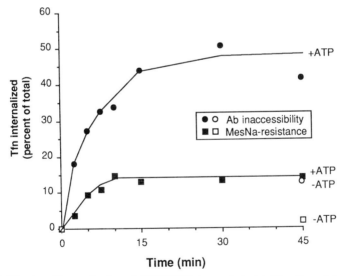

FIG. 2. Kinetics of internalization of [^{125}I]BSST into semiintact A431 cells as determined by Ab-inaccessibility and MESNa-resistance assays. Semiintact cells were prepared and incubations performed at 37° in the presence of 4 μg/ml [^{125}I]BSST, cytosol, and either an ATP-regenerating system (●, ■) or an ATP-depleting system (○, □) as described in text. Internalized [^{125}I]BSST was determined either by the Ab-inaccessibility assay (●,○) or the MESNa-resistance assay (■, □) as described in text. Data are expressed as the percentage of total cell-associated [^{125}I]BSST internalized. Backgrounds corresponding to 3.7 and 3.8% for Ab inaccessibility and MESNa resistance, respectively, have been subtracted from the results.

logical characterization of Tfn internalization into semiintact A431 cells demonstrated that shallow pits convert to deeply invaginated coated pits in the absence of ATP.[7,9] Studies in intact cells demonstrated that acquisition of Ab inaccessibility was much less sensitive to ATP depletion than acquisition of MESNa resistance.[8] In addition, a significant increase in deeply invaginated pits was observed in ATP-depleted cells.[8] These and other results[3] support our interpretation that Ab inaccessibility measures the formation of deeply invaginated coated pits in addition to coated vesicle budding whereas the acquisition of MESNa resistance specifically measures coated vesicle budding. Although both these assays give identical results in intact cells, the differences in efficiencies obtained using semiintact cells suggest that coated vesicle budding is much less efficient than coated pit formation *in vitro*. This interpretation is also supported by the morphological studies of others.[7,9]

Acknowledgments

This work was supported by grants to S. L. Schmid from the National Institutes of Health (GM42445 and CA27489) and from the Lucille P. Markey Charitable Trust. S.L.S. is a Lucille P. Markey Scholar. E. Smythe was supported by a NATO/SERC Postdoctoral Fellowship.

[9] M. Pypaert and G. Warren, this volume [21].

[23] Movement from Trans-Golgi Network to Cell Surface in Semiintact Cells

By STEPHEN G. MILLER and HSIAO-PING H. MOORE

Introduction

The trans-Golgi is a key compartment involved in trafficking between cellular organelles. Newly synthesized proteins to be released into the extracellular space, incorporated into the plasma membrane, or localized to lysosomes all traverse the same compartments until they reach the trans-Golgi. At the trans-Golgi network, these proteins are sorted from one another and packaged into unique classes of vesicles that are targeted to distinct post-Golgi destinations. In some specialized cell types, proteins destined for the cell surface are also sorted. Cells that release proteins in response to extracellular stimuli, such as neurons and endocrine cells, sort their secretory proteins into either the constitutive or the regulated pathway. Likewise, polarized cells [such as epithelial Madin–Darby canine kidney (MDCK) cells] sort proteins into vesicles that deliver cargos directly to either the apical or basolateral plasma membrane. In addition to proteins exiting the Golgi via the biosynthetic route, some proteins return to the trans-Golgi from post-Golgi compartments. For example, mannose 6-phosphate receptors continuously recycle to the trans-Golgi from prelysosomes.[1]

In this chapter, we describe a method that utilizes semiintact Chinese hamster ovary (CHO) cells to reconstitute vesicular traffic from the trans-Golgi region to the plasma membrane. The assay utilizes a bacterial cytolysin, streptolysin O (SL-O), to introduce large aqueous pores selectively into the plasma membrane. The general application of SL-O to permeabilize cells has been reviewed.[2] Because this method produces minimal damage to the cell surface, it is suitable for studying various transport pathways between the trans-Golgi and the plasma membrane. This approach has been successfully utilized by our group to study intracellular transport in CHO cells,[3] AtT-20 cells,[4] PC12 cells,[5] and by Gravotta et al.[6] to study transport in MDCK cells.

Preparation of Semiintact Cells with Streptolysin O

Cell Culture

This chapter describes methods for reconstituting constitutive transport between the trans-Golgi and the plasma membrane. Chinese hamster ovary cells are routinely used for this purpose for several reasons. First, CHO cells secrete newly synthesized proteins exclusively by the constitutive pathway in a nonpolarized fashion, therefore this pathway can be characterized in isolation. Second, the biochemistry of vesicular transport from the endoplasmic reticulum to the cis-Golgi, from the cis- to the medial-Golgi, and from a late endosomal compartment to the trans-Golgi network have been characterized using *in vitro* assays in CHO cells (see chapters by Rothman, Balch, and Pfeffer in this volume); thus, the results can be directly compared with other exocytic and endocytic transport steps.

[1] J. R. Duncan and S. Kornfeld, *J. Cell Biol.* **106**, 617 (1988).
[2] G. Ahnert-Hilger, W. Mach, K. J. Fohr, and M. Gratzl, *Methods Cell Biol.* **31**, 63 (1989).
[3] S. G. Miller and H.-P. H. Moore, *J. Cell Biol.* **112**, 39 (1991).
[4] S.-C. Chao and H.-P. H. Moore, unpublished observations, 1991.
[5] L. Carnell and H.-P. H. Moore, unpublished observations, 1992.
[6] D. Gravotta, M. Adesnik, and D. Sabatini, *J. Cell Biol.* **111**, 2893 (1990).

Chinese hamster ovary cells are maintained as monolayer cultures in Ham's F10 medium supplemented with 10% (v/v) fetal calf serum at 37° in a humidified atmosphere containing 5% CO_2. Cells to be used in suspension transport assays are plated at a density of $2-3 \times 10^6/10$-cm dish and grown for 48 hr before use. This results in a semiconfluent monolayer containing $1-1.5 \times 10^7$ cells/10-cm plate. One 10-cm plate typically provides sufficient cells for approximately 50 transport assays. For assays using attached cells, 12-well plates are routinely used. Depending on the sensitivity of the assays used, 24-, 48-, or 96-well plates are also suitable. For 12-well plates, 2×10^5 cells are plated in each well and used after 48 hr, resulting in approximately 1×10^6 cells/well.

Reconstitution, Storage, and Activation of Streptolysin O

Streptolysin O (M_r 69,000) is obtained from culture supernatants of various strains of β-hemolytic streptococci. Streptolysin O introduces a heterogeneous population of large (up to 20 nm) aqueous pores into the plasma membrane of eukaryotic cells by complexation with cholesterol.[7] Streptolysin is available from several sources, but we have found that commercial preparations of SL-O vary considerably in their ability to permeabilize eukaryotic cells. We have had good success with the product distributed by Burroughs-Wellcome (Research Triangle Park, NC). Streptolysin O from this source is provided as a lyophilized preparation containing buffer salts that is isotonic and buffered to pH 6.5 when reconstituted at 2 units (U)/ml with distilled water. GIBCO/BRL (Grand Island, NY) currently markets an SL-O preparation that is lyophilized in the absence of buffer salts, which may be more convenient for specific applications. One must be aware that the SL-O provided by most manufacturers is a crude preparation and potentially contains contaminating enzymatic activities that may interfere with transport.

We generally reconstitute the lyophilized SL-O preparation at a concentration of 5 U/ml with distilled water and immediately freeze small (50 to 100-μl) aliquots in liquid nitrogen and store them at $-80°$ until use. Streptolysin O that has been stored in this manner is stable for at least 2 months. Streptolysin O activity is inhibited by oxidation and is commonly "activated" with dithiothreitol (DTT) immediately before use. To activate, an aliquot is thawed rapidly in a 37° water bath, brought to 20 mM DTT using a 0.1 M stock solution, and incubated in an ice/water bath for 15 min. Any material not used immediately after "activation" is discarded.

---

[7] S. Bhakdi, J. Tranum-Jensen, and A. Sziegoleit, Infect. Immun. 47, 52 (1985).

Optimization of Perforation Conditions

Several factors are important in optimizing cell permeabilization while retaining competence for transport. As mentioned above, the ability of different commercial preparations to permeabilize cells is highly variable. Therefore, it may be difficult to compare directly the concentrations of SL-O reported in the literature by individual investigators. In addition to the concentration of SL-O, the time, temperature, and ratio of SL-O to cells all affect the degree of perforation. The concentration of SL-O required also varies considerably among different cell lines. We have found that the concentration of SL-O required to permeabilize > 90% of cells can vary by one to two orders of magnitude depending on the cell type. For example, 0.05 U/ml of SL-O is sufficient to permeabilize > 90% of CHO cells while up to 3.0 U/ml is required to permeabilize the same fraction of PC12 cells.[5] In addition, it has been reported that the effective size of the pores introduced by SL-O may vary depending on the concentration used.[8] Therefore, at low concentrations of SL-O, the plasma membrane may be permeable to small molecules (e.g., ions, nucleotides) but not to large proteins, while at higher concentrations large molecules are freely exchanged. Thus, at intermediate concentrations of SL-O, transport may be dependent on the addition of ATP but independent of added cytosol.

We use a method that involves "prebinding" of SL-O to the surface of cells at 4°, washing away excess SL-O, and warming to 37° to effect permeabilization.[9] This method may require higher concentrations of SL-O to produce the same degree of permeabilization compared to direct addition of SL-O without a prebinding step. The prebinding method has several advantages. First, commercial preparations of SL-O are not pure and prebinding allows the removal of potentially problematic contaminants. Second, some commercial SL-O preparations contain high concentrations of salt at the concentrations required for permeabilization of some cell types. Prebinding allows one to perform the permeabilization step under controlled conditions of pH and ionic composition.

Permeabilization of Cells Attached to Plastic Dishes

Chinese hamster ovary cells are seeded in 12-well plates at a density of $2-3 \times 10^5$ cells/well and grown for 48 hr before use. The medium is aspirated and the cells are washed three times with 2 ml of ice-cold buffer A [20 mM Na-2-hydroxyethylpiperazine-N'-2-ethanesulfonic acid

[8] L. Buckingham and J. L. Duncan, *Biochim. Biophys. Acta* **279**, 115 (1983).
[9] F. Hugo, J. Reichweiss, M. Arvand, S. Krämer, and S. Bhakdi, *Infect. Immun.* **54**, 641 (1986).

(HEPES), pH 7.2, 110 mM NaCl, 5.4 mM KCl, 0.9 mM Na$_2$HPO$_4$, 10 mM MgCl$_2$, 2 mM CaCl$_2$, 1 g/liter glucose]. The final wash is aspirated and 500 μl of 0.2 U/ml activated SL-O, preequilibrated in ice-cold buffer A, is added to each well. The plate is incubated on an ice/water bath for 10 min to allow binding of SL-O. Excess unbound SL-O is then removed by aspirating and washing twice with 1 ml of ice-cold buffer B [20 mM HEPES–KOH, pH 7.4, 100 mM potassium glutamate, 40 mM KCl, 5 mM MgCl$_2$, 5 mM ethylene glycol-bis(β-amino ethyl ether)-N,N,N',N'-tetraacetic acid (EGTA)]. Permeabilization is effected by aspirating the final wash, adding 1 ml of prewarmed buffer B (at 37°), and floating the dish on a 37° water bath for 3–5 min. In experiments where the steps of interest occur early during the transport reaction, an ATP-depleting system (4 mM glucose and 30 U/ml hexokinase) is included in the buffer to minimize vesicle formation during SL-O treatment. Permeabilization is terminated by aspirating the medium from each well, adding 1 ml of ice-cold buffer B, and transferring to an ice/water bath.

Permeabilization of Cells in Suspension

Chinese hamster ovary cells are grown to approximately 80% confluency on a 10-cm dish. To resuspend, cells are washed twice with 10 ml of Ca^{2+}/Mg^{2+}-free phosphate-buffered saline (PBS) followed by incubation in 10 ml of Ca^{2+}/Mg^{2+}-free PBS containing 5 mM ethylenediaminetetraacetic acid (EDTA) for 5 min at 37°. The suspended cells are then transferred to a 15-ml centrifuge tube and pelleted by centrifugation at 1000 rpm in a clinical centrifuge for 4 min at room temperature. The supernatant is aspirated, and the cells are resuspended in 1 ml of buffer A by gentle trituration and subsequently counted. The cell suspension is transferred to a 1.5-ml polypropylene microcentrifuge tube, pelleted briefly (5 sec) in a microcentrifuge, then gently washed twice with 1 ml of ice-cold buffer A by resuspension and brief centrifugation. The cells are finally resuspended in ice-cold buffer A at a density of 1 × 10^4 cells/μl and kept in an ice/water bath until use. It is important that the cells are not subjected to excessive shearing during these steps.

To produce semiintact cells, a 0.2 to 0.3-ml aliquot of the cell suspension is gently mixed with an equal volume of ice-cold buffer A containing 0.08–0.1 U/ml activated SL-O. The tube is returned to an ice/water bath for 10 min to allow binding of SL-O to the cells. Excess unbound SL-O is then removed by centrifugation in a microcentrifuge for 5 sec at 4°. The cells are washed twice with 1 ml of ice-cold buffer B, and finally resuspended at 1 × 10^4 cells/μl. An ATP-depleting system (4 mM glucose and 30 U/ml hexokinase) may also be included at this point if desired. Permeabilization is effected by shifting the cell suspension to a 37° water bath for 3–5 min. At the end of the permeabilization 1 ml of ice-cold buffer B is added, the cells are recovered by brief centrifugation, and finally resuspended at 1×10^4 cells/μl and kept on ice until use.

Assessment of Degree of Permeabilization

Several methods have been employed to determine the extent and degree of permeabilization by SL-O.[2] The simplest method is examination using phase-contrast light microscopy; permeabilized cells are generally less refractile than intact cells. Permeabilized cells can be visualized by staining with propidium iodide followed by examination with a fluorescence microscope. For attached cells, 0.5 ml of 5 μg/ml propidium iodide (M_r 668) in buffer B is added to each well of monolayer of cells in a 12-well plate and incubated for 5 min at 4° For cells in suspension, a 10-μl aliquot of permeabilized cells is mixed with 40 μl of 5 μg/ml propidium iodide, incubated for 5 min at 4°, spotted onto a glass slide, and covered with a glass cover slip. The nuclei of permeabilized cells will appear bright red/orange when viewed under fluorescein optics while intact cells remain unstained. Alternative methods, such as staining with ethidium bromide or trypan blue, can also be used to assess permeability to small molecules.

Permeability to larger molecules can be assessed by several methods. The loss of a cytoplasmic marker, such as lactate dehydrogenase (M_r 135,000), has been used to monitor permeabilization with SL-O.[10] Gravotta et al.[6] examined the release of trichloroacetic acid-precipitable radioactivity from [^{35}S]methionine-labeled cells to measure the release of cytoplasmic proteins from SL-O-permeabilized MDCK cells. In addition, accessibility of the cytoplasmic compartment to antibodies provides a means of estimating the fraction of cells accessible to large proteins.[3,11]

Labeling and Quantitation of Sulfated Glycosaminoglycans

Background

Two methods are routinely used to monitor transport from the trans-Golgi to the cell surface. To follow a membrane-bound molecule, the cells are infected with a temperature-sensitive mutant of vesicular stomatitis virus, VSV ts-045, and subsequently incubated at 20° to accumulate the viral membrane glycoprotein (VSV G protein) at the trans-Golgi network

---

[10] T. W. Howell and B. Gomperts, Biochim. Biophys. Acta 927, 177 (1987).
[11] G. Ahnert-Hilger, M. F. Bader, S. Bhakdi, and M. Gratzl, J. Neurochem. 52, 1751 (1989).

(TGN).[12] The movement of the viral membrane protein to the cell surface can then be assayed by indirect immunofluorescence or by surface immunoprecipitation.[3] Tryptic cleavage of the influenza virus hemagglutinin can also be used to measure transport, as has been shown in SL-O-permeabilized MDCK cells.[6] Alternatively, a soluble bulk-flow tracer such as glycosaminoglycan chains may be used to measure transport. We have found that measurement of the secretion of sulfated glycosaminoglycans (GAGs) provides an easier and more quantitative method for measuring constitutive secretion than using viral membrane proteins. The ability to perform multiple assays rapidly makes this method ideal for analyzing column fractions or other experiments requiring large numbers of transport reactions. Fifty *in vitro* transport reactions can be performed and quantitated within about 6 hr. In addition, the assay can be performed without any specialized reagents and eliminates the potential complication introduced by viral infection. However, because this method measures the release of a soluble molecule, it may give a false signal due to rupture of intracellular organelles or transport vesicles (although methods using cleavage or immunoprecipitation of viral proteins suffer from the same problems). Therefore, the results obtained from assays using sulfated GAG transport should be verified by other means, such as indirect immunofluorescence using vesicular stomatitis virus G protein or influenza virus hemagglutinin.[3,6]

Incubation of tissue culture cells with the membrane-permeant, 4-methylumbelliferyl-β-D-xyloside ("xyloside"), results in the synthesis of large quantities of free GAG chains in the Golgi lumen.[13] These free GAG chains are heavily sulfated during synthesis and therefore can be readily labeled by pulse-labeling of xyloside-treated cells with $^{35}SO_4$. The newly synthesized sulfated GAG chains are rapidly secreted in a soluble form by the constitutive pathway.[14] A rapid, quantitative, precipitation/filtration assay[15] can be employed to quantitate sulfated GAG release from large numbers of samples. The assay takes advantage of the high charge density of sulfated glycosaminoglycans for precipitation with cetylpyridinium chloride. The insoluble complex is then captured on membrane filters and counted in a scintillation counter.

[12] G. Griffiths, S. Pfeffer, K. Simons, and K. Matlin, *J. Cell Biol.* **101,** 949 (1985).
[13] N. B. Schwartz, L. Galligani, P.-L. Ho, and A. Dorfman, *Proc. Natl. Acad. Sci. USA* **71,** 4047 (1974).
[14] T. L. Burgess and R. B. Kelly, *J. Cell Biol.* **99,** 2223 (1984).
[15] S. D. Luikart, J. L. Sackrison, and C. V. Thomas, *Blood* **66,** 866 (1985).

Biosynthetic Labeling of Glycosaminoglycans with $^{35}SO_4$

Prior to labeling, the cells are starved for sulfate and equilibrated with xyloside to initiate the synthesis of glycosaminoglycan chains. For transport assays using cells in suspension, the labeling is done in 1.5-ml microcentrifuge tubes after resuspension of cells. The cells are washed twice with buffer A by brief centrifugation/resuspension and incubated for 30 min at 37° in 400 μl of buffer A containing 500 μM xyloside (approximately 25,000–35,000 cells/μl). A 0.5 M (100×) stock solution of 4-methylumbelliferyl-β-D-xyloside is prepared in dimethyl sulfoxide and stored at −20°. At the end of the starvation period, 100 μl of buffer A containing 100–200 μCi/ml of carrier-free $^{35}SO_4$ (25–40 Ci/mg) is added and the reaction is shifted to 37° for 120 sec. Labeling is terminated by adding 750 μl ice-cold buffer A supplemented with 4 mM nonradioactive Na$_2$SO$_4$ and shifting to an ice/water bath. Excess $^{35}SO_4$ is removed by washing twice with 1 ml of ice-cold buffer A containing 4 mM Na$_2$SO$_4$ by brief centrifugation and gentle resuspension. The labeled cell suspension is kept in an ice/water bath until permeabilization with SL-O as described above.

Quantitation of $^{35}SO_4$-Labeled Glycosaminoglycans

The precipitation/filtration assay of Luikart *et al.*[15] is used to quantitate $^{35}SO_4$-labeled glycosaminoglycan chains in cell extracts or medium from transport reactions. The assay takes advantage of the high charge density of the heavily sulfated glycosaminoglycans to form an insoluble precipitate with cetylpyridinium chloride. A commercially available preparation of chondroitin sulfate is used as a carrier and the precipitate is captured by vacuum filtration through 0.45-μm pore size membrane filters. Samples from *in vitro* transport reactions may be frozen at −20° before carrying out the precipitation/filtration assay. This assay is linear over at least a 70-fold range of concentrations of glycosaminoglycan chains. The assay is also relatively insensitive to the presence of free $^{35}SO_4$ in the amounts typically present in transport reactions.

Media samples or cell extracts are first digested with a nonspecific protease (pronase E; Sigma, St. Louis, MO) to reduce background. One hundred microliters of a 6-mg/ml solution of pronase E is added to each 500 μl of medium or extract sample (final concentration, 1 mg/ml pronase E) and the digestion is carried out for at least 2 hr at 37°. One hundred microliters of 15% (w/v) cetylpyridinium chloride [CPC; 2% (w/v) final concentration] and 50 μl of 2 mg/ml chondroitin sulfate (100 μg) are added to each sample to precipitate sulfated glycosaminoglycans. The samples are incubated further for 30 min at 37°. Precipitated sulfated glycosaminoglycans are collected by vacuum filtration through 2.4-cm diameter, 0.45-μm membrane filters (GN-6 Metricel or equivalent; Gelman Sciences, Ann Arbor, MI) using a vacuum manifold. Each filter is washed four times with 5 ml of 1% CPC, 25 mM Na_2SO_4 and the filters are transferred to glass scintillation vials and dried in a vacuum oven. Four milliliters of scintillation fluid is added to each vial and the samples are counted in a scintillation counter.

In Vitro Transport from Trans-Golgi Region to Plasma Membrane

Attached Cells

Attached cells grown in 12-well plates are metabolically labeled and permeabilized with SL-O as described above. They are then incubated in buffer B for 30–60 min at 4° to allow loss of cytoplasmic molecules. Transport is then carried out in a total volume of 250 μl of buffer B containing 350–400 μg of gel-filtered bovine brain cytosol, and either an ATP-regenerating system (1 mM ATP, 2 mM creatine phosphate, 40 U/ml creatine kinase) or an ATP-depleting system (2 mM glucose, 30 U/ml hexokinase). Transport buffer is added to the cells at 4° and transport is initiated by shifting the plates to 37°. We typically carry out the reactions by floating the plates on a temperature-controlled water bath. The reaction is terminated by cooling to 4°. The medium is transferred to 1.5-ml microcentrifuge tubes, 250 μl of ice-cold PBS is used to wash each well, and this is combined with the medium sample. The cells are solubilized by adding 100 μl of 50 mM Tris (pH 8.0), 150 mM NaCl, 2 mM $MgCl_2$, 1% (v/v) Triton X-100 to each well and incubating for 5 min at 37°. After solubilization, 400 μl of ice-cold buffer B is added to each well and the cell extracts are transferred to microcentrifuge tubes. ^{35}S-Labeled GAG chains in the medium and extract samples are quantitated as described above.

Cells in Suspension

Transport reactions are carried out in a total of 50 μl in 1.5-ml polypropylene microcentrifuge tubes. Permeabilized, [35]SO_4-labeled CHO cells are first incubated in buffer B for 30–60 min at 4° to allow loss of cytoplasmic molecules. Ten microliters of cells (approximately 10^5 cells) is then incubated in a total volume of 50 μl of buffer B containing 75 μg of gel-filtered bovine brain cytosol, and either an ATP-regenerating system (1 mM ATP, 2 mM creatine phosphate, 40 U/ml creatine kinase) or an ATP-depleting system (2 mM glucose, 30 U/ml hexokinase). The ATP-regenerating and ATP-depleting systems are made up as 20× stock solutions immediately prior to addition to the reaction. Reactions are initiated by shifting to 37°, and terminated by adding 450 μl of ice-cold buffer B and shifting to an ice/water bath. The cells are pelleted by centrifugation for

2 min in a microcentrifuge at 4° and the supernatant fraction, containing released ^{35}S-labeled GAG chains, is transferred to a fresh tube. The cell pellet is solubilized by adding 100 μl of 50 mM Tris (pH 8.0), 150 mM NaCl, 2 mM MgCl$_2$, 1% (v/v) Triton X-100 and incubating for 5 min at 37°. After solubilization 400 μl of ice-cold buffer B is added to the cell

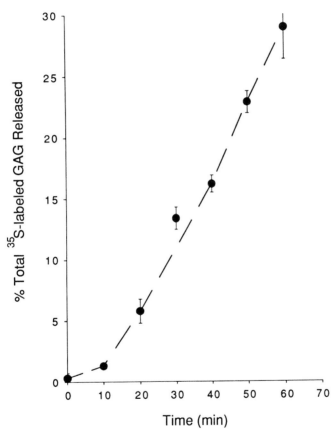

FIG. 1. Time course of ATP-dependent release of ^{35}S-labeled glycosaminoglycans from SL-O-permeabilized CHO cells. Chinese hamster ovary cells in suspension were equilibrated with xyloside, pulse-labeled with [35]SO$_4$, and permeabilized with SL-O (see text for details). Transport reactions were performed for the indicated times in the presence of cytosol and either an ATP-depleting or ATP-regenerating system as described in the text. Transport reactions were terminated by shifting the tube to 4°. The ATP-dependent release of ^{35}S-labeled glycosaminoglycans, expressed as a percentage of the total synthesized during the 2-min labeling period, is plotted as a function of time at 37°. Data are plotted as the mean ± standard deviation of duplicate samples.

extract and ^{35}S-labeled GAG chains in the medium and extract samples are quantitated as described above.

The time course of ATP-dependent ^{35}S-labeled GAG chain secretion from SL-O-permeabilized cells is shown in Fig. 1. Release of ^{35}S-labeled GAGs in the absence of ATP reaches a maximum level of 5–15% after 10–20 min of incubation. In contrast, ATP-*dependent* release of labeled GAG chains begins after a lag of approximately 5–10 min and reaches a

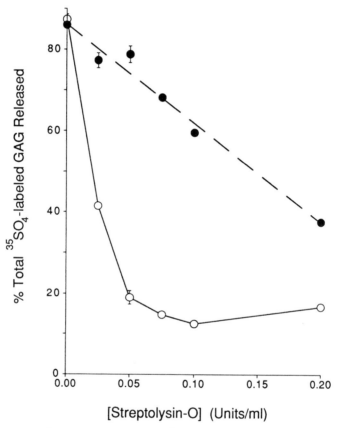

[Streptolysin-O] (Units/ml)

FIG. 2. Extent of ATP-dependent release of ^{35}S-labeled glycosaminoglycan as a function of the concentration of SL-O. Chinese hamster ovary cells were equilibrated with cyloside, pulse-labeled with [35]SO$_4$, and permeabilized for 3 min with the indicated concentrations of SL-O. Transport reactions were carried out at 37° in the presence of cytosol and either an ATP-depleting (O) or ATP-regenerating (●) system for 60 min as described in the text. The amount of ^{35}S-labeled glycosaminoglycans released, expressed as a percentage of the total synthesized during the 2-min labeling period, is plotted versus the concentration of SL-O used. Data are plotted as the mean ± standard deviation of duplicate samples.

maximal level of 30–50% of total ^{35}S-labeled GAG synthesized by 60 min. As described above, the concentration of SL-O must be optimized for maximal transport. As shown in Fig. 2, low concentrations of SL-O result in high levels of ATP-independent ^{35}S-labeled GAG release, suggesting incomplete permeabilization. At higher concentrations of SL-O, transport in the presence of ATP begins to decrease while ATP-independent release stabilizes at a constant low level (typically 5–15% of total ^{35}S-labeled GAG synthesized). At very high SL-O concentrations (>0.2 U/ml for CHO cells), ^{35}S-labeled GAG release is identical in the presence or absence of ATP. The low level of ATP-independent GAG release at high concentrations of SL-O indicates that the trans-Golgi and transport vesicles are not being lysed by SL-O.

Under optimized permeabilization and incubation conditions, transport from the trans-Golgi to the cell surface in SL-O-permeabilized cells requires cytosolic factors supplied by a crude cytosol fraction[3,6] (Fig. 3). We generally prepare cytosol from either bovine or rat brain because these sources provide an abundant and inexpensive source for cytosol preparation. Fresh tissue is frozen and stored at −80° before cytosol preparation. All steps are carried out at 4°. Brain tissue (typically 10 g) is thawed, minced, and homogenized in 2 vol of buffer C (25 mM HEPES–KOH, pH 7.2, 0.1 M potassium glutamate, 1 mM dithiothreitol, 0.1 mM phenyl-methylsulfonyl fluoride, 10 μg/ml leupeptin, 1 μM pepstatin, and 0.5 mM 1,1-phenanthroline) using a motor-driven glass/Teflon homogenizer. The crude homogenate is subjected to centrifugation at 800 rpm for 20 min in a Sorvall (Boston, MA) SS-34 rotor. The supernatant is recovered and further fractionated by centrifugation at 50,00 rpm for 90 min in a Beckman (Palo Alto, CA) 50.2 Ti rotor. The supernatant from this step (crude cytosol fraction) is then applied to a column of Sephadex G-25 (1.5 × 50 cm) equilibrated in 25 mM HEPES–KOH, pH 7.2, 125 mM potassium glutamate, 2 mM dithiothreitol. The protein peak eluting in the void volume of the column is pooled, aliquoted into 100-μl fractions, flash-frozen in liquid nitrogen, and stored at −80°. Cytosol stored in this fashion is stable for at least several months at −80°. Immediately before use, an aliquot is thawed quickly in a 37° water bath, then shifted to an ice/water bath.

Discussion

Many intracellular transport steps have been reconstituted *in vitro.* Steps involving the plasma membrane as either the donor or the acceptor membrane are more problematic for reconstitution than those involving only intracellular membranes. This is because commonly used methods for

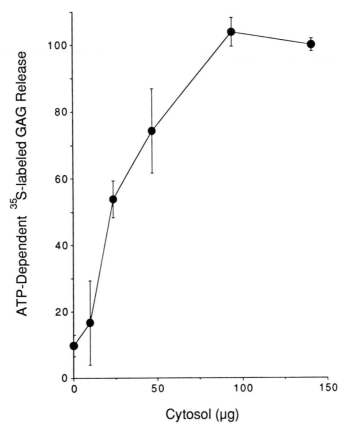

Fig. 3. Release of ³⁵S-labeled glycosaminoglycan from SL-O-permeabilized CHO cells as a function of added cytosol. CHO cells in suspension were equilibrated with cyloside, pulse-labeled with $^{35}SO_4$, and permeabilized with SL-O. Transport reactions were performed at 37° in the presence of indicated amounts of bovine brain cytosol and either an ATP-regenerating or ATP-depleting system as described in text. Transport was terminated after 60 min by shifting to 4°. The ATP-dependent release of ³⁵S-labeled glycosaminoglycans, expressed as a percentage of the total synthesized during the 2-min labeling period, is plotted versus the amount of exogenous cytosol present in the reaction. Release of ³⁵S-labeled glycosaminogly-cans in the absence of ATP is not dependent on the concentration of cytosol. Data are plotted as the mean ± standard deviation of duplicate samples.

preparing semiintact cells, such as digitonin, saponin, nitrocellulose, or osmotic shock, often damage the cell surface to such an extent that they are no longer competent for vesicle production and consumption. In contrast to these methods, semiintact cells prepared using SL-O allow efficient transport from the trans-Golgi to the plasma membrane. This general method has been used to study the fusion of regulated secretory vesicles in a number of cell types (for review see Ref. 2). Here we have extended the use of this reagent to study the formation, targeting, and fusion of a variety of exocytic vesicles in several cell lines.

Two methods are used for preparing semiintact cells, one using cells attached to plastic tissue culture dishes and the other using cells in suspension. Each technique produces semiintact cells that are competent for *in vitro* transport from the trans-Golgi to the plasma membrane. The use of cells in suspension offers several advantages over attached cells, making it the preferred technique when using CHO cells. With cells in suspension, it is easier to perform large numbers of transport assays in a single experiment. In addition, kinetic experiments are simpler to perform because individual tubes can be shifted to ice or aliquots can be removed from a single reaction at different times. Moreover, we have found that the efficiency of transport is generally higher in cells in suspension compared to attached cells. We have observed that transport is not consistently dependent on exogenous cytosol when attached cells are used. In contrast, when cells in suspension are permeabilized under optimal conditions, transport consistently requires the addition of exogenous cytosol. The use of attached cells may be required for other cell types that are more fragile in suspension. For example, we have better success reconstituting regulated secretion in attached AtT-20 cells.

We use two independent methods to monitor movement from the trans-Golgi to the plasma membrane: the transport of vesicular stomatitis virus G glycoprotein (VSV G protein) to the plasma membrane and the secretion of ^{35}S-labeled GAGs.[3] The use of ^{35}S-labeled GAG as a marker for secretion has a number of advantages over methods involving VSV G protein. The extent of arrest of VSV G protein in the trans-Golgi network at 20° is not complete and the amount of G protein distributed in compartments other than the TGN may vary in different cell types; this makes it difficult to perform kinetic experiments that require synchronous transport from the trans-Golgi. In contrast, ^{35}S-labeled GAGs are synthesized and secreted with similar kinetics in several cell lines that we have examined.[4] The ability to label sulfated GAGs with short (30–120 sec) pulses of [35]SO_4 allows one to monitor the synchronous movement of this marker through the secretory pathway.

Quantitation of ^{35}S-labeled GAG secretion utilizes only commercially available reagents, is less complicated, and is significantly faster than monitoring the appearance of VSV G protein at the cell surface. Assays using VSV G protein transport require several days to perform, whereas experiments utilizing ^{35}S-labeled GAG chains are completed in several hours. Due to these factors, the ^{35}S-labeled GAG assay is amenable to titration or kinetic experiments that require large numbers of experimental conditions to be analyzed in duplicate or triplicate. The linearity and sensitivity of the ^{35}S-labeled GAG assay also allows one to perform kinetic measurements with high precision. The development of these methods should greatly facilitate the molecular dissection of components involved in final stages of secretion.

Acknowledgments

This work was supported by grants from the National Institutes of Health (GM35239), the Markey Charitable Trust, and the American Cancer Society (CD-64184). S.G.M. is a Merck Postdoctoral Fellow of the Helen Hay Whitney Foundation.

Section III

Identification of Transport Intermediates

[24] Isolation and Characterization of Functional Clathrin-Coated Endocytic Vesicles

By Philip G. Woodman and Graham Warren

Introduction

The transferrin cycle is well characterized, and serves as a model of receptor-mediated endocytosis.[1] Halotransferrin binds to cell surface receptors located in clathrin-coated pits. These pits invaginate and pinch off to form endocytic, clathrin-coated vesicles, which carry the transferrin into the cell. Removal of at least part of the clathrin coat enables transferrin to be delivered to the endosome by a membrane fusion event. Once exposed to the low pH of the endosome, iron dissociates from transferrin and the apoprotein is recycled back to the plasma membrane by vesicular transport.

For the transferrin cycle to work efficiently, vesicles must recognize and fuse with their target membrane. The biochemical specificity that underlies this selection has allowed us to reconstitute fusion of endocytic vesicles in a cell-free system. The assay for vesicle fusion requires the preparation of "donor" endocytic vesicles containing ^{125}I-labeled transferrin. These are mixed with "acceptor" endocytic vesicles, containing internalized anti-transferrin antibody, in the presence of a cytosol fraction and an ATP-regenerating cocktail. Vesicle fusion permits the formation of a radiolabeled immunocomplex, which is isolated on *Staphylococcus aureus* cells after solubilization of the vesicle membrane. Measurement of the fusion of endocytic vesicles within crude preparations has been described in detail elsewhere.[2] However, further analysis of the interaction between cytosolic and membrane components during the fusion reaction requires use of purified membranes as substrates. We have chosen to isolate a donor preparation of clathrin-coated vesicles,[3] because this is the best characterized endocytic compartment. In addition, the unique composition of coated vesicles simplifies both purification and identification. Here, we describe the isolation of functional clathrin-coated vesicles. Isolation is monitored by the ability to fuse with acceptor endocytic vesicles.

[1] R. D. Klausner, G. Ashwell, J. van Renswoude, J. B. Harford, and K. R. Bridges, *Proc Natl. Acad. Sci. USA* **80,** 2263 (1980).
[2] P. G. Woodman and G. Warren, *Methods Cell Biol.* **31,** 197 (1989).
[3] P. G. Woodman and G. Warren, *J. Cell Biol.* **112,** 1133 (1991).

Isolation of Donor Clathrin-Coated Vesicles

Cells

All membrane and cytosol fractions used in this study are prepared from A431 cells, a human cell line rich in transferrin receptors.[4] A431 cells are grown in Dulbecco's modified Eagle medium (DMEM) supplemented with 10% (v/v) fetal calf serum and 100 U/ml of penicillin and streptomycin, in an atmosphere of 95% air/5% CO_2. After trypsinization cultures are split 1:5 every 48 hr. All tissue culture media and supplements were bought from Northumbria Biologicals, Ltd, Northumberland, UK.

Radiolabeling

Human transferrin is radiolabeled with iodine-125 according to the method of Fraker and Speck.[5] Dissolve Iodogen (1,3,4,6-tetrachloro-$3\alpha,6\alpha$-diphenylglycoluril; Pierce Chemical Co., Rockford, IL) to 0.5 mg/ml in chloroform. Evaporate 20 μl in a glass tube under a stream of nitrogen. Add 30 μl sodium phosphate buffer, pH 7.2, containing 100 μg human transferrin. To this add 2.5 mCi (25 μl) Na^{125}I (16 mCi/μg; Amersham, UK). Incubate for 15 min on ice, then stop the reaction by addition of 166 mg/ml unlabeled KI (50 μl) and 2.5 μg/ml sodium metabisulfite (50 μl). Remove the free iodine by gel filtration over a BioGel P-6 column (Bio-Rad Laboratories, Richmond, CA), prewashed with sodium phosphate buffer containing 0.1 mg/ml bovine serum albumin (BSA) and equilibrated with BSA-free phosphate buffer, followed by dialysis against 400 vol of the same buffer. This method should achieve a specific activity of 2–3 $\times$ 10^7 counts per minute (cpm)/μg transferrin.

Isolation of Coated Vesicles

Donor coated vesicles are prepared using a modification of the method of Pearse,[6] and isolation is monitored by the cell-free assay for vesicle fusion (see below). Care must be taken to avoid pelleting the membranes, because this leads to aggregation and loss of activity. The preparation is carried out at pH 6.6 throughout, to stabilize the clathrin coat. For a standard preparation, grow A431 cells to near confluence on four 24 $\times$ 24 cm tissue culture dishes, supplied by GIBCO (Paisley, Scotland). For each dish, wash cells with ice-cold Dulbecco's phosphate-buffered saline (PBS), then incubate on a slowly rocking platform at 4° with 15 ml bind-

---

[4] C. R. Hopkins and I. S. Trowbridge, *J. Cell Biol.* **97,** 508 (1983).
[5] P. J. Fraker and J. C. Speck, *Biochem. Biophys. Res. Commun.* **80,** 849 (1978).
[6] B. M. F. Pearse, *Proc. Natl. Acad. Sci. USA* **79,** 451 (1982).

ing medium [BM; DMEM containing 20 mM N-2-hydroxyethylpipera-zine-N'-2-ethanesulfonic acid (HEPES), pH 7.4, and 0.2% (w/v) BSA] containing 1.5 μg/ml ^{125}I-labeled transferrin. After 2 hr, wash the cells three times in PBS, and incubate for 2 min at 31° with 50 ml prewarmed BM. The short period for internalization should ensure that the greatest proportion of internalized transferrin is in coated vesicles. Wash the dish four times in 50 ml ice-cold vesicle buffer [140 mM sucrose, 0.5 mM MgCl$_2$, 1 mM EGTA, 20 mM 2-(N-morpholino)ethanesulfonic acid (MES), 70 mM potassium acetate, pH 6.6]. Drain for 2 min at 4° to remove excess buffer, then scrape the cells from the dish with a rubber policeman. Suspensions from four plates should be combined (approximately 3–4 ml), supplemented with dithiothreitol (DTT; 1 mM) and pro-tease inhibitors [1 μg/ml antipain, 1 μg/ml chymostatin, 1 μg/ml pepstatin, 2 μg/ml E64, 40 μg/ml phenylmethylsufonyl fluoride (PMSF), all stored in a 1000× concentrate in dimethyl sulfoxide (DMSO)], and broken in a stainless-steel homogenizer. Cells are passed 10 times through a 0.2540-in. bore containing a 0.2530-in. ball.[7] These conditions should break 80–90% of cells with little damage to nuclei, as assessed by trypan blue staining and microscopy. Prepare a postnuclear supernatant by centrifuging at 500 g_{av} for 5 min at 4°. Polyribosomes, potential contaminants, are disassembled by incubating the extract with ribonuclease A (50 μg/ml; Worthington Enzymes, Ltd., Freehold, NJ) for 30 min at 4°. Centrifuge at 7000 g_{av} for 30 min at 4°, and then apply the supernatant (approximately 3 ml) to a 10-ml continuous deuterium oxide (D$_2$O) rate sedimentation gradient of 10–90% (w/v) D$_2$O in vesicle buffer, containing 1 mM DTT throughout. Centrifuge for 30 min at 45,000 g_{av} in an SW40 rotor (Beckman Instru-ments, Inc., Palo Alto, CA) and collect 1-ml fractions from the bottom by tube puncture. The coated vesicles sediment slowly, and should remain in the top 5 ml of the gradient. This should be monitored by assaying 50-μl samples for vesicle fusion activity (see below). Pool the peak fractions, dilute to 18 ml in vesicle buffer containing 1 mM DTT, and apply to the top of an equilibrium density gradient (20 ml) of 2% (w/v) Ficoll/9% (w/v) D$_2$O–20% (w/v) Ficoll/90% (w/v) D$_2$O in vesicle buffer containing 1 mM DTT. (Before use, the Ficoll should be dissolved in water, dialyzed extensively to remove low molecular weight contaminants, and lyophi-lized.) Centrifuge for 16 hr at 80,000 g_{av} in a Beckman SW28 rotor and remove twenty 1-ml fractions from the bottom by using a fine capillary tube attached to a peristaltic pump. Samples (50 μl) from each fraction should be analyzed for vesicle fusion activity. As indicated below, to dilute

[7] W. E. Balch, W. G. Dunphy, W. A. Braell, and J. E. Rothman, *Cell* **39**, 405 (1984).

out the viscous Ficoll and vesicle isolation buffer, samples should be diluted at least 1:4 into the fusion assay mix.

Analysis of Donor Coated Vesicle Preparations

Endocytic Vesicle Fusion Activity

Because fusion occurs only between endocytic compartments,[2] only those fractions supporting fusion activity should contain clathrin-coated vesicles. ^{125}I-labeled transferrin migrates as two peaks on the equilibrium gradient (Fig. 1). Only the lower peak possesses fusion activity and, therefore, contains endocytic vesicles. Observed efficiency of fusion is dependent on the concentration of acceptor membranes; with an excess of acceptor membranes, up to about 40% of total ^{125}I-labeled transferrin should be precipitated in an ATP-dependent manner. The peak migrates to the same position as placental clathrin-coated vesicles isolated on similar gradients. The unique properties of clathrin-coated vesicles permit several independent methods of confirming that vesicles in this peak are coated, and of assessing the purity of the preparation. Analysis can include detection of coated vesicle proteins by Western blotting with anti-(coat protein)

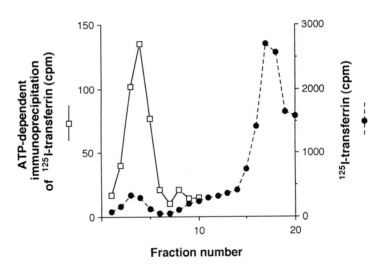

FIG. 1. Isolation of endocytic coated vesicles. Fractions (1 ml) from the D_2O/Ficoll density gradient are sampled (100 μl) for ^{125}I-labeled transferrin and endocytic vesicle fusion activity. Fusion activity is expressed as the ATP-dependent immunoprecipitation of ^{125}I-labeled transferrin.

antisera, electron microscopy, sodium dodecyl sulfate-polyacrylamide gel electrophoresis (SDS-PAGE), and susceptibility of the preparation to the action of the clathrin-uncoating ATPase.[8]

Western Blotting

To obtain sufficient material to detect coat proteins by Western blotting, supplement the normal donor vesicle preparation with twelve 24 × 24 cm dishes of untreated cells. After collecting the fractions from the equilibrium gradient, sediment vesicles by diluting 10-fold in vesicle buffer and centrifuging for 2 hr at 100,000 g_{av} in a Beckman SW40 rotor. Wash the pellets carefully with vesicle buffer, and resuspend in sample buffer [1 M sucrose, 200 mM Tris-HCl, pH 6.5, 5 mM ethylene diamine tetraacetic acid (EDTA), 0.04% (w/v) bromphenol blue, 10 mM DTT, 4% (w/v) SDS], boil for 5 min, cool, add 5 mM iodoacetamide, and electrophorese overnight (70 V, constant voltage) on a 10% (w/v) polyacrylamide gel. Transfer to nitrocellulose, according to the method of Towbin, for 2 hr at 1.5 A. Incubate the nitrocellulose on a rocking platform in Tris/salt buffer (200 mM Tris-HCl, pH 7.4, 150 mM NaCl) containing 0.2% (w/v) polyoxyethylene sorbitan monolaurate (Tween 20) and 0.1% (w/v) fish skin gelatin. Incubate with the appropriate dilution of an anti-(coat protein) antiserum (e.g., anti-clathrin light chain) in the same buffer for 90 min at room temperature, then wash three times in 1 hour with Tris/salt containing 0.05% (w/v) Tween 20, followed by Tris/salt without Tween 20. Incubate for 60 min at room temperature with ^{125}I-labeled protein A [0.1 μCi/ml in Tris/salt containing 5% (w/v) BSA]. Wash three times in Tris/salt containing 0.05% (w/v) Tween 20, dry, and expose to photographic film.

SDS-PAGE

The unique protein composition of coated vesicles (clathrin heavy chain, M_r 180,000; clathrin light chain, M_r 34,000–36,000; adaptins, M_r approximately 100,000, 50,000, and 16,000) makes SDS-PAGE an ideal method of assessing the purity of a preparation. Because the quantity of material isolated is very small, the best method of detection is to add carrier untreated, metabolically labeled cells to the preparation. Wash one 24 × 24 cm dish of semiconfluent A431 cells three times with sterile PBS, and add 80 ml of labeling medium [DMEM containing 0.75 mg/liter methionine (make up from an MEM-Selectamine kit, supplied by GIBCO

[8] D. M. Schlossman, S. L. Schmid, W. A. Braell, and J. E. Rothman, *J. Cell Biol.* **99**, 723 (1984).

Laboratories, Grand Island, NY) and supplemented with 10% (v/v) dialyzed fetal calf serum, and 5 mCi [^{35}S]methionine]. Label overnight at 37° in an atmosphere of 95% air/5% CO_2. Wash the cells four times with vesicle buffer, scrape from the dish, and combine with cells containing labeled transferrin before homogenization. Follow the isolation procedure, and process the fractions of peak fusion activity for SDS-PAGE as described for Western blotting. Run the 10% gel overnight at 70 V, constant voltage, then fix the gel in 10% acetic acid, 20% methanol for 1 hr at room temperature. Discard the fixative, replace with Amplify (Amersham), and incubate for a further 1 hr. Dry the gel and expose to film. The autoradiograph (for example, see Fig. 2, lane c) should clearly show clathrin heavy chain and the adaptin proteins. A Coomassie blue-stained gel of a placental coated vesicle preparation is shown for comparison (Fig. 2, lane d). Clathrin light chains are less easily distinguished, because they label poorly with [^{35}S]methionine but can be visualized after longer exposure (Fig. 2,

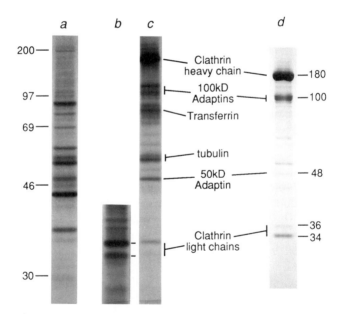

FIG. 2. SDS-PAGE of coated vesicles. Autoradiograph of an SDS-PAGE of a postnuclear supernatant (lane a) and coated vesicle preparation (lane c) from metabolically labeled cells. Clathrin light chains are distinguished on longer exposure of the film (lane b). A Coomassie blue-stained gel of a placental coated vesicle preparation is shown for comparison (lane d). [Reproduced from the *Journal of Cell Biology* **112**, 1133–1141 (1991).]

lane b). A major contaminant is tubulin (M_r 55,000). This can be reduced by addition of 10 μg/ml colchicine before the final density gradient.

Electron Microscopy

Clathrin-coated vesicles are readily identified by electron microscopy, owing to their polygonal protein coat. Again, it is advisable to include carrier untreated cells in the preparation. Use 15 dishes of untreated cells and 1 dish of cells labeled with ^{125}I-labeled transferrin. Dilute the peak fractions from the D$_2$O/Ficoll gradient 1:9 in vesicle buffer, then centrifuge for 2 hr at 100,000 g_{av} in a Beckman SW40 rotor. Wash the pellet twice in vesicle buffer, then fix at room temperature in 3% (w/v) glutaraldehyde, in the same buffer. Rinse the pellet three times in 0.1 M sodium cacodylate, pH 7.4, then postfix in 1% (w/v) osmium tetroxide and 1.5% (w/v) potassium ferrocyanide for 30 min at 4°. Wash the pellet three more times in sodium cacodylate, then dehydrate, using standard procedures, in graded ethanol. Embed in Epon and cut 70-nm sections. Stain the sections with alcoholic uranyl acetate and Reynold's lead citrate. A typical preparation is shown in Fig. 3.

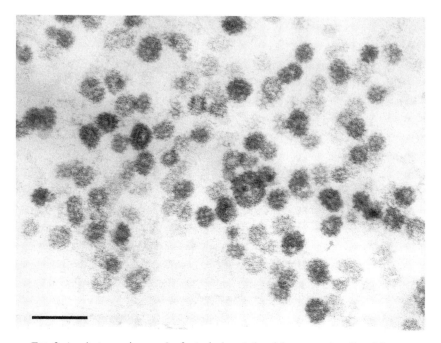

FIG. 3. An electron micrograph of a typical coated vesicle preparation. Bar: 0.2 μm.

Uncoating ATPase

Clathrin is removed from coated vesicles by an uncoating ATPase.[8] Therefore the most direct method of showing that all of the ^{125}I-labeled transferrin is within coated vesicles is to incubate the preparation with uncoating ATPase. The clathrin coat confers a high density to the vesicle, permitting purification on D_2O/Ficoll density gradients. ATP-dependent removal of the coat will lower the density of the vesicle, which will migrate to a higher position on a similar D_2O/Ficoll equilibrium gradient. No change in density will result from incubations lacking uncoating ATPase, or performed without ATP, or at pH 6.6. For each sample, use 25 μl donor coated vesicles. Add ATP-regenerating or -depleting cocktails (25 μl), as for the fusion assay. Add 10 μg uncoating ATPase, prepared from brain or placenta as described by Schlossman *et al.*[8] Dilute to 250 μl in HEPES buffer/1 mM DTT, or vesicle buffer/1 mM DTT, containing protease inhibitors, and incubate for 15 min at 37°. Stop the reaction by diluting to 2 ml in ice-cold vesicle buffer/1 mM DTT. Each sample is analyzed on a 10-ml gradient of 2% (w/v) Ficoll/9% (w/v) D_2O–20% (w/v) Ficoll/90% (w/v) D_2O, in vesicle buffer containing 1 mM DTT. Centrifuge for 16 hr at 80,000 g_{av} in a Beckman SW40 rotor, and collect fractions from the bottom using a capillary tube and peristaltic pump. Figure 4 demonstrates the uncoating ATPase-dependent change in vesicle density.

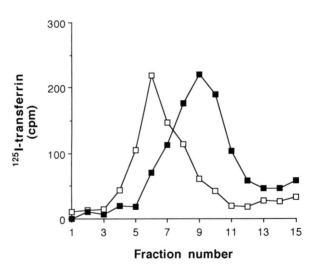

FIG. 4. Action of uncoating ATPase. Vesicles labeled with ^{125}I-labeled transferrin are incubated with MgATP, with (■) or without (□) uncoating ATPase. Samples are loaded onto D_2O/Ficoll gradients (10 ml) and fractionated (0.5 ml) from the bottom.

Vesicle Fusion Assay

Materials

Creatine phosphate (CP) and creatine phosphokinase (CPK) are obtained from Boehringer Mannheim (Indianapolis, IN). Sheep anti-transferrin antiserum is supplied by the Scottish Antibody Production Unit (Carluke, Scotland) and heat inactivated by incubating for 30 min at 56°. All other reagents are supplied by Sigma Chemical Co. (St. Louis, MO) unless specified.

Acceptor Membrane Preparations

For a standard acceptor membrane preparation grow cells to near confluence on four 24×24 cm tissue culture dishes. For each plate, wash the cells four times in ice-cold PBS and incubate on a rocking platform at 4° for 1 hr with 10 μg/ml unlabeled human transferrin in 25 ml BM. Wash the cells a further four times with ice-cold PBS, then incubate for 1 hr with 1 ml sheep anti-transferrin antiserum diluted to 25 ml with BM. Wash the cells four times with PBS, and incubate in a water bath for 5 min at 37° with 50 ml of prewarmed BM. Wash the cells four times with approximately 50 mi ice-cold HEPES buffer (140 mM sucrose, 20 mM HEPES–KOH, 70 mM potassium acetate, pH 7.2) to cool rapidly. Drain the dishes, then scrape the cells, add DTT and protease inhibitors, and homogenize as for donor coated vesicles. Centrifuge the homogenate for 5 min at 500 g_{av} at 4°, and apply the supernatant to the tops of two discontinuous sucrose gradients (it is important not to overload the gradients; extracts from not more than two dishes of cells should be applied to each gradient, to prevent membrane aggregation) of 2 ml 40% (w/v) sucrose in HEPES buffer containing 1 mM DTT overlaid with 8 ml 20% (w/v) sucrose in the same buffer. Centrifuge for 2 hr at 155,000 g_{av} in a Beckman SW40 rotor and recover the crude membrane preparation at the 20%/40% interface by tube puncture. Membranes can be snap-frozen and stored in 100-μl aliquots in liquid nitrogen. Preparations contain approximately 2–3 mg/ml protein.

Cytosol Fractions

Wash four 24×24 cm dishes of near-confluent A431 cells four times with HEPES buffer, drain the dishes, and scrape off the cells. Add DTT to 1 mM, together with protease inhibitors, and homogenize as described above. Centrifuge the extract for 30 min at 400,000 g_{av} in a Beckman TL100 benchtop ultracentrifuge and carefully remove the supernatant. Apply 3.5 ml to a 10-ml BioGel P-6 desalting column, equilibrated with ice-cold HEPES buffer containing 1 mM DTT, and collect the fractions of peak protein concentration. Combine, then freeze and store in liquid nitrogen. Protein concentrations should be 5–10 mg/ml.

Fusion Assay Conditions

The fusion assay conditions are similar to those described previously,[2] but modified to account for the need to dilute the donor coated vesicle preparations (see below). A typical incubation contains 50 μl donor coated vesicles (normally 2000–5000 cpm) and 2 mg/ml cytosol (final concentration). Add 25 μl unlabeled transferrin (1 mg/ml in HEPES buffer) to prevent immunoprecipitation of any ^{125}I-labeled transferrin present outside the sealed coated vesicles, and 50 μl of acceptor membranes. The incubation should include an ATP-regenerating cocktail (added in 25 μl as a 10× concentrate of 10 mM MgATP, 50 mM CP, 80 IU/ml CPK), and the total incubation is diluted to 250 μl with HEPES buffer/1 mM DTT. Include a control sample with an ATP-depleting cocktail (25 μl) of 500 IU/ml hexokinase in 50 mM glucose, to give a background ATP-independent signal. This value is typically 2% or less of that obtained in the standard incubation. Incubate for 2 hr at 37°, then dilute to 1 ml with immunoprecipitation buffer [0.1 M Tris-HCl, pH 8.0, 0.1 M NaCl, 5 mM MgCl$_2$, 1% (w/v) Triton X-100, 0.5% (w/v) SDS, 1% (w/v) sodium deoxycholate, 0.1% (v/v) BSA]. Add 20 μl S. aureus cells (Calbiochem, San Diego, CA), washed three times in immunoprecipitation buffer, and incubate on ice for a further 1 hr. Pellet the cells at low speed in a "microcentaur" microfuge (Measuring & Scientific Equipment, London, England) for 4 min at room temperature and carefully remove the supernatant with a syringe needle. Repeat the washing procedure, then count the pellet for radioactivity. Samples are normally counted for 10–30 min.

Future Prospects

This chapter describes the isolation of a functional transport intermediate. It should be possible to use this in combination with inhibitors of vesicle fusion to isolate intermediates in the fusion pathway. For example, incubation with a specific inhibitor of fusion may lead to the association of cytosolic proteins with the vesicle membrane, to form a fusion intermediate that cannot be consumed. Reisolation of these vesicles should result in enrichment of these proteins.

[25] Use of Two-Stage Incubations to Define Sequential Intermediates in Endoplasmic Reticulum to Golgi Transport

By H. W. Davidson and W. E. Balch

Export of protein from the endoplasmic reticulum (ER) involves the formation (fission) of carrier vesicles from the ER, and their vectorial targeting and fusion to the cis-Golgi compartment. Each of these steps in protein transport is likely to require biochemically distinct components. In this chapter we describe an assay that efficiently reconstitutes the transport of vesicular stomatitis virus (VSV) G protein between the ER and the cis-Golgi compartment. To investigate the time at which individual components are required during the course of a single round of transport we utilize two-stage incubations. We also describe the principles related to the practice and interpretation of such assays.

Vesicular Stomatitis Virus G Protein Transported as Synchronous Wave after Temperature Shift

Transport of VSV G protein between the ER and the cis-Golgi compartment involves a protocol that results in the export of a synchronous wave of protein from the ER. As illustrated in Fig. 1A, transport between the ER and the Golgi proceeds through at least three distinctive kinetic phases. The first phase is a lag period (generally requiring 20 min at 32°). It is involved in vesicle formation and probably targeting. During this lag period no oligosaccharide processing of VSV G protein can be detected, indicating that the protein has not been delivered to the lumen of the cis-Golgi compartment. The lag period is followed by a period in which there is a progressive increase in the amount of processed VSV G protein, indicating delivery to the cis-Golgi compartment (Fig. 1A). Because processing is virtually simultaneous with exposure of VSV G protein to the resident mannosidases and glycosidases of the cis-Golgi compartment, this step defines the progressive fusion of vesicles with the Golgi. Finally, there is a plateau after which no additional processing is observed, presumably measuring the maximum capacity of the assay to support transport between the ER and the cis-Golgi.

To define the temporal requirements for protein factors involved in vesicle formation, targeting, and fusion, two different protocols can be employed. In the first protocol, inhibitors can be added during the time course of the reaction to prevent the function of one or more key compo-

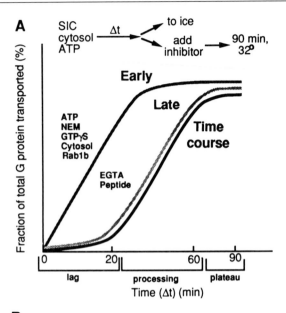

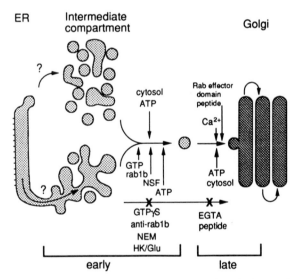

Fig. 1. Transport of VSV G protein between endoplasmic reticulum and cis-Golgi compartment.

nents during the course of transport. This is the method of choice for irreversible inhibitors. With reversible inhibitors, an additional approach may be applied in which a reversible inhibitor used to accumulate a transport intermediate is removed and the components required for further transport examined. These two protocols are described below.

Time of Addition Experiments: Two-Stage Incubations to Determine Early vs Late Function in Transport

In the first protocol, which we refer to as a "time of addition" experiment, specific or general inhibitors believed to block the function of one or more components are added to complete cocktail containing semiintact cells, cytosol, and ATP during the first stage at increasing times of incubation as illustrated in Fig. 1A [top flow diagram (Δt)]. After the addition of the inhibitor, incubations are continued in its presence in a second stage for the duration of the assay (generally a total time of 90 min). The second-stage incubation allows VSV G protein that was mobilized past the step sensitive to inhibition during the first stage to continue to the cis-Golgi compartment for processing in the second stage.

Results obtained from such two-stage incubations will determine the potential role of the component(s) in vesicle formation (and possibly targeting) or fusion. If a component is required for vesicle formation, the component is likely to be required during the lag period. In this case, if the inhibitor is added at zero time in the first stage, vesicle formation will not occur and no processing will be observed during the second stage, because VSV G protein cannot exit the ER. However, after 20–30 min of incubation in the first stage in the absence of the inhibitor, a time period in which export from the ER is complete but little delivery to the Golgi can detected (Fig. 1), an early-acting component will no longer be required. At this time, addition of the inhibitor will have only a marginal effect on an early-acting component. Further incubation for 90 min in the second stage will result in the efficient transport of VSV G protein to the cis-Golgi compartment. For early-acting components, when the total protein transported (that observed after a total of 90 min of incubation) is plotted for each time of addition of the inhibitor (Δt), a curve is generated that shows no lag period and precedes the standard time course by 15–25 min (Fig. 1; compare the early curve to the time course curve). Moreover, the total extent of transport observed after 90 min in the presence of inhibitor plateaus at a correspondingly earlier time point (Fig. 1, early curve). This idealized curve is diagnostic of components required for an early transport step. As shown in Fig. 1B, using this protocol we have shown that formation of functional vesicles requires ATP and cytosol, requires the function of two known proteins, Rab1b and N-ethylmaleimide-sensitive fusion protein (NSF), uncharacterized soluble and membrane-associated factors, and is sensitive to the general chemical inhibitor GTPγS.[1-9]

In contrast to components required during vesicle formation in the lag period, components are also required for a late, vesicle fusion step. The activity of these components generates a different diagnostic curve when the time of addition experiment described above is performed. In this case, if an inhibitor blocks the step immediately preceding fusion (and processing by Golgi glycosidases and mannosidases is rapid relative to the fusion step), then addition of the inhibitor at each time point (Δt) in the first stage will stop the reaction abruptly, preventing further processing. Because inhibition of a late-acting component is equivalent to transferring the incubation to ice, further incubation for 90 min during the second stage results in no additional processing. For late-acting components, when the total protein transported (that observed at 90 min) is plotted for each time of addition of the inhibitor (Δt), a curve is generated that is identical to the time course (Fig. 1A; compare late curve to the time course curve). In the extreme case, a membrane-permeant inhibitor may block the function of the Golgi processing enzymes. This possibility can be resolved by assaying the effect of the inhibitor on solubilized enzymes. Such inhibitors would obviously not provide useful information concerning transport components involved in late fusion steps. As illustrated in Fig. 1B, using this type of protocol we have shown that vesicle fusion requires Ca^{2+} (0.1 μM), ATP, uncharacterized soluble and membrane-associated components, and is sensitive to a synthetic peptide analog homologous to the effector domains found in the rab gene family of small GTP-binding proteins.[1-9]

In the above protocol there are several variables that need to be considered. First, it is important to establish that processing in the Golgi compartment per se is not the rate-limiting step in the assay. If processing is slow relative to transport, the significance of early versus late becomes

[1] H. Plutner, A. D. Cox, R. Khosravi-Far, S. Pind, J. Bourne, R. Schwaninger, C. J. Der, and W. E. Balch, *J. Cell Biol.* **115**, 31 (1991).
[2] R. Schwaninger, C. J. M. Beckers, and W. E. Balch, *J. Biol. Chem.* **266**, 13055 (1991).
[3] H. Plutner, R. Schwaninger, S. Pind, and W. E. Balch, *EMBO J.* **9**, 2375 (1990).
[4] C. J. M. Beckers, H. Plutner, H. W. Davidson, and W. E. Balch, *J. Biol. Chem.* **265**, 18298 (1990).
[5] C. J. M. Beckers, M. R. Block, B. S. Glick, J. E. Rothman, and W. E. Balch, *Nature (London)* **339**, 397 (1989).
[6] C. J. M. Beckers, and W. E. Balch, *J. Cell Biol.* **108**, 1245 (1989).
[7] W. E. Balch, *J. Biol. Chem.* **164**, 16965 (1989).
[8] C. J. M. Beckers, D. Keller, and W. E. Balch, *Cell* **50**, 523 (1987).
[9] W. E. Balch, *Curr. Opinions Cell Biol.* **2**, 634 (1990).

difficult to interpret because even a late-acting fusion protein would yield an apparent early phenotype in such two-stage incubations. As indicated previously, processing of VSV G protein by α-1,2-mannosidase I (Mann I) in the cis-Golgi compartment[4] or by N-acetylglucosamine transferase I (Tr I) in the medial-Golgi compartment is not rate limiting.[10]

Second, it is important determine the half-time required for the function of an inhibitor. Some inhibitors, such as N-ethylmaleimide (NEM; a sulfhydryl alkylating reagent), are chemical reagents that can modify proteins during incubation on ice. These work very rapidly ($t_{1/2}$ of seconds). Pretreatment on ice has the added advantage that it allows the investigator to eliminate any variables related to kinetics of inhibition per se during incubation at temperatures that support transport. On the other hand, some inhibitors have a slow half-time [such as GTPγS ($t_{1/2}$ of minutes)] and will not inhibit on ice. These inhibitors block transport and may serve as competitive or noncompetitive substrates through incorporation into one or more key components during the cycling of transport components. For example, GTPγS efficiently competes for GTP, blocking transport at the step presumably requiring GTP hydrolysis.[6] The half-time of inhibition will strongly influence the observed slope and point of origin of the time of addition curves. Determination of the half-time of inhibition for some inhibitors is easily accomplished by two-stage incubations. In this case, a complete reaction cocktail is incubated in the presence of the inhibitor from time zero. After increasing time (Δt), the inhibitor is washed out by pelleting of the semiintact cells (5 sec at 10,000 g in a microfuge), followed by a subsequent incubation of the cells in the second stage for 90 min in complete cocktail lacking the inhibitor. A parallel incubation in the absence of the inhibitor serves to control for variables other than those directly affected by the inhibitor. Determination of the half-time required for inhibition is applicable principally to inhibitors that are irreversible and can be readily washed out. In instances where it is not possible to obtain a $t_{1/2}$ of inhibition, interpretation of the participation of a component in an early or late step is more difficult.

In general, time of addition experiments have been useful for determination of the temporal role of nucleotides through use of nucleotide analogs, and the temporal role of cytosolic and membrane components through the use of either the chemical inhibitors described above,[2-6] neutralizing antibodies that inhibit components of the transport machinery,[1] or augmentation of incubations depleted of known components by addition of the purified protein.[5] Because many of the components required for

[10] W. E. Balch, W. G. Dunphy, W. A. Braell, and J. E. Rothman, *Cell* **39**, 405 (1984).

vesicular trafficking are likely to function in the context of molecular complexes that undergo maturation and/or recycling for subsequent rounds of transport, these experiments are likely to inform the investigator of only one step in which a particular protein may participate in a complete round of vesicle fission and fusion. It is also important to emphasize that while the idealized curves shown above serve to delineate components potentially participating in the early vesicle formation vs late vesicle fusion steps, curves of intermediate values are likely to be obtained. These may reflect the properties of the inhibitors per se (such as efficiency of inhibition and $t_{1/2}$ of inhibition) or may reflect the participation of a component in an intermediate step in transport, such as targeting. The latter step is presently a difficult step to identify in semiintact cells, although current evidence tentatively suggests that targeting occurs rapidly relative to fusion.[4]

Two-Stage Assays to Accumulate Transport Intermediates

A second approach that is complementary to the above experiments and provides a more defined role for individual components is to accumulate VSV G protein in transport intermediates. This approach can be used to study either the components required to form the intermediate or those required for its targeting and/or fusion to the cis-Golgi compartment.

To identify components required to form a transport intermediate, semiintact cells are incubated in a first stage in the presence of a reversible inhibitor that is believed to block a later step in transport. Included in this first-stage incubation are the components being tested for their role in formation of the intermediate. Subsequently, the inhibitor is removed (by a brief pelleting of semiintact cells as described above) and the cells supplemented with a *complete* cocktail containing all the necessary components to support vesicle delivery to the cis-Golgi compartment. If a critical component required for vesicle formation is deleted in the first stage, no transport (processing of VSV G protein) will be observed in the second stage.

Alternatively, to identify components required for fusion of a transport intermediate to the cis-Golgi compartment, a transport intermediate is again generated in the first stage by incubation in the presence of the reversible inhibitor and a *complete* cocktail to ensure intermediate formation. In the second stage, cells are pelleted and reincubated in a cocktail containing components being tested for their putative role in later transport steps. If a particular component is required for delivery to the cis-Golgi compartment, its absence in the second-stage incubation would preclude vesicle fusion and processing of VSV G protein.

Summary

Identification of the temporal requirement for components through the use of two-stage incubations is valuable in dissecting the overall transport reaction into steps relevant to vesicle fission and those related to vesicle fusion. In the context of semiintact mammalian cells in which a functional vesicle intermediate has not been detected, components playing a role in targeting are presently difficult to identify. However, the two-stage incubations are particularly powerful when either the donor or acceptor compartments can be manipulated independently, as is the case for intra-Golgi transport using enriched Golgi fractions or in the case of ER-to-Golgi transport in perforated yeast, in which a vesicle intermediate can be physically isolated.

[26] Use of *sec* Mutants to Define Intermediates in Protein Transport from Endoplasmic Reticulum

By Michael F. Rexach and Randy W. Schekman

Introduction

Vesicle-mediated protein transport from the endoplasmic reticulum (ER) to the Golgi apparatus requires as many as 25 "transport proteins," which orchestrate the intermediate stages of vesicle budding, targeting, and fusion. The dissection of intermediates in this process is facilitated by the availability of specific inhibitors: point mutations in any of the genes that encode a transport protein may render the gene product temperature sensitive for function, resulting in a conditional block at an intermediate stage in ER-to-Golgi protein transport. Electron microscopic analysis of *Saccharomyces cerevisiae* strains that carry point mutations in *sec*[1] and *ypt1*[2] genes showed a differential requirement for the gene products in the formation or consumption of 60-nm vesicles.[3,4] Are these vesicles intermediates in the transfer of proteins from the ER to the Golgi apparatus? Which intermediate stage of vesicle formation is blocked, the generation of the vesicle or its scission from the ER membrane? What stage of vesicle consumption is blocked, the targeting of the vesicle to the Golgi membrane

[1] P. Novick, C. Field, and R. Schekman, *Cell* **21**, 205 (1980).
[2] D. Gallwitz, C. Donath, and C. Sander, *Nature (London)* **306**, 704 (1983).
[3] C. Kaiser and R. Schekman, *Cell* **61**, 723 (1990).
[4] J. Becker, T. Tan, H. Trepte, and D. Gallwitz, *EMBO J.* **10**, 785 (1991).

or the fusion of lipid bilayers? These questions are difficult to address using living cells, but *in vitro* analysis is providing the answers.

Endoplasmic reticulum to Golgi protein transport can be reconstituted *in vitro* using perforated yeast spheroplasts.[5] In the following pages we will discuss how to use the *sec* mutants to define intermediates in transport vesicle budding, targeting, and fusion. This *in vitro* mutant analysis resulted in the identification of transport vesicles that function in the transfer of core-glycosylated yeast α factor precursor from the ER to the Golgi apparatus.[6]

Review of *in Vitro* Transport Reaction

Protein transport from the ER was reconstituted in gently lysed yeast spheroplasts (perforated cells) using ^{35}S-labeled prepro-α-factor (ppαf) as a marker secretory protein.[5] The marker is posttranslationally translocated into the lumen of the ER during a 15-min incubation at 10° (stage I). Once in the ER lumen, pro-α-factor is glycosylated with three N-linked core carbohydrate chains. Membranes are separated from untranslocated precursor by centrifugation, then incubated at 20° with a cytosol fraction, an ATP-regenerating system, and GDPmannose (stage II). During this stage, core-glycosylated pro-α-factor (core-gpαf) is transported to the Golgi apparatus where it is further modified with "outer-chain" mannose residues in α-1,6-linkage. All glycosylated forms of pro-α-factor (gpαf) bind to the plant lectin concanavalin A (Con A) whereas all outer chain-modified forms of gpαf bind to antibodies specific for α-1,6-mannose linkages.[7,8] Transport efficiency is expressed as the ratio of radiolabeled gpαf precipitated with "outer chain" antibodies to total gpαf precipitated with Con A. We previously showed that transport reaches 25% efficiency, requires cytosol, ATP, GTP hydrolysis, Ca^{2+}, Sec23p, Sec12p, Sec18p, and Ypt1p, and exhibits a 6- to 10-minute lag phase.[5,6,9]

---

[5] D. Baker, L. Hicke, M. Rexach, M. Schleyer, and R. Schekman, *Cell* **54**, 335 (1988).
[6] M. Rexach and R. Schekman, *J. Cell Biol.* **114**, 219 (1991).
[7] A. Franzusoff and R. Schekman, *EMBO J.* **8**, 2695 (1989).
[8] T. Graham and S. Emr, *J. Cell Biol.* **114**, 207 (1991).
[9] D. Baker, L. Wuestehube, R. Schekman, and N. Segev, *Proc. Natl. Acad. Sci USA* **87**, 355 (1990).

Experimental Preparation

In Vitro Transport

Materials

Gently lysed yeast spheroplasts: Spheroplasts are prepared as described by Baker *et al.*[5,10] The strains used in this study were RSY255 (*MATα ura3-52 leu2-3, -112*), RSY281 (*MATα sec23-1 ura3-52 his4-619*), RSY271 (*MATα sec18-1 ura3-52 his4-619*), sf226 (*MATα sec12-4*), and RSY453/TSB4B[2] (*MATα ypt1ᵗˢ LEU2 leu2 his3*). All strains were grown at 24° in YPD medium to early log phase (approximately 4 OD$_{600}$/ml; 1 OD$_{600}$ unit is ~ 10⁷cells). YPD medium contained 1% (w/v) Bacto yeast extract, 2% (w/v) Bacto peptone (Difco Laboratories, Inc.), and 5% (w/v) glucose.

Cytosol fractions: Cells are grown at 24° in YPD medium to early log phase. Approximately 2000 OD$_{600}$ units of cells is harvested by centrifugation and washed twice by dilution in reaction buffer B88 [20 m*M* *N*-2-hydroxyethylpiperazine-*N'*-2-ethanesulfonic acid (HEPES), pH 6.8, 150 m*M* potassium acetate, 250 m*M* sorbitol, and 5 m*M* magnesium acetate]. Cell pellets are resuspended in a Corex 30-ml glass tube, with 1 ml 20 m*M* HEPES, pH 6.8, 5 m*M* magnesium acetate, 50 m*M* potassium acetate 100 m*M* sorbitol, 1 m*M* ATP, 0.5 m*M* phenylmethylsulfonyl fluoride (PMSF), 1 m*M* dithiothreitol (DTT). Glass beads (4 g) are added and the cells are lysed by eight 30-sec periods of agitation in a vortexer at full speed. Samples are mixed with 1.5 ml of 20 m*M* HEPES, pH 6.8, 5 m*M* magnesium acetate, 2 *M* KCl, 400 m*M* sorbitol, 1 m*M* ATP, 0.5 m*M* PMSF, 1 m*M* DTT, and vortexed four times, each for 30 sec at full speed. The homogenate is clarified initially by centrifugation at 3000 *g* for 5 min at 4° and the supernatant (3 ml) is further clarified by centrifugation at 100,000 *g* for 30 min at 4°. Cytosol is desalted by filtration in a Sephadex G-25-column (15 ml, fine) that is equilibrated in B88 and 1 m*M* ATP. The eluted protein peak is pooled and distributed in 50- to 60-μl fractions that are frozen in liquid nitrogen for storage at −80°. Protein concentration, which ranges from 12 to 16 mg/ml, is measured by the Bradford assay and compared to a bovine serum albumin (BSA) standard

[³⁵S]Methionine-labeled prepro-α-factor in yeast S100 translation lysate: Prepare lysate as described in Baker *et al.*[5]

ATP regeneration mix (10×): 10 m*M* ATP, 400 m*M* creatine phos-

---

[10] D. Baker and R. Schekman, *Methods Cell Biol.* **31,** 127 (1989).

phate, 2 mg/ml creatine phosphokinase, and 500 μM GDPmannose (in B88). The ATP, creatine phosphate, and GDPmannose are dissolved in B88 to near volume, the pH is adjusted to 6.8 with 1N KOH, then the creatine phosphokinase is added and the volume adjusted. GDPmannose is added to stimulate glycosylation[5]

Materials for immunoprecipitation: Concanavalin A- and protein A-Sepharose are obtained from Pharmacia (Piscataway, NJ). Anti-α-1,6 antibody is produced as described in Baker et al.[5] Immunoprecipitations and immunoprecipitate washes are performed as described in Baker et al.[5]

Methods

Stage I: Translocation

For each experiment an aliquot of frozen membranes (from 60 OD$_{600}$ units of cell equivalents) is thawed and washed three times by resuspension in B88 and brief ($\sim$ 10 sec) centrifugation in a Fisher microfuge (Fisher Scientific, Pittsburgh, PA). Membranes are then resuspended in an S100 yeast lysate that contains [^{35}S]methionine-labeled prepro-α-factor and an ATP-regenerating system in final volumes of either 160 or 80 μl. The final concentration of components in a 25-μl translocation reaction is 375–500 μg membranes [measured by the method of Lowry, modified to contain sodium dodecyl sulfate (SDS)], 90 μg of yeast S100 lysate (measured by the method of Bradford using BSA as a standard), 50 μM GDPmannose, 1 mM ATP, 40 mM creatine phosphate (CP), and 200 μ/ml creatine phosphokinase (CPK), all dissolved in B88. The mix is incubated for 15 min at 10° to allow posttranslational translocation, cooled to 4°, and washed with 1 ml B88 and centrifuged in a microfuge at 13,000 g for 30 sec at 4°. Membranes are then resuspended in 1 ml B88 and mixed by rotation at 4° for 7 min to allow "drainage" of cytosolic protein. A final pellet fraction is resuspended in B88 at a concentration of 150–200 μg/10 μl.

Stage II: Transport

Stage II incubations (25 μl) contain 10 μl of stage I membranes, 90 μg of cytosol (preincubated for 5 min at 29° to reduce wild-type thermolability), 50 μM GDPmannose, 1 mM ATP, 40 mM CP, 200 μg/ml CPK, all in B88. At the end of stage II incubations, the samples are chilled and treated with 250 μg/ml trypsin for 10 min at 4° and then with 250 μg/ml trypsin inhibitor for > 5 min at 4° to ensure that only membrane-bound gpαf is quantified. Sodium dodecyl sulfate is then added to a final concentration of 1% (w/v) and the tubes are heated at 95° for 7 min. Equal aliquots from each tube are treated with either Con A-Sepharose or protein A-Sepharose coupled with anti-α-1,6 antibody as described in Baker et al.[5] Pro-α-factor

constitutes the major radiolabeled protein. The radioactive immunoprecipitates are heated to 95° in 1% (w/v) SDS for 7 min and dissolved in
Universol ES scintillation fluid (ICN Biochemicals) for quantitation in a
scintillation counter.

Distinction between Vesicle Budding, Targeting, and Fusion

A differential sedimentation analysis that physically resolves ER, vesicle, and Golgi membranes is used to distinguish intermediates in vesicle
budding, targeting, or fusion.[6] An early stage prior to vesicle budding is
represented by [^{35}S]core-gpαf that cosediments with the ER. Vesicles detached from the ER, but not yet targeted to the Golgi apparatus, sediment
more slowly than either the donor or acceptor compartments. Vesicles
targeted to the Golgi sediment along with Golgi membrane but more
slowly than the ER membrane.

Figure 1 is a diagram of the *in vitro* transport reaction. The ER remains
associated with the broken cells and sediments rapidly (MSP, medium-
speed pellet) in a microcentrifuge while ER-derived vesicles are released
from the cells before they target and fuse with the Golgi apparatus, which

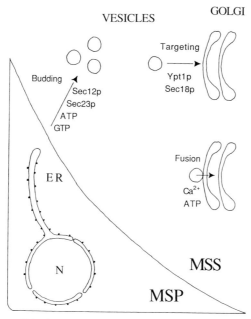

FIG. 1. A diagram of the *in vitro* transport reaction and the proposed role of Sec and Ypt1
proteins, GTP, and calcium in ER-to-Golgi transport. N, Nucleus; MSP, medium-speed
pellet; MSS, medium-speed supernatant.

also fractionates mostly in the supernatant (MSS) fraction. Vesicle and Golgi membranes in the supernatant fraction can be resolved from each other by velocity sedimentation in sucrose gradients. To establish the authenticity of this model, temperature-sensitive Sec proteins and chemical inhibitors were tested for their ability to block differentially these reactions. For convenience, the results are summarized in Fig. 1.

Sec Protein-Dependent, Temperature-Sensitive Transport

Rationale. We sought to reproduce temperature-sensitive Sec protein defects in the reconstituted reaction. To establish a "nonpermissive" temperature for transport in mutant lysates we determined the highest temperature that sustains efficient transport in wild-type lysates. We found that cytosol prepared by agitation of cells with glass beads in low osmotic support followed by exposure to high salt to extract peripherally associated proteins had the most activity at high temperatures: 29° is the highest temperature that will sustain efficient transport in wild-type extracts using this particular cytosol preparation.

Methods. Gently lysed spheroplasts and cytosol are prepared from all the ER-to-Golgi *sec* mutant strains and the *ypt1*[ts] strain grown at 24°. To test the thermolability of *in vitro* transport, stage II reactions are incubated at 20 or 29° in a water bath. At the end of a 75-min incubation the samples are chilled and processed as described above. For the *sec12* reaction, wild-type cytosol is used because Sec12p is an integral membrane protein.[11] For the *ypt1*[ts] reaction, wild-type membranes are used because *ypt1*[ts] membranes are deficient at 20°. Reactions deficient at all temperatures may be inactive as a result of a secondary defect.

Results. *sec12, sec23, sec18,* and *ypt1*[ts] reactions are 5- to 10-fold more thermolabile than wild-type reactions (Fig. 2). All other strains show a thermolability profile comparable to wild type (not shown). Rapid thermal inactivation may be important to obtain thermosensitivity *in vitro.*

Although the method described above is the simplest to study the thermolability of Sec proteins, there are other ways to inactivate thermolabile proteins. For example, the cytosol fraction may be preincubated at high temperatures before being added to a reaction.

Comments. Which stage in transport is blocked by the temperature-sensitive proteins? Electron microscopic analysis suggests that the Sec12 and Sec23 proteins participate in the formation of transport vesicles, whereas the Sec18 and Ypt1 proteins are needed for vesicle consumption.[3,4] Having developed assays that distinguish vesicle budding, targeting, and fusion, we tested the requirement of these proteins in each subreaction.

---

[11] A. Nakano, D. Brada, and R. Schekman, *J. Cell Biol.* **107,** 851 (1988).

	SEC+	sec12	sec18	sec23	ypt1ts
Transport efficiency at 20°	25%	25%	22%	27%	26%
Transport efficiency at 29°	23%	7%	12%	8%	16%
Thermolability	8%	72%	46%	70%	39%

FIG. 2. Sec protein-dependent, temperature-sensitive transport. Stage II reactions containing wild-type or mutant fractions were incubated during 75 min at 20 or 29° as indicated. Chilled reactions were treated with trypsin and then trypsin inhibitor. The protease-treated reactions were heated in 1% SDS and the glycosylated pro-α-factor quantified as described in the section Experimental Preparation. For the wild-type reaction, wild-type membranes and cytosol were used. For the *sec12* reaction, *sec12* membranes and wild-type cytosol were used; we used wild-type cytosol because Sec12p is an integral membrane protein. For the *sec23* reaction, *sec23* membranes and cytosol were used. For the *sec18* reaction, *sec18* membranes and cytosol were used. For the *ypt1ts* reaction, wild-type membranes and *ypt1ts* cytosol were used; we used wild-type membranes because *ypt1ts* membranes are deficient at all temperatures. Thermolability represents the percentage inhibition of transport at 29° relative to 20°: {100-[(α-1,6-precipitable counts generated at 29°/α-1,6-precipitable counts generated at 20°) × 100]}.

Vesicle Budding Assay

Chase of 10° Intermediate

Rationale. The vesicle budding assay is based on a peculiarity of gently lysed yeast spheroplasts: during a transport reaction the ER, marked by translocation activity, remains attached to the perforated cells and sediments (MSP) during a brief (~ 30 sec) centrifugation in a microcentrifuge while ER-derived transport vesicles are released from the cells and remain in the medium-speed supernatant fraction (MSS)[6] (see Fig. 1). Transport vesicle budding is measured by the release of [35S]core-gpαf-containing vesicles from perforated cells. The MSS fraction is treated with trypsin and then trypsin inhibitor to degrade any core-gpαf released by rupture of ER membranes. Vesicle release reaches a plateau after 35 min of incubation, and requires cytosolic proteins and ATP at an optimum temperature range between 20 and 29°.[6]

The ER-to-Golgi transport reaction described here measures the chase of a 10° intermediate that is formed during the translocation reaction (stage I). At 10° [35S]ppαf is translocated efficiently into the lumen of the ER while transport is inhibited.[5] Sedimentation analysis shows that at the end of a translocation reaction core-gpαf accumulates within a compartment that cosediments with the ER membrane. The 10° intermediate represents an early, possibly prebudding stage (see text above).

Methods. A 200-μl stage II mix is prepared and aliquoted in 25-μl portions into 0.5-ml microfuge tubes. The tubes are incubated at 0° for the zero-time point control, or 20° for intervals; each tube represents one time point. Reactions are terminated by placing each tube at 0° until the last time point is taken. After completion of the time course each tube is fractionated by centrifugation for 37 sec in a Fisher microfuge. A 15-μl medium-speed supernatant (MSS) fraction is taken from the meniscus. The remaining MSS is aspirated and discarded. The medium speed pellet (MSP) fractions are resuspended to 25 μl, and 15 μl is processed. Each fraction is then treated with 250 μg/ml trypsin for 10 min at 0° and then with 250 μg/ml trypsin inhibitor for > 5 min at 0°. Sodium dodecyl sulfate is added to a final concentration of 1% (w/v) and the tubes are heated at 95° for 7 min. Equal aliquots from each tube are treated with either Con A-Sepharose or protein A-Sepharose coupled with anti-α-1,6 antibody. The washed radioactive immunoprecipitates are heated to 95° in 1% (w/v) SDS for 7 min and dissolved in Universol ES scintillation fluid for quantitation in a scintillation counter.

When performing vesicle budding reactions it is important not to disturb the stage II mix after the incubations have started. Although the ER remains part of a rapidly sedimenting particulate fraction, it is fragile during incubations at 29°, and to a lesser extent at 20°. The ER is sheared from broken cells by repeated pipetting of reactions incubated in the presence of ATP and cytosol. This treatment results in the release of up to 33% of the ER as judged by translocation activity (M. Rexach, unpublished results, 1990).

Results. The results of a typical budding assay are depicted in Fig. 3. The production of slowly sedimenting vesicles containing gpαf (Fig. 3, top), and depletion from the pellet fraction (Fig. 3, bottom), are monitored during a stage II incubation of wild-type components at 20°. Of the total core-gpαf within the ER at time zero, 45% is released into the supernatant fraction within transport vesicles, in a reaction that is linear at early time points and plateaus after 35 min. Of the core-gpαf released, 40% is transferred to an early Golgi compartment, where it receives outer chain carbohydrate. This reaction initiates 6–10 min after the appearance of the first vesicles in the supernatant fraction. At the end of transport reactions, 65–85% of the Golgi-modified gpαf accumulates within a membrane compartment located outside of the broken cells. Based on the observation that outer chain growth is incomplete *in vitro*[7] we believe that outer chain gpαf accumulates within an early Golgi compartment, possibly because the requirements for transport to a distal Golgi compartment are not met by the conditions of our incubation.

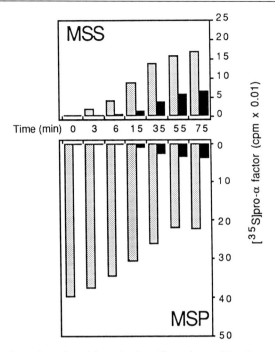

FIG. 3. Pro-α-factor is packaged into slowly sedimenting vesicles that are released from broken cells *en route* to the Golgi apparatus. Stage I membranes containing [³⁵S]gpαf were mixed with 90 μg of cytosol, an ATP-regenerating system, and GDPmannose. Portions (25 μl) were distributed into 0.5-ml microcentrifuge tubes and incubated at 20° for various lengths of time (stage II). The zero time point remained at 0° throughout. Chilled samples were then fractionated into medium-speed pellet (MSP) and medium-speed supernatant (MSS) fractions. Each fraction was treated with trypsin and then trypsin inhibitor. The protease-treated reactions were heated in 1% (w/v) SDS and the glycosylated pro-α-factor quantified as described in the section Experimental Preparation. The zero time point of the MSS (always less than 10% of maximum signal) was subtracted as background from the other MSS fractions to obtain the values shown. Gray bars, core glycosylated; solid bars, outer chain modified.

Comments. The two forms of gpαf appear in the supernatant fraction with different kinetics, consistent with a slowly sedimenting vesicle intermediate containing core-gpαf being released from the perforated cells *en route* to the Golgi apparatus. Vesicle budding proceeds with no lag phase while the overall transport reaction has a delay of 6–10 min before outer chain gpαf is detected.

Sec Protein Requirements for Vesicle Budding

Rationale. Based on morphological data mentioned above, we expect to find that *sec12* and *sec23* lysates are thermolabile for the release of core-gpαf-containing vesicles, whereas *sec18 and ypt1ts* lysates are thermolabile for the transfer of core-gpαf from the vesicles to the target Golgi compartment. Medium-speed centrifugation is used to monitor the appearance of slowly sedimenting membranes that contain core and outer chain forms of gpαf in lysates prepared from *Sec+, sec12, sec23, sec18,* and *ypt1ts*.

Results. The kinetics of release of membrane-enclosed gpαf (Fig. 4, circles; left hand side, vesicle budding) and formation of outer chain gpαf (Fig. 4, triangles; right-hand side, vesicle fusion) are monitored at 20 and 29°. Release of membrane-enclosed gpαf into the supernatant fraction is more rapid at 29° than at 20° in wild-type lysates (Fig. 4A, left). No defect in the release is detected in incubations containing *sec18* membranes and cytosol (Fig. 4D, left) or *ypt1ts* cytosol (Fig. 4E, left). *sec12* and *sec23* lysates show a considerable reduction, 75 and 60%, respectively, in the final extent of release of gpαf at 29° relative to 20° (Fig. 4B and C, left). The initial rate of gpαf release in *sec12* and *sec23* lysates is not reduced, perhaps reflecting a lag in the inactivation of the mutant protein. Unfortunately, preincubation of mutant or wild-type stage II membranes at 29° under conditions of no budding (e.g., no ATP) causes membranes to lose transport activity.

The formation of outer chain gpαf proceeds more rapidly at 29° than at 20° in wild-type lysates (Fig. 4A, right). In *sec12* and *sec23* lysates this transport is reduced by 75%, in *sec18* by 50%, and in *ypt1ts* by 40% (Fig. 4B–E, right).

Comments. Taken together, these results suggest that Sec12p and Sec23p are required to generate a slowly sedimenting membrane that contains core-gpαf whereas Sec18p and Ypt1p are required for the transfer of this species to the compartment that assembles outer chain carbohydrate.

Distinction between Block in Vesicle Targeting or Fusion

Rationale. Sucrose gradient separation is used to test whether the core-gpαf-containing vesicles that accumulate in the MSS fractions of *sec18* and *ypt1ts* reactions are attached (targeted) or not to the Golgi membrane. The vesicle and Golgi membranes are distinguished by their content of core-gpαf in vesicles (Con A-precipitable [35S]gpαf corrected by subtraction of outer chain antibody-precipitable [35S]gpαf) or outer chain gpαf in Golgi membranes.

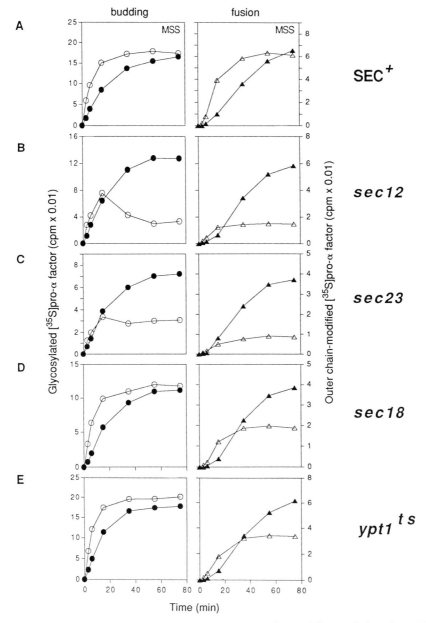

FIG. 4. Temperature-sensitive Sec proteins block vesicle budding or fusion. Stage II reactions containing wild-type or mutant fractions were performed at 20° (solid symbols) or 29° (open symbols) as indicated, or at 0° for the zero time point control. Membranes released into the supernatant fraction (MSS) were treated with trypsin and processed as before. The release of gpαf is plotted in the left-hand panels and the appearance of outer chain gpαf in the MSS is plotted in the right-hand panels. The zero time point of each set of MSS samples was subtracted as background from the corresponding MSS fractions to obtain the values shown. In all cases, uninhibited transport (formation of outer chain gpαf from core-gpαf) was ~25% efficient.

Methods. Large (150 μl) supernatant fractions from incubations of wild-type, *sec18,* and *ypt1^{ts}* lysates at 29° are obtained as described above. Each fraction is cooled and loaded on top of a sucrose gradient with a log-linear distribution of 15 to 45% (w/w) sucrose in B88. The gradients are then centrifuged at 32,000 rpm for 2 hr at 4° in an SW50.1 Beckman (Palo Alto, CA) rotor. Gradient fractions are collected from the top into forty 150-μl fractions. The density of every other fraction is measured in a Zeiss refractometer and expressed as the percentage (w/w) sucrose. The [^{35}S]gpαf content of every other fraction is determined by Con A- and anti-α-1,6-mannose/protein A precipitation as before.

The log-linear sucrose gradient is prepared just prior to use by overlaying at room temperature 0.4 ml of 55% (w/v) sucrose, 0.5 ml of 40% sucrose, 0.5 ml of 30% sucrose, 1.2 ml of 25% sucrose, 1.2 ml of 20% sucrose, and 1.2 ml of 15% sucrose in a 5.3-ml Ultraclear Beckman thin-walled tube. This step gradient is centrifuged for 3 hr at 32,000 rpm in an SW 50.1 rotor at 4° to create a smooth sucrose gradient.

Results. Membranes collected in the supernatant fraction of a 29° incubation of a wild-type lysate sediment fast in a 15–45% (w/w) sucrose gradient. Roughly one-half of the gpαf released remains core-glycosylated and accumulates within membranes that sediment slightly slower (open circles) than membranes containing outer chain gpαf (closed triangles)(Fig. 5A). The sedimentation profile of membranes collected in the supernatant fractions of *sec18* and *ypt1^{ts}* reactions incubated at 29° is compared to that of wild-type membranes in Fig. 5B and C. Outer chain-modified gpαf, produced because the mutant blocks are not completely restrictive (see Fig. 4), accumulates within membranes that sediment near the bottom of the gradient as in wild type. The majority of the gpαf released, however, remains core-glycosylated and accumulates within a membrane compartment that sediments slowly; thus the vesicles and the Golgi membrane can be physically separated.

Comments. The results of sedimentation analysis show that a slowly sedimenting vesicle, distinct from the ER and the Golgi compartment in which outer chain is assembled, accumulates when the function of Ypt1p or Sec18p is inactivated. Thus, the slowly sedimenting vesicles appear to be an intermediate compartment between the ER and the Golgi.

Are these slowly sedimenting vesicles true intermediates in ER-to-Golgi transport, or are they fragments of the ER that lack translocation activity? To establish the authenticity of the intermediate compartment one must show that on removal of the inhibitory block the vesicles transfer their contents to the Golgi compartment. Transport from an intermediate compartment should be faster, more efficient, and should require only a subset of the components necessary for transport from the ER.

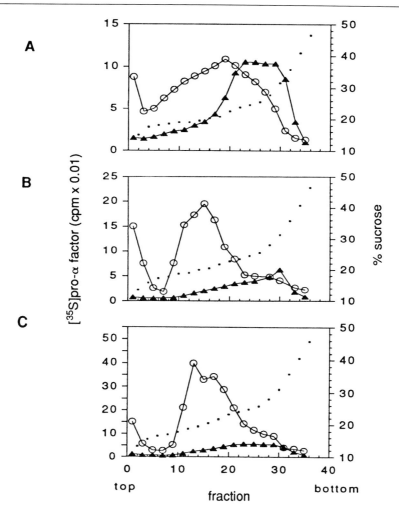

FIG. 5. Endoplasmic reticulum-derived vesicles and the Golgi apparatus are resolved by velocity sedimentation in sucrose gradients. (A) *Sec⁺;* (B) *sec18;* (C) *ypt1ᵗˢ.* Scaled-up MSS fractions obtained from stage II reactions that were blocked in vesicle targeting were analyzed by velocity sedimentation in sucrose gradients. Fractions were collected from the top. The density and the gp*α*f content of every other fraction were quantified as described in the section Experimental Preparation. Open circles represent core-gp*α*f. Filled triangles represent outer chain-modified gp*α*f. The black dots represent percentage (w/w) sucrose. Cytosolic proteins did not enter the gradient and remained in the top three fractions.

Vesicle Targeting/Fusion Assay

Chase of Ypt1p-Requiring Intermediate

Rationale. To develop a vesicle targeting/fusion assay a reversible transport block is used to accumulate vesicles that retain their ability to target and fuse with the Golgi membrane. Vesicles that accumulate in *sec18* and *ypt1^{ts}* reactions remain functional and fuse with a cis-Golgi compartment when incubated at 20° in the presence of fresh membrane, cytosol, and ATP (M. Rexach, unpublished results, 1991). However, to test the requirement for Sec12p and Sec23p in the targeting/fusion reaction, an alternative targeting block may be used. Antibody Fab fragments against Ypt1p block *in vitro* ER-to-Golgi transport at 20° [5] and, like Ypt1p^{ts}, cause the accumulation of transport vesicles that have not targeted to the Golgi membrane.[6] Completion of transport, measured by formation of outer chain gpαf, is detected when the core-gpαf-containing vesicles accumulated in the presence of Ypt1p antibody Fab fragments are incubated at 20° with fresh cytosol, fresh membranes, and required ATP (Fig. 6A).

Methods. A large (200 μl) stage II incubation using wild-type or mutant lysates is incubated at 20° for 60 min in the presence of 400 μg/ml anti-Ypt1p Fab fragments. At the end of stage II incubations, the reactions are chilled and fractionated by centrifugation in a Fisher microfuge at 13,000 g for 1 min at 4°. A 150-μl MSS fraction is taken from the meniscus and further centrifuged at 88,000 g for 15 min at 4° to generate a high-speed pellet (HSP) fraction devoid of soluble Fab fragments. The HSP is then resuspended in a small volume of B88 and mixed with an ATP-generating system, cytosol, and fresh membranes to the same concentrations as in a stage II incubation. The mix is incubated at 20° for 60 min (stage III). Completed stage III incubations are chilled on ice, mixed with SDS to 1% (w/v), and heated to 95° for 7 min. Equal aliquots from each tube are processed as described for stage II reactions.

Results. Optimum stage III transport from the Ypt1p Fab intermediate is achieved at 20° in the presence of fresh cytosol and membranes, and requires ATP (Fig. 6A). Formation of outer chain gpαf is more rapid and efficient (34 vs 26%) starting from this intermediate than in a typical stage II incubation starting from the 10° budding intermediate (Fig. 7).

Comments. The Ypt1p Fab intermediate is functionally distinct from the 10° intermediate. First, the typical 6 to 10-min transport lag phase seen in stage II incubations is absent in stage III incubations (Fig. 7), indicating that the vesicles have progressed past the most rate-limiting step in transport. Second, transport is more efficient from the vesicle intermediate than from the ER (Fig. 7), consistent with the vesicles being a later intermediate in the pathway.

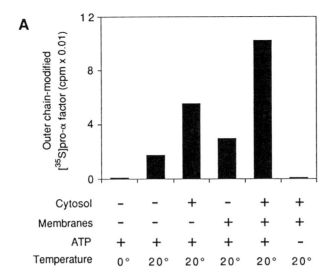

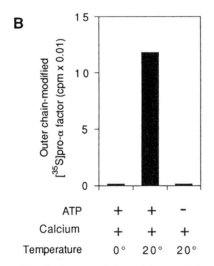

FIG. 6. Reversible blocks: completion of transport from (A) the Ypt1p Fab and (B) Ca²⁺-requiring intermediates. Large stage II reactions blocked with anti-Ypt1p Fab fragments or EGTA/Mn²⁺ were fractionated in a microcentrifuge. Each supernatant fraction, which contained the intermediate compartment, was further centrifuged at 88,000 g for 15 min at 4° to generate a high-speed pellet fraction (HSP) devoid of soluble proteins and inhibitors. The pellet fractions were resuspended in B88 for the Ypt1–Fab intermediate, or in EGTA/ Mn²⁺ buffer for the Ca²⁺-requiring intermediate, and mixed with fresh cytosol, membranes, and ATP-regeneration system as indicated. For the tubes containing no ATP, apyrase (0.25 U) was added to hydrolyze residual ATP; these tubes were supplemented with 50 μM GDPmannose, normally added in the ATP-regeneration mix. Transport intermediates were chased for 75 min at 20°. A tube incubated at 0° served as a negative control for transport during stage III and reflected background transport from stage II. Transport was quantified by precipitation and scintillation counting. The value obtained from the 0° incubation was subtracted from the other values as background. The Con A-precipitable counts averaged 3249 cpm for the Ypt1p-requiring intermediate chase and 2714 cpm for the Ca²⁺-requiring intermediate.

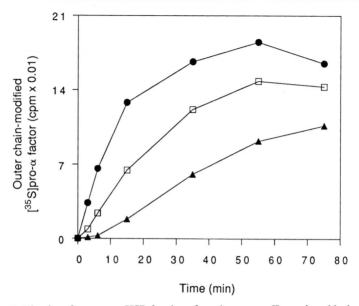

Fɪɢ. 7. Kinetics of transport. HSP fractions from large stage II reactions blocked with anti-Ypt1p Fab fragments or EGTA/Mn^{2+} were generated. The pellet fractions were resuspended in the presence of fresh cytosol, membranes, and an ATP-regeneration system, for the case of the Ypt1 Fab intermediate, or in the presence of an EGTA/Mn^{2+} buffer with Ca^{2+} and an ATP-regeneration system for the case of the Ca^{2+}-requiring intermediate. Transport intermediates were chased for various intervals at 20°. A tube incubated at 0° served as a negative control for transport during stage III and reflected background transport from stage II. Transport was quantified by precipitation and scintillation counting. The value obtained from the 0° incubation was subtracted from the other values as background. The Con A-precipitable counts averaged 3839 cpm for the Ypt1p-requiring intermediate (□) chase and 3555 cpm for the Ca^{2+}-requiring intermediate (●). A time course of transport from the ER/10° intermediate (▲) is plotted for reference; Con A-precipitable counts averaged 3941 cpm. To present a direct comparison of transport efficiencies, the time courses plotted were normalized to 4000 Con A-precipitable cpm, which standardized the maximum core-gpαf available for transport from each intermediate.

Which step is most rate limiting in transport? Although a lag phase follows transport from the 10° intermediate, no lag is detected in vesicle budding or consumption of the Ypt1p-requiring vesicle intermediate. Thus, a significant rate-limiting step in transport occurs after the vesicles have budded, but prior to docking to the Golgi membrane. Perhaps the vesicles undergo a "maturation" process in preparation for targeting. Vesicle uncoating or the recruitment of proteins required for targeting could be rate limiting.

Sec Protein Requirements for Vesicle Targeting/Fusion

Rationale. To further distinguish transport vesicles from ER fragments the consumption of the vesicle intermediates is assessed for Sec protein requirements. Ypt1p Fab vesicle intermediates are accumulated in Sec^+, *sec12, sec23,* and *sec18* lysates at 20°. The thermolability of the chase reaction (stage III) is tested by incubating reactions at 20 or 29°. Wild-type reactions are somewhat thermolabile in these incubations (25% inhibited at 29°) and thus not a good control to compare mutant thermolability. Nevertheless, thermolability may be tested by preincubating either the HSP, or the mix of required membrane and cytosol fractions, at 20 or 29° to inactivate the thermolabile mutant proteins prior to the chase reaction.

Methods. The HSP fraction, or the mix of cytosol, membrane, and ATP, is preincubated for 10 min at 20 or 29°. The fractions are chilled, the complete stage III mix assembled, and the chase reaction conducted at 20° for 60 min. After completion the samples are chilled and processed as before.

Results. Chase of the vesicle intermediate is not affected by incubation of Sec^+ or *sec12* fractions at 29° prior to a stage III reaction, but is reduced by 50% in *sec18* and 18% in *sec23* fractions (Fig. 8). The reduction seen for *sec18* is similar to the effect of *sec18* on the overall stage II reaction (Fig. 2). The reduction observed in *sec23* is only a small fraction of the effect seen on the overall stage II reaction (Fig. 2). The *ypt1*[ts] reaction is not significantly thermolabile (not shown), but addition of fresh anti-Ypt1p Fab to a stage III wild-type reaction reduces transport by 84%.

Comments. Consumption of the vesicle intermediate requires the action of Ypt1p and Sec18p, while Sec12p and Sec23p are not required for transport from this intermediate compartment. The vesicle intermediate is functionally distinct from the ER, which requires all four proteins to execute transport.

Vesicle Fusion Assay

Chase of Ca²⁺-Requiring Intermediate

Chase of Ca^{2+}-Requiring Intermediate

Rationale. An assay that specifically measures membrane fusion requires vesicles that are targeted/attached to the Golgi membrane. Chelation of Ca^{2+} during a transport reaction results in the accumulation of core-gpαf in a membrane compartment that cosediments with Golgi membranes in sucrose velocity gradients.[6] Addition of Ca^{2+} and ATP allows completion of transport from this intermediate. We interpret these results to imply a Ca^{2+}-dependent fusion of vesicles to the cis-Golgi compartment, although it remains a possibility that the Ca^{2+}-requiring step involves outer chain glycosylation after a fusion event.

	SEC^+	sec12	sec18	sec23	SEC^+ +Ypt1p Fab
Ypt1p-requiring intermediate (%) inhibition	2%	4%	50%	18%	84%
Ca^{2+}-requiring intermediate (%) inhibition	3%	0%	0%	0%	0%

FIG. 8. Chase of Ypt1p- and Ca^{2+}-requiring transport intermediates in the presence of Sec mutant proteins or Ypt1p Fab fragments. Ypt1p-requiring intermediate/HSP fractions were generated in wild-type and mutant lysates at 20° and were mixed with ATP, cytosol, and fresh membranes from each mutant, respectively. The mix of required cytosol, membranes, and ATP was pretreated for 10 min at 20 or 29° prior to mixing with the HSP fractions. The reactions were performed at 20° for 65 min and processed as described before. For the antibody inhibition, the Ypt1p Fab intermediate/HSP fraction was generated in wild-type lysates and mixed with ATP, cytosol, fresh membranes, and 400 μg/ml of Fab fragments, and incubated for 75 min at 20°; percentage transport efficiency was compared to an equivalent reaction without added Fab fragments. Background stage II transport was subtracted in all cases. Percentage inhibition represents thermolability in the mutant lysates (see Fig. 2) or percentage decrease in transport efficiency caused by the presence of antibody. Ca^{2+}-requiring intermediate/HSP fractions were generated in wild-type and mutant lysates at 20° and resuspended with $EGTA/Mn^{2+}$ buffer containing ATP. The tubes were incubated at 20 or 29° for 20 min and then calcium (250 μM final concentration) was added and the reactions were further incubated at 20 or 29° for 55 min. Chilled reactions were processed as described before. For the antibody inhibition, the Ca^{2+}-requiring intermediate/HSP fraction was generated in wild-type lysates and mixed with ATP, Ca^{2+}, and 400 μg/ml of Fab fragments, and incubated for 75 min at 20°; percentage transport efficiency was compared to an equivalent reaction without added Fab fragments. Background stage II transport was subtracted in all cases. Percentage inhibition represents thermolability in the mutant lysates (see Fig. 2) or percentage decrease in transport efficiency caused by the presence of antibody.

Methods. A large (200 μl) stage II incubation composed of wild-type or mutant components is incubated at 20° for 60 min in the presence of 5 mM ethylene glycol-bis(β-aminoethyl ether)-N,N,N',N'-tetraacetic acid (EGTA)/600 μM $MnCl_2$. The concentration of free ions in this EGTA buffer was estimated to be < 1 nM Ca^{2+} and ~ 10 nM Mn^{2+} using a modified version of the computer program described by Robertson and Potter.[12] At the end of stage II incubations, the reactions are chilled and fractionated by centrifugation in a Fisher microfuge at 13,000 g for 1 min at 4°. A 150-μl MSS fraction is taken from the meniscus and further centrifuged at 88,000 g for 15 min at 4° to generate a high-speed pellet (HSP) fraction. The HSP is then resuspended in a small volume of EGTA buffer (B88, 5 mM EGTA, 600 μM $MnCl_2$, and 250 μM $CaCl_2$) and

---

[12] S. Robertson and J. Potter, *Methods Pharmacol.* **5**, 63 (1984).

mixed with an ATP-regenerating system. The concentration of EGTA, $MnCl_2$, and $CaCl_2$ must be kept constant after any additions. The concentration of free ions in this EGTA buffer was estimated to be ~63 nM Ca^{2+} and ~10 nM Mn^{2+}. The mix is then incubated at 20° for 60 min (stage III). Completed stage III incubations are chilled on ice, mixed with SDS to 1% (w/v), and heated to 95° for 7 min. Equal aliquots from each tube are processed as described for stage II reactions.

Results. Optimum transport from the Ca^{2+}-requiring intermediate is achieved at 20° in the presence of Ca^{2+} and ATP (Fig. 6B). Addition of fresh cytosol and membranes does not stimulate transport. Formation of outer chain gpαf is more rapid and efficient (40 vs 34%) starting from this intermediate than from the Ypt1p-requiring intermediate (Fig. 7).

Comments. The Ca^{2+}-requiring intermediate is functionally distinct from the Ypt1p-requiring intermediate. First, transport from the Ca^{2+}-requiring intermediate does not require added cytosolic protein or membranes; the HSP fraction contains all components sufficient for the fusion reaction. Second, transport from the Ca^{2+}-requiring intermediate is more rapid and efficient than from the Ypt1p-requiring intermediate, consistent with it being a later intermediate in the pathway.

Sec Protein Requirements for Vesicle Fusion

Rationale. To further distinguish the vesicle targeting and fusion intermediates, transport from the Ca^{2+}-requiring intermediate is tested for Sec protein requirements. Calcium ion-requiring intermediates are accumulated in *Sec+, sec12, sec23,* and *sec18* lysates at 20°, and stage III reactions are conducted at 20 or 29°. No significant thermolability is detected in any of the lysates. However, considering the observed delay in the inactivation of the mutant proteins and the fast rate of transport from this intermediate, we tested the effect of preincubations at 20 or 29°. Preincubations to inactivate mutant protein are performed in the absence of calcium to prevent passage through the Ca^{2+}-requiring step. After preincubation, Ca^{2+} is added and completion of transport allowed to proceed at either 20 or 29°.

Methods. The HSP fractions containing the Ca^{2+}-requiring intermediate are obtained as described above. The HSP is resuspended in a small volume of EGTA buffer without Ca^{2+} (B88, 5 mM EGTA, and 600 μM $MnCl_2$) and mixed with an ATP-regenerating system. The mix is then incubated at 20 or 29° for 20 min (stage III). At the end of preincubations, calcium is added to a final concentration of 250 μM and the incubations continued at the respective temperatures for another 60 min. Completed incubations are chilled on ice, mixed with SDS to 1% (w/v), and processed as before.

Results. Chase of the Ca^{2+}-requiring intermediate is not affected by preincubation of any of the lysates or by addition of anti-Ypt1p Fab fragments (Fig. 8).

Comments. The late, fusion intermediate does not require the action of Sec12, Sec18, Sec23, or Ypt1 proteins to complete transport. This reaction allows a distinction to be drawn between proteins required for vesicle targeting and those required for membrane fusion.

Summary

In this chapter we have discussed the methodology used to identify and characterize three intermediates in protein transport from the ER that represent stages of transport vesicle budding, targeting, and fusion. The intermediates are obtained using a variety of transport inhibitors: low-temperature incubations, addition of chemicals, or inactivation of Sec protein function using temperature-sensitive mutants or specific antibodies. In all cases, the transport block imposed by the inhibitor is reversible, permitting assessment of the requirements for transport from each intermediate stage. Based on the differential requirements for Sec protein function, as well as the distinct kinetics and efficiency of transport from each intermediate, we demonstrate that each transport intermediate is functionally distinct.

[27] Purification of Golgi Cisternae-Derived Non-Clathrin-Coated Vesicles

By Tito Serafini and James E. Rothman

Introduction

Non-clathrin-coated [or coat protein-coated (COP-coated)] vesicles[1-3] are thought to act within the Golgi apparatus and other portions of the constitutive secretory pathway as "bulk flow" carriers.[3-5] These vesicles can be produced *in vitro* in a cell-free system that reconstitutes intercisternal

[1] L. Orci, M. Ravazzola, M. Amherdt, D. Louvard, and A. Perrelet, *Proc. Natl. Acad. Sci. USA* **82**, 5385 (1985).

[2] G. Griffiths, S. Pfeiffer, K. Simons, and K. Matlin, *J. Cell Biol.* **101**, 949 (1985).

[3] L. Orci, B. S. Glick, and J. E. Rothman, *Cell* **46**, 171 (1986).

[4] F. T. Wieland, M. L. Gleason, T. A. Serafini, and J. E. Rothman, *Cell* **50**, 289 (1987).

[5] A. Karrenbauer, D. Jeckel, W. Just, R. Birk, R. R. Schmidt, J. E. Rothman, and F. T. Wieland, *Cell* **63**, 259 (1990).

protein transport[6-8] (described elsewhere in this volume[9]). The nonhydrolyzable nucleotide analog GTPγS inhibits the acceptor function of the Chinese hamster ovary (CHO) Golgi membranes present in the reaction, and the key observation that transformed the vesicles from a totally electron microscopic phenomenon into a possibly biochemical, purifiable one was the accumulation of the vesicles attached to Golgi cisternae in the presence of the inhibitor.[10]

Electron microscropy experiments revealed that the vesicles could be removed from CHO Golgi cisternal membranes by raising the ionic strength of the wash buffer (T. Serafini, V. Malhotra, and J. E. Rothman, unpublished observations, 1987). Recovery of intact vesicles from these salt washes was first accomplished using rat liver Golgi as a large-scale membrane source (T. Serafini and J. E. Rothman, unpublished observations, 1987). Together with the use of bovine brain cytosol as a large-scale source of soluble factors for the *in vitro* production of the vesicles, this observation was the starting point for development of a large-scale purification.

The non-clathrin-coated vesicles were originally purified in a bootstrapping effort, as no markers existed that served to identify the vesicles specifically during a purification.[11] The purification was developed employing the following assumptions: (1) the coated vesicles should be much smaller than the parental Golgi cisternae; (2) the vesicles, bearing a protein coat and thus a higher protein-to-lipid ratio than the parental Golgi cisternae, should have an equilibrium density greater than that of the parental membranes; (3) the coated vesicles, being membranous, should "float up" under centrifugation conditions that result in no net movement of soluble, non-membrane-associated proteins; (4) as transport intermediates, the vesicles should possess within them transported secretory proteins to serve as biochemical markers (an assumption previously employed successfully in the purification of late secretory vesicles accumulating *in vivo* in a yeast *sec* mutant[12]); (5) the vesicles should not be present in mock vesicle preparations (i.e., from *in vitro* reactions lacking GTPγS); and (6) any preparation meeting biochemical criteria for containing vesicles should possess mor-

[6] W. E. Balch, W. G. Dunphy, W. A. Braell, and J. E. Rothman, *Cell* **39**, 405 (1984).
[7] W. E. Balch, E. Fries, W. G. Dunphy, L. J. Urbani, and J. E. Rothman, this series, Vol. 98 [4].
[8] W. E. Balch and J. E. Rothman, *Arch. Biochem. Biophys.* **240**, 413 (1985).
[9] J. E. Rothman, this volume [1].
[10] P. Melançon, B. S. Glick, V. Malhotra, P. J. Weidmann, T. Serafini, M. L. Gleason, L. Orci, and J. E. Rothman, *Cell* **51**, 1053 (1987).
[11] V. Malhotra, T. Serafini, L. Orci, J. C. Shepherd, and J. E. Rothman, *Cell* **58**, 329 (1989).
[12] N. C. Walworth and P. J. Novick, *J. Cell Biol.* **105**, 163 (1987).

phologically identifiable non-clathrin-coated vesicles as judged by electron microscopy. Using rabbit liver Golgi as a membrane source and bovine brain cytosol as a source of soluble factors, coated vesicles produced *in vitro* were purified to morphological homogeneity from salt washes with GTPγS-inhibited membranes by a procedure utilizing differential centrifugation and an equilibrium float-up centrifugation. Corresponding to the morphologically purified vesicles were many proteins as observed by sodium dodecyl sulfate-polyacrylamide gel electrophoresis (SDS-PAGE), likely vesicle-specific species having molecular weights of 160K and 95K to 105K. These proteins, however, were neither the only nor the major proteins present in the preparation.

However, with these likely vesicle-specific markers and knowledge of the buoyant density of the coated vesicles (1.18 g/ml), an improved purification procedure was developed by maximizing the purity and yield of these proteins at that density.[13] This procedure is detailed in this chapter. The chief revisions of the new procedure were (1) the initial use of CHO Golgi membranes as opposed to the "dirtier" tissue-derived membranes, (2) elimination of a sucrose cushion in the initial membrane collection, (3) inclusion of a low-salt wash before the high-salt wash to remove contaminants but retain coated vesicles associated with cisternae, (4) nonpelleting of vesicles once they had been removed from the Golgi cisternal membranes, and (5) the use of a steeper isopycnic gradient, into which the vesicles were centrifuged from above rather than from below.

Four major proteins of the vesicle-containing preparations produced using this improved method were shown to be coat proteins, or COPs: α-COP (160K), β-COP (110K), γ-COP (98K), and δ-COP (61K). Furthermore, β-COP was shown to be identical to a previously identified protein[14] whose predicted protein sequence showed homology with the adaptin family of clathrin-coated vesicle coat proteins,[15] thus showing the two types of vesicular carriers to be more fundamentally similar than previously thought. The COPs are found together in a large, cytosolic complex termed the "coatomer," as it is presumably the protomer for coat formation.[16] The purification of this complex from bovine brain cytosol is described elsewhere in this volume.[17] More recent data have shown the existence of at least three other coat proteins, one of which is identical to the GTP-binding protein ADP-ribosylation factor (ARF).[18]

[13] T. Serafini, G. Stenbeck, A. Brecht, F. Lottspeich, L. Orci, J. E. Rothman, and F. T. Wieland, *Nature (London)* **349**, 215 (1991).

[14] V. J. Allen and T. E. Kreis, *J. Cell Biol.* **103**, 2229 (1986).

[15] R. Duden, G. Griffiths, R. Frank, P. Argos, and T. E. Kreis, *Cell* **64**, 649 (1991).

[16] M. G. Waters, T. Serafini, and J. E. Rothman, *Nature (London)* **349**, 248 (1991).

[17] M. G. Waters, C. J. M. Becker, and J. E. Rothman, this volume [31].

[18] T. Serafini, L. Orci, M. Amherdt, M. Brunner, R. A. Kahn, and J. E. Rothman, *Cell* **67**, 239 (1991).

The following details the procedures for preparing the CHO or rabbit liver Golgi membranes and bovine brain cytosol used as the starting materials for the *in vitro* production of the vesicles, for assaying the best concentration of cytosol to use to obtain a maximal inhibition of acceptor Golgi membrane function, and for preparing the COP-coated vesicles themselves.

Preparation of Bovine Brain Cytosol

This procedure is based on one originally developed in our laboratory by Wattenberg,[19] which as modified (by T. Serafini and P. Melançon) provides a much higher concentration of protein (and thus active transport factors) and eliminates the need for gel filtration of the cytosol (because low molecular weight components are inhibitory in the *in vitro* intra-Golgi transport system[8]). The procedure yields approximately 250 ml of an orange-colored cytosol at a protein concentration of 20–30 mg/ml; this high protein concentration permits repeated freeze-thaws with minimal effect on activity.

Solutions

Brain sauce (2 liters):
320 mM Sucrose,
25 mM Tris-HCl (pH 7.4)

Breaking buffer (2 liters):
500 mM KCl,
250 mM sucrose,
25 mM Tris-HCl (pH 8.0),
2 mM EGTA,
0.5 mM 1,10-phenanthroline (stock at pH 5),
1 mM dithiothreitol (DTT),[19a]
2 μg/ml protinin,[19a]
0.5 μg/ml leupeptin,[19a]
2 μM pepstatin A [stock in dimethyl sulfoxide (DMSO)],[19a]
1 mM phenylmethylsulfonyl fluoride (PMSF) (stock in 2-propanol)[19a]

Dialysis buffer (prepare three 30-liter volumes):
50 mM KCl,
25 mM Tris-HCl (pH 8.0),
1 mM DTT[19b]

---

[19] B. W. Wattenberg and J. E. Rothman, *J. Biol. Chem.* **261**, 2208 (1986).
[19a] Add immediately before use.
[19b] Add immediately before use, and omit where indicated.

Method

1. Collect bovine brains fresh at the abattoir, and keep on ice in brain sauce while in transit. (All subsequent steps should be carried out on ice or at 4°.) Six brains should provide enough workable material for the procedure.

2. Remove the meninges and clotted blood from the cerebra, and remove as much white matter as is reasonably possible. The meninges can be peeled more easily from large, undamaged cerebra. Keep the pieces as cold as possible during handling.

3. Weigh out two 500-g quantities of tissue. Add a small volume of breaking buffer (complete except for PMSF) and chop the tissue coarsely with a pair of scissors; rinse the pieces with breaking buffer and place one 500-g quantity into a 1200-ml capacity blender (e.g., a Waring commercial blender). Fill the blender to capacity with breaking buffer, add PMSF, and homogenize with two 30-sec bursts (leaving 30 sec between bursts). Repeat for the other 500 g and pool the homogenates.

4. Centrifuge the homogenate at $g_{av} = 9000\ g$ for 60 min and recover the supernatant. Centrifuge this supernatant at $g_{av} = 150,000\ g$ [e.g., at 45,000 rpm in a Beckman (Palo Alto, CA) 45 Ti rotor using Quick-Seal tubes] for 90 min and again recover the supernatant.

5. Dialyze the supernatant using Spectra/Por 2 (Spectrum Medical Industries, Los Angeles, CA) dialysis tubing (6.4 ml/cm) twice for a minimum of 2 hr each time against 30 liters of dialysis buffer. A white, flocculant precipitate should form. (The second dialysis can proceed overnight.)

6. Centrifuge the dialyzed material as in step 5, and recover the supernatant, being careful not to recover any of the more loosely pelleted material.

7. In the cold room, add ammonium sulfate to 60% saturation slowly over 20–30 min; stir an additional 30 min. The ammonium sulfate need not be pulverized, although large aggregates should be dispersed before addition.

8. Recover the precipitate by centrifugation as in step 4 for 30 min. Redissolve the protein in dialysis buffer (saved from the second dialysis of step 6) to approximately one-seventh the volume originally dialyzed in step 6; use a Dounce homogenizer with a B pestle. The solution should be somewhat turbid.

9. Dialyze as before against 30 liters of dialysis buffer (using the tank from the second dialysis of step 6). This dialysis may proceed overnight; otherwise, 2 hr will suffice. Then dialyze a second time for at least 2 hr against 30 liters of dialysis buffer without DTT.

10. Centrifuge the dialyzed cytosol at $g_{av} = 100,000\ g$ for 90 min (e.g., at 35,000 rpm in a 45 Ti rotor using polycarbonate bottles). Recover the

supernatant; mix, aliquot (25 and 5 ml), freeze in liquid nitrogen, and store at −80°.

Preparation of Chinese Hamster Ovary Golgi-Enriched Membrane Fraction

The method for preparing CHO Golgi membranes is given elsewhere in this volume.[9]

Preparation of Rabbit Liver Golgi-Enriched Membrane Fraction

This method is an adaptation (by T. Serafini and Y. Goda) of the method of Tabas and Kornfeld.[20] Their choice of buffers yields Golgi membranes having better GTPγS-sensitive acceptor activity than other methods (e.g., that of Leelavathi *et al.*[21]; T. Serafini and J. E. Rothman, unpublished observations, 1987). Originally, rat livers were utilized, but the larger size of rabbit livers allows one to obtain enough tissue more easily. The method yields 10–15 ml of a tan-colored membrane preparation at a protein concentration of 2–3 mg/ml.

Solutions

Ethylenediamine tetraacetic acid (EDTA, pH 7–8), 100 mM
Tris-HCl (pH 7.4), 10 mM
S/T (0.5 M): 0.5 M sucrose/10 mM Tris-HCl (pH 7.4)
S/T, 1.0 M
S/T, 1.1 M
S/T, 1.25 M
Sucrose, 2.0 M

Method

1. Remove the liver from a euthanized rabbit, and cut away the gall bladder and connective tissue (be careful not to puncture the gallbladder). Collect at least 100 g of tissue lacking extensive amounts of connective tissue. Rabbits should be 5–6 lb; one rabbit yields approximately 80–100 g tissue. (We use female New Zealand White rabbits.) Put the tissue on ice (all subsequent steps should be carried out on ice or at 4°).

2. To a given mass of tissue add 3.75 times the mass in ml of 0.5 M S/T. Add EDTA to 5 mM (volume of EDTA solution added equals 0.25 times tissue mass). Mince the tissue using a pair of scissors.

[20] I. Tabas and S. Kornfeld, *J. Biol. Chem.* **254**, 11655 (1979).
[21] D. E. Leelavathi, L. W. Estes, D. S. Feingold, and B. Lombardi, *Biochim. Biophys. Acta* **211**, 124 (1970).

3. Transfer the tissue to a square bottle and homogenize twice, 30 sec each time, using a Brinkmann (Westbury, NY) Polytron homogenizer with PTA 20S probe tip on setting 4. Between bursts, remove the connective tissue that will accumulate on the shearing surfaces of the homogenizer. Thorough homogenization is essential to a good recovery of Golgi membranes.

4. Centrifuge the homogenate at $g_{av} = 650$ g [e.g., 2500 rpm in a Du Pont (Wilmington, DE) SA-600 rotor using Oak Ridge bottles] for 10 min, and recover the postnuclear supernatant through layered cheesecloth, being careful not to recover any of the loosely packed brown pellet.

5. Overlay 8 ml of 1.25 M S/T with 30 ml of this supernatant in a Beckman SW 28 Ultra-Clear tube. (Use of a large-bore pipette and an air-driven pipetting aid greatly facilitates this and other layering steps.) Using two SW 28 rotors, one may process 360 ml of postnuclear supernatant. Centrifuge at 25,000 rpm in a SW 28 rotor ($g_{av} = 83,000$ g) for 90 min.

6. After centrifugation, the crude smooth membranes will be found in a milky brown band present at the 0.5/1.25 M sucrose interface. Remove the white layer of triglycerides found at the top of the tube by absorbing into two bunched Kimwipes, and recover the crude smooth membranes using a long Pasteur pipette. Recover only two pipette volumes (3–4 ml) per tube. Poor membrane recovery at this step is often a cause of poor final recoveries of Golgi membranes.

7. Adjust the sucrose in the pooled crude smooth membranes to 1.2 M using either 2.0 M sucrose or 10 mM Tris-HCl (pH 7.4). Determine the sucrose concentration using a refractometer (using a table to convert to molarity from refractive index), and adjust according to Eqs. (1) and (2). To raise sucrose concentration by adding 2.0 M sucrose

$$V_{to\,add} = \frac{V_{original}(1.2\ M - [original])}{(2.0\ M - 1.2\ M)} \tag{1}$$

To lower sucrose concentration by adding 10 mM Tris-HCl (pH 7.4),

$$V_{to\,add} = \left(\frac{V_{original}[original]}{1.2\ M}\right) - V_{original} \tag{2}$$

8. Place the crude smooth membranes into SW 28 tubes, using no less than 13–15 ml/tube (usually three or four tubes if one has started with 100 g of tissue). Overlay with 10 ml 1.1 M S/T, then with 10 ml 1.0 M S/T, and then finally 4–5 ml 0.5 M S/T. Centrifuge at 25,000 rpm as before for 2.5 hr.

9. Collect the Golgi-enriched fraction at the 0.5/1.0 M interface. If a large layer (a "felt") or smaller pieces are present, recover them using a

Pasteur pipette in the smallest possible volume, and then recover the remainder using side puncture of the tube slightly above the interface, recovering no more than 3.5 ml/tube (a small piece of Scotch tape applied to the tube at the site of puncture will prevent leaking during and after recovery). While a felt or at least some pieces should be present, if they are not, merely use side puncture. Use a 20-gauge needle on a 3-ml syringe for side puncture, and remove the needle before expelling the collected membranes into a collecting tube. Swirl by hand to disperse large clumps, aliquot (1-ml volume in screw-capped tubes), freeze in liquid nitrogen, and store at $-80°$.

Assay for GTPγS Inhibition of Acceptor Function

The use of the *in vitro* intra-Golgi transport assay to measure the cytosol dependence of GTPγS-dependent inhibition of transport (and concomitant COP-coated vesicle accumulation) with CHO donor and acceptor membranes is based on the procedure presented elsewhere in this volume,[9] and is accomplished by using one-half of the normal amount of membranes in a reaction while varying the cytosol concentration.[10] However, rabbit liver Golgi, because it is so distinct in its properties from the CHO Golgi membranes, cannot simply be substituted for the CHO wild-type acceptor in such reactions while still obtaining meaningful results applicable to a situation in which the rabbit liver Golgi membranes are the sole membranes present in a reaction. A two-step assay was therefore developed to measure the cytosol dependence of the GTPγS-dependent inhibition of acceptor function in liver tissue-derived Golgi membranes (T. Serafini and J. E. Rothman, unpublished results, 1987).

In the first part of the assay, rabbit liver Golgi membranes ($50-120$ μg/ml) are incubated in the presence of GTPγS ($20 \mu M$) with varying concentrations of bovine brain cytosol ($0-12$ μg/ml) present during the incubation. The reaction mix is otherwise standard (with balanced KCl concentrations in the different samples), except for the omission of the radiolabeled sugar nucleotide ([^{3}H]UDPGlcNAc) required to measure intra-Golgi transport. In the second part of the assay, GTP (and extra Mg^{2+}, both to 2 mM final), [^{3}H]UDPGlcNAc, and donor membranes are added along with balancing cytosol by the dilution of the first assay samples with an equal volume of second, matched reaction mixes. In other words, the first stage of the assay inhibits the acceptor activity of the rabbit liver Golgi membranes in a cytosol-dependent fashion, while the second stage of the assay measures that inhibition by assaying any normal transport that can still occur under identical cytosol concentrations for each assay sample. The donor membrane and [^{3}H]UDPGlcNAc levels are normal during this stage of the reaction; only the volume of the assay samples (100 μl) is twice the standard level. The assay samples are incubated in the second stage for 30 min and processed as normal. The GTP present during the second stage of the assay prevents any further inhibition by GTPγS during the remainder of the assay.[10] Figure 1 shows the cytosol dependence of GTPγS-dependent inhibition obtained in such a two-stage assay.

Purification of Non-Clathrin-Coated Vesicles

In the reaction mixes below, the cytosol and membrane concentrations we have found to yield effective inhibition of acceptor function are given. All proteinaceous components are added after other reagents have been mixed, and the membranes are added last, together with the GTPγS. The reaction mixes are assembled on ice, with the membranes being quickly thawed for use in a 37° water bath. We incubate the reactions directly in the glass Corex tubes subsequently used for centrifugation after incubation (the long glass tubes allow for efficient heating and cooling of the solution).

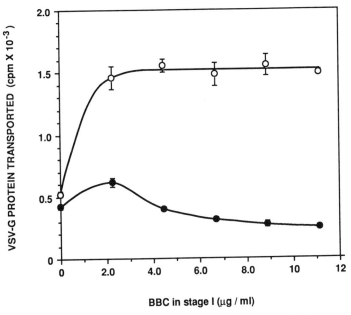

FIG. 1. Results of a two-stage transport assay to determine the optimal concentration of bovine brain cytosol (BBC) required for inhibition of rabbit liver Golgi acceptor function (and, by assumption, the optimal production of vesicles). The details of the assay procedure are given in text. As is observed in this case, inhibition (the difference between the two curves) does not increase very much beyond 5.0 mg/ml. (O) − GTPγS; (●) + GTPγS.

Solutions

Reaction mix for CHO Golgi membranes (60 ml):

20 mM KCl (total after addition of cytosol),

25 mM N-2-hydroxyethylpiperazine-N'-2 ethanesulfonic acid (HEPES)–KOH (pH 7.0),

2.5 mM magnesium acetate,

0.2 M sucrose (total after addition of membranes),

125 μM DTT,

50 μM ATP,

250 μM UTP,

5 mM creatine phosphate,

8.0 IU/ml creatine kinase,

25 μg/ml RNase A,

2.4 mg/ml bovine brain cytosol,

50–100 μg/ml CHO Golgi membranes,

20 μM GTPγS

Reaction mix for rabbit liver Golgi membranes (30 ml):

Same as above, except for the following:

250 μM DTT,

4.8 mg/ml bovine brain cytosol,

50–120 μg/ml rabbit liver Golgi membranes

Low-salt stripping buffer (LSSB):

50 mM KCl,

25 mM HEPES–KOH (pH 7.2),

2.5 mM magnesium acetate,

0.2 M sucrose

High-salt stripping buffer (HSSB):

Same as low-salt stripping buffer, except for the following:

250 mM KCl

##% stripping buffers (##% SB):

Same as high-salt stripping buffer, except at the indicated percentage (w/w) sucrose

Method

1. After adding the membranes and GTPγS, quickly incubate the reaction for 15 min at 37°. Chill in an ice/water bath for 10 min after incubation. All subsequent steps should be carried out on ice or at 4°.

2. Collect the membranes by centrifugation for 30 min at g_{av} = 11,400 g [e.g., 10,500 rpm in a Sorvall SA-600 rotor, k (clearing factor) $\approx$ 1200].

3. Remove the supernatant by aspiration, and resuspend the membrane pellet. To resuspend, after transferring the pellet to a microcentrifuge tube in a total of 600 μl LSSB [400 μl to transfer, 200 μl to rinse the centrifuge tube(s)] using a Pipetman P-1000, set the Pipetman to 400 μl and homogenize the pellet by working the Pipetman rapidly 20–30 times while holding the blue tip almost but not against the microcentrifuge tube bottom. Vortex briefly, and incubate on ice for 15 min.

4. Collect the membranes by microcentrifugation for 10 min, and resuspend the membrane pellet as before in 600 μl HSSB. Incubate on ice 15 min.

5. Remove the larger contaminating membranes by microcentrifugation for 10 min. Carefully recover the supernatant (600 μl), transfer to a new tube, and microcentrifuge again for 15 min.

6. Carefully recover only 545 μl of supernatant ("crude COP-coated vesicles"), place into a new tube, and adjust to 20% (w/w) sucrose by the addition of 169 μl of 55% SB. Layer this on an initially discontinuous sucrose gradient formed by layering 714 μl of each of the following in a Beckman SW 55 tube: 50% SB, 45% SB, 40% SB, 35% SB, 30% SB, and 25% SB.

7. Centrifuge the gradient for 18 hr at $g_{av} = 100,000$ g (e.g., in a Beckman SW 55 rotor at 32,500 rpm).

8. Fractionate the isopycnic gradient from the bottom using a capillary tube attached to tubing feeding through a peristaltic pump. Collect approximately twenty 260-μl fractions at the rate of 260 μl/min. The pump tubing outlet should be moved by hand due to the small volumes involved.

The percentage sucrose (w/w) composition of the fractions should be determined by refractometry; the COP-coated vesicles band at 41.5% sucrose (1.18 g/ml), and are usually found in the three fractions centered on this concentration. The sodium dodecyl sulfate polyacrylamide gel electrophoresis (SDS-PAGE) protein band patterns (in a 7.5% acrylamide gel) of the fractions from isopycnic gradients of both CHO- and rabbit liver Golgi-derived COP-coated vesicle preparations are shown in Fig. 2 along with the measured sucrose composition of the gradient fractions. The four high molecular weight COPs (α-COP, 160K; β-COP, 110K; γ-COP, 98K; and δ-COP, 61K) are indicated; the coated vesicles band tightly in fractions 7–9 in each case. In fact, if gradient construction and fractionation are carefully performed, refractometry is unnecessary, as the fractionation is quite reproducible from preparation to preparation. Electron microscopy of fixed vesicles shows their "fuzzy" coat morphology, which is distinct from the geometric basketwork of clathrin coats (Fig. 3).

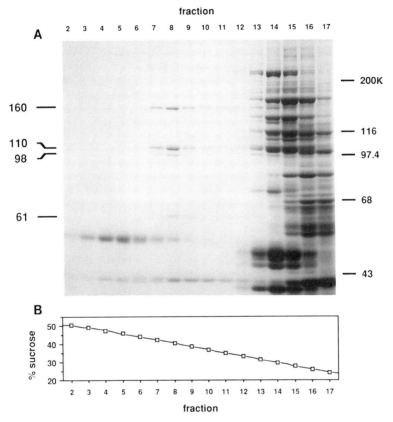

FIG. 2. Protein composition of Golgi-derived COP-coated vesicles. (A) Coomassie Brilliant blue (CBB)-stained SDS-PAGE gel (7.5% acrylamide) showing fractions 2–17 of the isopycnic gradient from a vesicle preparation using CHO Golgi membranes as the membrane source. Proteins in the fractions were trichloroacetic acid (TCA) precipitated and electrophoresed as described in Ref. 13. The four high molecular weight COPs are indicated on the left by their molecular weights: α-COP, 160K; β-COP, 110K; γ-COP, 98K; and δ-COP, 61K. Molecular weight standards are indicated on the right. (B) Percentage sucrose (w/w) composition of the fractions in (A) as determined by refractometry. (C) CBB-stained SDS-PAGE gel showing fractions from a coated vesicle preparation using rabbit liver Golgi membranes as the membrane source. The COPs are indicated as in (A). (D) Percentage sucrose composition of the fractions in (C).

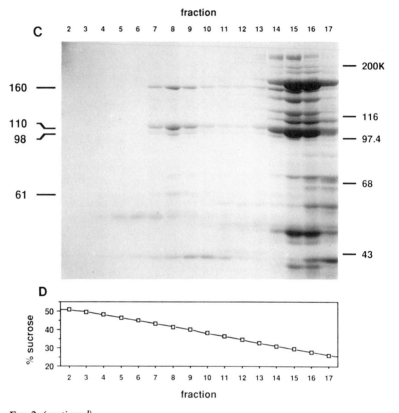

FIG. 2. *(continued)*

Concluding Remarks

The availability of antibody reagents specific for the COPs (such as the monoclonal antibody M3A5[14]) should lead to a more improved vesicle generation protocol, because the above-described method was based on an inhibition of acceptor function as an assay for vesicle production. This inhibition, while correlating with COP-coated vesicle accumulation, does not necessarily reflect the degree of vesicle accumulation or eventual recovery. Parameters now determined based on the inhibition of acceptor function (such as time of incubation, cytosol and membrane concentrations, GTPγS concentration) will necessarily be redefined by assaying for vesicle production directly through the use of these reagents. As many

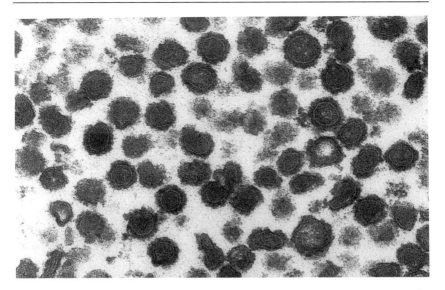

FIG. 3. Purified COP-coated vesicles as visualized by electron microscopy. Note the "fuzzy" coat morphology. (These particular vesicles were isolated using the original procedure.)[11] Vesicle-containing gradient fractions were fixed by glutaraldehyde in suspension and processed as previously described.[3,10,11] Magnification: ×70,160.

low-abundance proteins are present in the vesicles derived from both CHO and rabbit liver membrane sources[13,18] (and thus are transport machinery candidates), improved purification methods will be necessary to utilize fully the COP-coated vesicles to unravel the biochemistry of intracellular transport.

Acknowledgments

We wish to thank our longtime collaborator, Lelio Orci (Département de Morphologie, Institut d'Histologie et d'Embryologie, Université de Genève, Genève, Switzerland), for performing the electron microscopy of Fig. 3, and for his essential role in the development of the first coated vesicle purification procedure. We also wish to thank Friedericke Freymark, Lyne Pâquet, and Barbara Devlin for their technical assistance; Vivek Malhotra, Paul Melançon, and Yukiko Goda for their collaboration; and the many past and present members of the Rothman laboratory for their advice and criticism during the development of these procedures.

[28] Purification of *N*-Ethylmaleimide-Sensitive Fusion Protein

By MARC R. BLOCK and JAMES E. ROTHMAN

Elucidation of the molecular mechanisms involved in vesicle budding, targeting, and fusion that occur during protein transport has been one of the major challenges of the last decade in the field of cellular biology. *In vitro* reconstitutions of most of the different steps of the endocytic and exocytic pathways (this volume, Section I) have enabled a biochemical study of the transport machinery. However, vesicular transport requires numerous protein components acting in concert. The removal of a single essential component along one purification step results in the complete loss of transport activity. The task of purifying transport components is greatly facilitated when one component can be eliminated at a time, transforming the overall assay into an assay specific for the single missing component. In a cell-free system that reconstitutes transport within the Golgi stack[1-4] Glick and Rothman[5] made the crucial discovery that *N*-ethylmaleimide (NEM) selectively allowed the measurement of a single component even in the crudest fractions. This led to the first purification of a component of the transport machinery, named *N*-ethylmaleimide-sensitive fusion protein (NSF), according to its function (Block *et al.*[6]). Herein, we describe the detailed procedure for purification of NSF from Chinese hamster ovary (CHO) cells, together with the main guidelines for the purification of this component from other sources.

Assay for NSF Activity

One of the peculiarities of NSF activity is that it can be found in soluble (cytoplasmic) and membrane-associated forms in CHO cells. However, little activity was present in the original preparation of the cytosol fractions, due to the high instability of soluble NSF. This allowed the original discovery that the mild treatment of vesicular stomatitis virus (VSV)-infected 15B CHO and uninfected CHO Golgi membranes with NEM (15

---

[1] E. Fries and J. E. Rothman, *Proc. Natl. Acad. Sci. USA* **77**, 3870 (1980).
[2] W. E. Balch, W. G. Dunphy, W. A. Braell, and J. E. Rothman, *Cell* **39**, 405 (1984).
[3] W. E. Balch, B. S. Glick, and J. E. Rothman, *Cell* **39**, 525 (1984).
[4] L. Orci, B. S. Glick, and J. E. Rothman, *Cell* **46**, 171 (1986).
[5] B. S. Glick and J. E. Rothman, *Nature (London)* **326**, 309 (1987).
[6] M. R. Block, B. S. Glick, C. A. Wilcox, F. T. Wieland, and J. E. Rothman, *Proc. Natl. Acad. Sci. USA* **85**, 7852 (1988).

min at 0°) was sufficient to eliminate most of the activity in the standard transport assay (Balch et al.[2]). Nevertheless, a residual transport activity was still present after NEM treatment. To decrease this background, the cytosol used in the NSF assay (8 mg/ml of protein) is now prepared as described in Balch and Rothman[7] with minor modifications: it is first desalted on a BioGel P-6DG column (Bio-Rad, Richmond, CA), and further incubated at 37° for 20 min. In these conditions, NSF has a half-time of inactivation of 3–5 min and is readily destroyed in the preparation. A mixture of equal volumes of acceptor and donor Golgi membranes in 1 M sucrose, 10 mM Tris-HCl, pH 7.4, is incubated at 0° for 15 min with 1 mM NEM (final concentration) added from a fresh stock solution (50 mM). Then dithiothreitol (DTT) is added to 2 mM from a 0.1 M stock. This mixture of NEM-treated Golgi membranes may be refrozen in liquid nitrogen and stored at −80° until use.

Incubation mixtures of 50 μl contain NEM-treated Golgi membranes (10 μl), 5 μl of NSF-free cytosol, 10 μM palmityl-CoA, 50 μM ATP, 2 mM creatine phosphate, creatine kinase (7.3 international units/ml), 250 μM UTP, and 0.4 μM UDP[^{3}H]GlcNAc (0.5 Ci) in an assay buffer containing 25 mM N-2-hydroxyethylpiperazine-N'-2-ethanesulfonic acid (HEPES)–KOH (pH 7.0), 15 mM KCl, 2.5 mM magnesium acetate, and 0.2 M sucrose (derived from the Golgi fractions). The NSF fraction to be tested (up to 20 μl) is added last. After incubation at 37° for 1 hr, the VSV G protein is immunoprecipitated at 4° for at least 6 hr or alternatively at 30° for 2 hr as described in Balch et al.[3] Assays of all fractions must be performed in a predetermined linear range (0–0.2 mg/ml for crude NSF preparations).

Biological Starting Material for NSF Purification

Initially, Glick and Rothman[5] found that NSF could be removed from CHO Golgi membranes by adding small amounts of ATP. Actually, ATP not only induces NSF to stay in a soluble form, but it readily stabilizes it. Indeed, at 37°, the half-time of thermal inactivation of NSF in solution is only 3 min without ATP, but it rises to 45 min when 300 μM ATP or ADP is added. This finding has been crucial for the success of the purification. Therefore in the purification procedure all our buffers are supplemented with 0.5 mM ATP when not stated otherwise. This addition allows a purification procedure lasting more than 24 hr at 4° with a minimal loss of activity.

Even though NSF can be conveniently ATP extracted from CHO Golgi

[7] W. E. Balch and J. E. Rothman, *Arch. Biochem. Biophys.* **240**, 413 (1985).

membranes, this starting material results in an insufficient yield of NSF to be practical. Alternatively, we took advantage of the fact that NSF can be removed from CHO membranes with a moderate ionic force of 100 mM KCl without ATP to purify NSF from the crude supernatant (cytosol) fraction of CHO cells. The cytosol preparation has been modified to optimize the amount of NSF activity present in the extract.

Routinely, a washed pellet of CHO cells (1 vol) is resuspended with 4 vol of a swelling buffer containing 20 mM piperazine-N,N'-bis(2-ethane sulfonic acid (PIPES)–KOH (pH 7.2) , 10 mM MgCl$_2$, 5 mM ATP and homogenized in a solution containing 100 mM KCl and 5 mM ATP, 5 mM dithiothreitol, 1 mM phenylmethylsulfonyl fluoride (PMSF), 0.5 mM o-phenanthroline, leupeptin (10 μg/ml), and 1 μM pepstatin. The cells are allowed to swell 20 min on ice and then are disrupted in a Waring blender at high speed for 30 sec. Subsequently, KCl is added slowly to the homogenate while stirring to achieve a final concentration of 0.1 M from a 2.5 M stock solution. After low-speed centrifugation at 800 g for 15 minutes, the postnuclear supernatant may be frozen in 45-ml aliquots in liquid nitrogen and stored at $-80°$. Just before use, this postnuclear supernatant must be rapidly thawed at 37° and spun at 45,000 rpm in a 45 Ti (Beckman, Palo Alto, CA) rotor for 90 min. The clear supernatant is used as the starting material for NSF purification. Typically, it has an NSF activity of 2500 cpm of [^{3}H]GlcNAc incorporated into VSV G protein per microgram of cytosolic protein.

Initial Purification Steps: Elimination of Bulk Proteins

Guidelines for Successful NSF Purification

Although NSF is stabilized with ADP or ATP, it remains a very unstable and fragile protein. Therefore, all the purification steps should be as short as possible and must be linked together whenever possible. All the biologically active fractions must be frozen in liquid nitrogen when not used and kept at $-80°$. The NSF activity can be further stabilized with 10% (v/v) glycerol or 1% (w/v) polyethylene glycol 4000 or 8000 (PEG), but it is destroyed by phosphate buffers and sulfate ions.

Fast buffer changes cannot be achieved by gel filtration: NSF activity is completely lost when this chromatography step is used in the absence of PEG. With PEG, NSF becomes more resistant to gel filtration but the yields remain quite low. Alternative desalting procedures, such as fast dialysis using the ultrathin Spectrapor 2 dialysis tubing, must be preferred. Desalting is complete within 2 hr of dialysis at 4° with one buffer change, resulting in a high recovery of the activity.

Concentrations of the fractions required before each analytical purification steps cannot be carried out with ammonium sulfate precipitation. Ultrafiltration on XM 300 membranes (Spectrum Medical Industries, Los Angeles, CA) was found to be a fast and convenient alternative and no loss of activity is usually observed. Surprisingly, this procedure results in little if any loss of the small molecular components and does not constitute an actual purification step.

Polyethylene Glycol Precipitation

This is the most crucial step of the whole purification procedure, and the major source of variability; therefore it must be performed with the maximum of care. Overprecipitation or incomplete precipitation will result in a poor final purification. Because a great deal of variability is observed with PEG precipitation, depending on the initial protein concentration, pH, and temperature,[8] all these parameters have been tested and standardized in our procedure. The clear high-speed supernatant from CHO extract (described above) containing NSF activity is diluted with a solution made of 100 mM KCl, 20 mM PIPES–KOH, pH 7.2, 10 mM MgCl$_2$, 5 mM ATP, 5 mM dithiothreitol, to adjust the protein concentration to 5 mg/ml. The temperature is set at 0°, and the pH precisely adjusted to 7.0 with 1 M KOH (pH electrode equilibrated at 20°). The appropriate volume of a 50% (w/v) PEG 4000 (Sigma, St. Louis, MO) must be added dropwise with vigorous stirring until a final concentration of 8% (w/v) is reached. Nevertheless, the total time of PEG addition should not exceed 1 min. The mixture is further incubated for 30 min at 0° with stirring. The precipitate is then pelleted for 15 min at 10,000 rpm in a JA-20 rotor (Beckman, Palo Alto, CA). The pellet is suspended in one-third of the initial volume in a solution of 100 mM KCl, 20 mM PIPES–KOH, pH 7.0, 2 mM MgCl$_2$, 2 mM dithiothreitol, and 0.5 mM ATP. The suspension is homogenized with a Dounce (Wheaton, Millville, NJ) homogenizer and then sonicated with three 10-sec bursts in a 3000-W Branson (Danbury, CT) water bath sonicator at 0°. Undissolved material is subsequently removed by a second 15-min centrifugation at 10,000 rpm in the JA-20 rotor. At this point, you may freeze the supernatant in liquid nitrogen or proceed.

Ion-Exchange Chromatography

The NSF activity does not bind to anionic or cationic exchange chromatography at pH 7 under our conditions, suggesting that this protein is

[8] K. C. Ingham, *Arch. Biochem. Biophys.* **186**, 106 (1978).

close to its isoelectric point. This fact was used to design a fast and efficient purification step. Most proteins do bind to either anionic or cationic exchanger at pH 7 and moderate ionic force while NSF activity runs in the flow through.

Routinely, DE-52 cellulose (Whatman, Clifton, NJ) is swollen in a solution of 1 M PIPES–KOH, pH 7.0, 0.5 M KCl and then poured into a column. The column volume is chosen to equal the sample volume to be treated. One-third of this volume of Sepharose fast-flow S resin (Pharmacia, Piscataway, NJ) is equilibrated with a 2 M KCl solution and poured into a column. Both columns are connected in series and equilibrated with a solution of 20 mM PIPES–KOH, pH 7.0, 100 mM KCl, 2 mM MgCl$_2$, 2 mM dithiothreitol, and 0.5 mM ATP until the pH and the conductivity of the eluate become identical to those of the starting buffer. It is noteworthy that ATP does not bind to the DE-52 column at the concentration of 100 mM KCl.

The resolubilized material from PEG precipitation is applied to the top of the DE-52 column. The protein concentration of the fractions is followed using the Bradford assay.[9] The flow-through protein peak is then concentrated 20-fold by ultrafiltration on an XM 300 membrane (Spectrum) at 4° with a pressure of 20 psi. At this point, some precipitation may occur in the concentrate and the insoluble material must be eliminated with a new centrifugation for 15 min at 25,000 g at 4°. The preparation may be frozen in liquid nitrogen at this point.

Analytical Purification Steps

Sedimentation on Glycerol Gradient

Even though NSF seems to be a large protein component, gel filtration does not give satisfactory results. Therefore, NSF is separated from the small proteins by velocity sedimentation through a glycerol gradient. Linear 10–35% (w/v) glycerol gradients (38 ml) are formed from the bottom of 40-ml Beckman Quick-Seal tubes with thin stainless steel tubing. Three gradients can be formed simultaneously using an automatic Beckman gradient maker. The buffer throughout the gradient is 20 mM HEPES–KOH (pH 7), 100 mM KCl, 2 mM dithiothreitol, 2 mM MgCl$_2$, 0.5 mM ATP. The concentrated DE-52 fast-flow S Sepharose flow-through peak is layered on the top of each gradient in 2-ml portions by using a peristaltic pump. The gradients are centrifuged in sealed tubes in a VTi 50 rotor (Beckman) at 50,000 rpm for 2.5 hr with a slow acceleration and decelera-

---

[9] M. M. Bradford, *Anal. Biochem.* **72**, 248 (1976).

tion program. The use of a vertical rotor rather than a swinging bucket rotor reduces the length of the run from 16 to 2.5 hr and results in a considerable improvement of the yield at this step. Typically, three gradients are required to process NSF derived from 450 ml of cytosol. Fractions (2 ml) are collected from the bottom of the tube. The reproducibility of the gradients made with the Beckman apparatus allows us to collect three gradients at a time using a three-way peristaltic pump and, subsequently, to reduce the overall duration of the preparation. The protein concentration and NSF activity (in 0.1-μl samples) of each fraction are measured. Only the most active fractions (Fig. 1) are pooled, yielding about 1.5 mg/ml of protein per milliliter. At this stage a protein doublet of 76 and 70 kDa on sodium dodecyl sulfate-polyacrylamide gel electrophoresis (SDS-PAGE) clearly follows the NSF activity peak. Nevertheless, the 70-kDa protein is neither a component nor a proteolytic fragment of NSF but a main contaminant of the preparation. It will be eliminated by the next purification step. Whenever the volume of the pooled fractions ex-

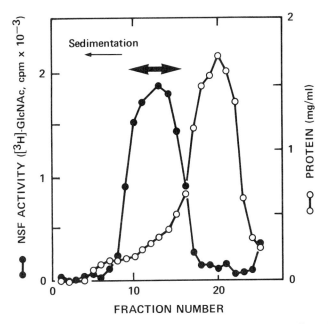

Fig. 1. Glycerol gradient step in the purification of NSF. A 10–35% (v/v) glycerol gradient was prepared and centrifuged as mentioned in the text. Then 0.1 and 5 μl of each 2-ml fraction were assayed for NSF activity (solid symbols) and protein concentration (open symbols), respectively. The double-headed arrow represents the fraction of the activity peak that is pooled and used for the next purification step.

ceeds 10 ml, the preparation is then concentrated to 8 ml or less by ultra-filtration on an XM 300 membrane (Spectrum). This partially purified NSF preparation can be conveniently stored frozen in liquid nitrogen at this stage.

Mono S Fast Protein Liquid Chromatography

The concentrated glycerol fractions are dialyzed in Spectrapor 2 tubing against 100 vol of a solution of 20 mM HEPES–KOH (pH 7.0), 10% (v/v) glycerol, 2 mM MgCl$_2$, 2 mM dithiothreitol, and 0.5 mM ATP for a total of 2 hr at 4° (with one buffer change). The dialysate is centrifuged for 10 min at 10,000 rpm in a JA-20 rotor, and 8-ml portions of the clear supernatant are injected into a 1-ml Mono S column (preequilibrated with the dialysis buffer) at a flow rate of 0.5 ml/min. After a 15-ml wash, NSF activity is eluted with a swallow 10-ml linear gradient of 0–100 mM KCl. Activity is mainly found in two 0.5-ml fractions corresponding to approximately 60 mM KCl. It is noteworthy that an initial KCl concentration of 10 mM in the sample to be injected (due to insufficient dialysis) hampers NSF binding to the column. The column is stripped with 2 M KCl before reuse. Two runs are necessary to process the entire preparation derived from 450 ml of cytosol. The NSF activity corresponds to an absorbance peak at 280 nm and SDS-PAGE analysis of the fractions reveals a single protein band of 76 kDa (Fig. 2). The NSF (0.15 mg/ml) can be frozen in aliquots in liquid nitrogen and stored at −80°. It can be repeatedly thawed and frozen in liquid nitrogen with little loss of activity.

Activity of the purified NSF fraction is completely eliminated by treatment with 1 mM NEM at 0°, and under the same conditions the 76-kDa polypeptide is labeled with [^{14}C]NEM. It is stimulated by long-chain acyl-CoA in a fashion similar to crude NSF. Monoclonal IgM directed at the 76-kDa polypeptide inhibited specifically the *in vitro* transport assay in Golgi[6] and many other *in vitro* assays related to other steps of protein transport,[10,11] suggesting a general use of this protein within the cell. One can conclude that the 76-kDa polypeptide contains the NSF activity required to promote transport.

Table I summarizes the purification procedure. The NSF is purified 1000-fold relative to cytosol with an overall yield of 12% in activity. This degree of purification may be an underestimate due to the inactivation of NSF at the different steps.

[10] C. J. M. Beckers, M. R. Block, B. S. Glick, J. E. Rothman, and W. E. Balch, *Nature (London)* **339**, 397 (1989).

[11] R. Diaz, L. S. Mayorga, P. G. Weidman, J. E. Rothman, and P. D. Stahl, *Nature (London)* **339**, 398 (1989).

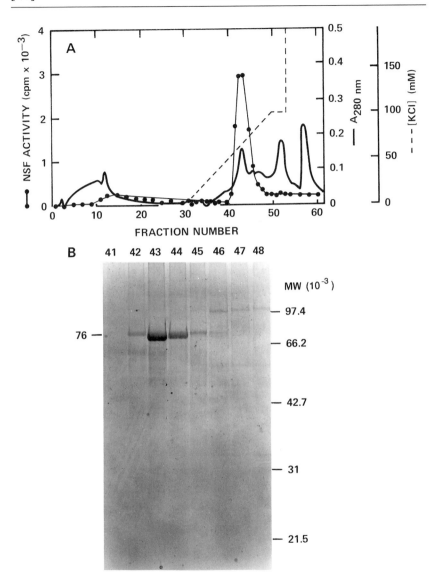

FIG. 2. Chromatography of NSF on a Mono S FPLC (fast protein liquid chromatography) column. (A) Elution profile. (●) NSF activity in 0.01-μl samples (the assay is saturated in fractions 43 and 44); —, absorbance at 280 nm, reflecting protein concentration; - - -, concentration of KCl. (B) SDS-polyacrylamide (10%, w/v) gel electrophoresis of the proteins present in the fractions containing NSF activity (Coomassie blue staining).

TABLE I
PURIFICATION OF NSF FROM CHO CELL CYTOSOL

Step	Yield of activity (% of step 1)	Relative specific activity normalized to step 1	Yield of protein (mg)
Cytosol	100	1	600
PEG 4000 precipitation	70	2.5	163
DE-52 flowthrough	70	12.5	54
Sepharose fast-flow S flow-through	35	29	11
Glycerol gradient	38	206	1.2
FPLC on Mono S	12	1000	0.14

Hints for Purification of NSF from Other Biological Sources

One can find NSF activity in many other biological sources, such as bovine brain cytosol, rat Golgi extract, and yeast cytosol.[12] However, in crude tissue extracts, the biological activity is often masked by a strong inhibitory activity that does not allow the use of our NSF assay. With a more purified preparation, such as rat Golgi extract, NSF can be easily detected and exhibits properties similar to CHO NSF, i.e., ATP stabilization of soluble NSF, NEM sensitivity (with some differences according to the species), and improved stability in buffers containing glycerol or PEG. Nevertheless, the extraction of NSF from the membranes in a soluble form may require slightly different conditions. For example, ATP alone is not sufficient to induce the release of rat NSF from Golgi membranes. However, it can be readily extracted using 0.5 mM KCl in a buffer containing 20 mM PIPES–KOH, pH 7.2, 2 mM MgCl$_2$, 2 mM dithiothreitol, 5 mM ATP, and protease inhibitors. With rat Golgi extract we have been able to follow the biological activity after velocity sedimentation on a glycerol gradient in a manner indistinguishable from CHO NSF. The SDS-PAGE revealed the presence of the 76/70-kDa polypeptide doublet following the activity peak (not shown). However, rat Golgi extract does not yield a sufficient amount of protein and is not a suitable preparation on which to perform the whole purification procedure. Because the procedure described above does not require the measurement of NSF activity until the fifth step of purification, the procedure can be achieved blindly up to this

[12] D. W. Wilson, C. A. Wilcox, G. C. Flynn, E. Chen, W.-J. Kuang, W. J. Henzel, M. R. Block, A. Ullrich, and J. E. Rothman, *Nature (London)* **339**, 355 (1989).

point using a crude whole-tissue extract as starting material (where no NSF activity can be detected). After the sedimentation on glycerol gradient, we found that most of the inhibitory activity is removed and that the NSF peak could be followed during the final purification steps. Therefore, it is likely that mammalian NSF in tissue extracts can be purified using our procedure with minor modifications.

[29] Expression and Purification of Recombinant N-Ethylmaleimide-Sensitive Fusion Protein from Escherichia coli

By Duncan W. Wilson and James E. Rothman

The N-ethylmaleimide (NEM)-sensitive fusion protein (NSF) is a homotetramer of 76 kDa, initially purified from Chinese hamster ovary (CHO) cells[1] and essential for the fusion of transport vesicles with their cognate acceptor membrane[2] at many stages in the secretory and endocytic pathways. Studies *in vitro* and *in vivo* have implicated mammalian NSF and Sec18p, the *Saccharomyces cerevisiae* homolog,[3] in endoplasmic reticulum (ER)-to-Golgi traffic,[4] in at least two stages of intra-Golgi transport,[1] and in endocytosis.[5] For an overview, see Wilson *et al.*[6] and references therein. Here we describe techniques for the expression and purification of large quantities of active, recombinant CHO cell NSF from the bacterium *Escherichia coli*. Availability of unlimited amounts of this protein, and the opportunity to prepare genetically modified derivatives, will greatly facilitate analysis of the biochemical and molecular properties of NSF.

[1] M. R. Block, B. S. Glick, C. A. Wilcox, F. T. Wieland, and J. E. Rothman, *Proc. Natl. Acad. Sci. USA* **85,** 7852 (1988).

[2] V. Malhotra, L. Orci, B. S. Glick, M. R. Block, and J. E. Rothman, *Cell* **54,** 221 (1988).

[3] D. W. Wilson, C. A. Wilcox, G. C. Flynn, E. Chen, W.-J. Kuang, W. J. Henzel, M. R. Block, A. Ullrich, and J. E. Rothman, *Nature (London)* **339,** 355 (1989).

[4] C. J. Beckers, M. R. Block, B. S. Glick, J. E. Rothman, and W. E. Balch, *Nature (London)* **339,** 397 (1989).

[5] R. Diaz, L. S. Mayorga, P. J. Weidman, J. E. Rothman, and P. D. Stahl, *Nature (London)* **339,** 398 (1989).

[6] D. W. Wilson, S. W. Whiteheart, L. Orci, and J. E. Rothman, *Trends Biochem. Sci.* **16,** 334 (1991).

Construction of Recombinant Expression System for NSF

We chose to express a derivative of NSF bearing a carboxy-terminal epitope "tag" because of the lack of anti-NSF antibodies suitable for immunoprecipitation or Western blotting. Cleavage of the DNA encoding NSF with the restriction endonuclease *Nhe*I excised the last six codons of the *NSF* gene (see Fig. 1). A synthetic oligonucleotide was used to restore these 6 codons and then extend the gene by 10 additional codons corresponding to the amino acids EQKLISEEDL (the sequence of the new 3' end of the *NSF* open reading frame was confirmed by DNA sequence analysis). The sequence EQKLISEEDL is recognized by a mouse monoclonal IgG$_1$ antibody termed 9E10, raised against peptides derived from the predicted amino acid sequence of human c-*myc*,[7] and which has previously proved successful as an epitope tag.[8] We refer to the recombinant NSF encoded by this gene as NSF/Myc. To test that the NSF/Myc protein is active we prepared cytosol from a *S. cerevisiae sec18-1* strain bearing the *NSF* or *NSF/myc* genes expressed from the *GAL1-10* promoter. Normally, cytosol from *sec18-1* mutant *S. cerevisiae* has no detectable NSF activity when tested in an NSF-dependent transport assay[3]; however, extracts from mutants expressing the *NSF* or *NSF/myc* genes contained levels of NSF activity comparable to wild-type *SEC18* yeast (data not shown). The "tagged" *NSF/myc* gene therefore has the capacity to encode an active protein in this eukaryote.

We then attempted to achieve expression of active protein in *E. coli,* in order to maximize the amount of NSF/Myc that could be recovered. An *Xba*I–*Xho*I DNA fragment, containing the entire *NSF/myc* gene, was ligated with the bacterial expression plasmid pTTQ18,[9] previously cleaved with the restriction enzymes *Xba*I and *Sal*I. The resulting plasmid, pTTQNSF/*myc,* is shown in Fig. 1. The *Xba*I restriction site is the point of translational fusion between the pTTQ18 vector and a linker contiguous with the CHO cell-derived DNA fragment bearing the *NSF* gene. As shown in Fig. 1, this fusion results in an amino-terminal extension of 22 amino acid residues (Met-Asn-Ser . . . Ser-Ala-Lys) prior to the methionine residue proposed by Wilson *et al.*[3] to be the true amino terminus of NSF on the basis of position and ATG context.

The *E. coli* strain NM522 (9) was transformed with pTTQNSF/*myc* and transformants selected on plates of Luria broth (LB; 10 g/liter tryptone, 5 g/liter yeast extract, 5 g/liter NaCl, pH 7.2) supplemented with 50

[7] G. I. Evan, G. K. Lewis, G. Ramsay, and M. J. Bishop, *Mol. Cell. Biol.* **5,** 3610 (1985).
[8] S. Munro and H. R. B. Pelham, *Cell* **48,** 899 (1987).
[9] M. J. R. Stark, *Gene* **51,** 255 (1987).

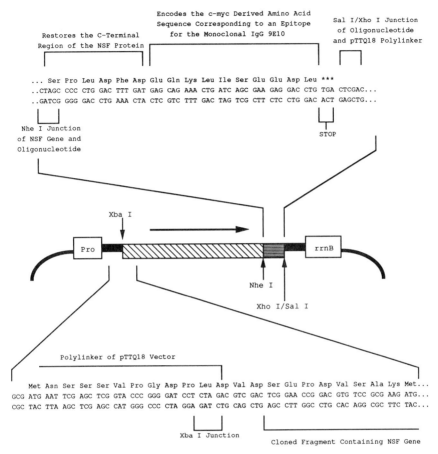

FIG. 1. Structure of the IPTG-inducible transcribed region of plasmid pTTQNSF/*myc*, used for expression of NSF/Myc in *E. coli*. Open bars, *tac* promoter (Pro) and ribosomal RNA transcriptional terminator *(rrnB)*; solid bars, polylinker regions derived from parent vector pTTQ18; hatched bar, DNA fragment encoding the *NSF* gene up to the *Nhe*I restriction site; horizontally striped bar, region derived from synthetic oligonucleotides. Direction of transcription is shown by a horizontal arrow above the *NSF/myc* gene. Note particularly the *Xba*I restriction site (point of fusion between DNA containing the *NSF/myc* gene and that from pTTQ18), the *Nhe*I restriction site (which marks the point at which the C terminus of NSF becomes encoded by a synthetic oligonucleotide), and the *Xho*I/*Sal*I junction (where the DNA encoding NSF/Myc rejoins that of vector pTTQ18).

μg/ml ampicillin (sodium salt), 5 g/liter dextrose (LBAD medium) and 2% (w/v) agar. pTTQNSF/*myc* transformants were grown using LBAD broth (except when inducing plasmid expression). Transformants rapidly became inviable when stored on LBAD-agar plates, even at 4°. Consequently, we store pTTQNSF/*myc* transformants frozen at −70° in LBAD containing 12.5% (w/v) glycerol.

Preparation of Bacterial Extracts with NSF Activity

To prepare bacterial extracts containing active NSF/*Myc*, 50 ml of LBAD broth is inoculated with cells scraped from the surface of a frozen aliquot of pTTQNSF/*myc* transformants. The culture is shaken overnight at 37°, then 10 ml is used to inoculate 1 liter of LBAD broth, prewarmed to 37°. This culture is shaken vigorously at 37° to ensure efficient aeration (typically, we shake 500 ml of broth in a 2-liter conical flask) until mid-log phase, when the optical density achieves 0.6, measured at a wavelength of 660 nm (under these conditions, this takes between 2.5 and 3 hr). Cells are pelleted at 6000 rpm for 10 min in a Sorvall (Norwalk, CT) GS3 rotor at room temperature and the pellets completely resuspended in 500 ml of prewarmed LB. The purpose of this wash is to remove dextrose from the cells, to ensure efficient induction. After pelleting again, the cells are resuspended in 500 ml of prewarmed LB supplemented with 50 μg/ml ampicillin and 0.5 mM isopropyl-β-D-thiogalactopyranoside (IPTG). We dispense 250 ml of culture to each of two 2-liter conical flasks and shake vigorously at 37° for 3 hr.

From this point, all procedures are at 4°, using prechilled bottles, solutions, and centrifuge rotors. Cells are pelleted as before, then washed by resuspension in 50 ml breaking buffer containing 100 mM N-2-hydroxyethylpiperazine-N'-2-ethanesulfonic acid (HEPES)–KOH, pH 7.8, 5 mM ATP, 1 mM ethylene glycol-bis(β-aminoethyl ether)-N,N,N',N'-tetraacetic acid (EGTA), 500 mM KCl, 5 mM MgCl$_2$, 1 mM dithiothreitol (DTT), 1 mM phenylmethylsulfonyl fluoride (PMSF). Cells are pelleted again, resuspended using 25 ml breaking buffer (yielding a final volume of approximately 35 ml), and disrupted by two passages through a prechilled 40K French pressure cell (SLM AMINCO Instruments, Urbana, IL), at 16,000 psi. The lysate is centrifuged for 30 min at 15,000 rpm in a Sorvall SA600 rotor and the low-speed supernatant (LSS) carefully decanted from the pellet. We typically recover 26–28 ml of LSS, at a protein concentration between 8 and 10 mg/ml.

At this point, NSF activity can be measured directly in the LSS (see the next section), and the polypeptide readily detected by Coomassie Brilliant blue staining when 10 μg of LSS is resolved by sodium dodecyl sulfate-polyacrylamide gel electrophoresis (SDS-PAGE) using a 7.5% (w/v) poly-acrylamide gel; NSF/Myc migrates as an 85K band comprising approximately 0.5% of total protein (estimated by eye). We have found it is most useful to compare the LSS extract of a pTTQNSF/*myc* transformant with that prepared from cells containing the pTTQ18 vector. In the latter extract, the 85K band is absent, as is NSF activity (see below). If 1 μg of LSS is electrophoresed as above, then Western blotted, the 85K band can be shown to cross-react with the 9E10 anti-*myc* monoclonal antibody (data not shown). No other *E. coli* protein has been observed to cross-react with the 9E10 antibody under these conditions.

NSF Activity Detected in Crude Extracts

N-Ethylmaleimide-sensitive fusion protein activity can be detected within the crude LSS extracts by assaying between 20 and 100 ng of LSS protein in a 25-μl NSF-dependent Golgi transport assay (see Block *et al.*[1] for assay conditions); however, transport activity is greatly inhibited if more than 1 μg of LSS is added to the assay. The LSS can be stored at $-70°$ following rapid freezing in liquid nitrogen, and on thawing shows little loss of activity. That the activity measured in this extract is indeed due to expression of active CHO cell NSF, and not to a nonspecific effect of the lysate, was confirmed in a number of ways. First, activity always copurified with the NSF/Myc polypeptide (see below). Second, activity is absent if LSS is prepared from *E. coli* that carry the pTTQ18 vector but not the *NSF*/*myc* gene (Fig. 2) or in LSS from pTTQNSF/*myc* transformants that have been "mock-induced" using 0.5% (w/v) dextrose rather than 0.5 mM IPTG (data not shown). Third, treatment of the LSS with 1 mM NEM or incubation at 37° in the absence of ATP, conditions known to inactivate CHO cell NSF, abolish NSF activity in these extracts (Fig. 3). Last, LSS-derived NSF activity is inhibited by the anti-CHO NSF monoclonal IgM antibody 4A6, which inhibits CHO cell NSF activity *in vitro*.[1] A control IgM monoclonal antibody fails to affect LSS NSF activity (see Fig. 4). Sensitivity to inhibition by 4A6 appears to be a highly specific characteristic of CHO NSF activity; 4A6 has no effect on the activity of Sec18p (data not shown), even though it shares 49% primary sequence identity with NSF (Wilson *et al.*[3]).

Purification of NSF/Myc from *Escherichia coli*

We have found that the chromatographic properties of bacterially expressed CHO cell NSF/Myc are very different from those of the protein obtained directly from CHO cells (for reasons discussed below). We have

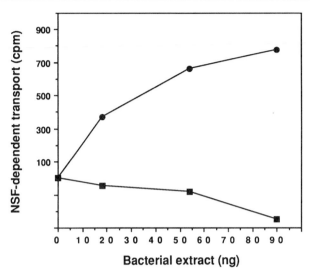

FIG. 2. NSF activity is found only in bacterial extracts if the *NSF/myc* gene is present in the expression vector. Low-speed supernatant extracts were prepared from IPTG-induced *E. coli* NM522 transformants (see text) and assayed in 25-μl NSF-dependent transport assays. Assays were in duplicate and the mean value plotted. A background (NSF/Myc-independent) signal of 343 counts per minute (cpm) has been deducted from the data. (●) *E. coli* transformed with pTTQNSF/*myc*; (■) *E. coli* transformed with pTTQ18 vector.

therefore developed a purification scheme that differs markedly from that described by Block *et al.*[1] The characteristics of the purification are listed in Table I, and the appearance of the active fractions following SDS-PAGE on a 7.5% (w/v) polyacrylamide gel is shown in Fig. 5.

First, an LSS is prepared exactly as described above, then clarified by centrifugation at 50,000 rpm for 1 hr in a Beckman (Palo Alto, CA) Ti 60 rotor. The high-speed supernatant (HSS) is recovered by decanting and pipetting, carefully avoiding the loose gelatinous membrane fraction that overlays a firmer, high-speed pellet. The protein concentration of the HSS is corrected to 6 mg/ml using breaking buffer, then a solution of polyethylene glycol 4000 [made to 50% (w/v) in water] slowly added to a final concentration of 10%, while stirring continuously at 4°. After stirring for a further 30 min the mixture is centrifuged at 5000 rpm for 20 min in a Sorvall SA600 rotor, the supernatant discarded, and the pellet gently resuspended in a Dounce homogenizer (Wheaton, Millville, NJ) in 10 ml of Q200 buffer [20 mM piperazine-N,N'-bis(2-ethanesulfonic acid) (PIPES)–KOH, pH 7.0, 2 mM MgCl$_2$, 2 mM DTT, 0.5 mM ATP, 10% (w/v) glycerol, 200 mM KCl].

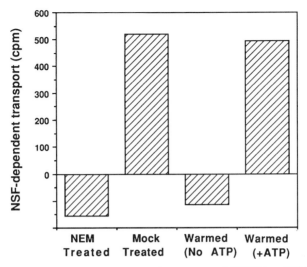

FIG. 3. Biochemical properties of bacterially expressed NSF/Myc. *N*-Ethylmaleimide sensitivity studies: LSS extracts were diluted 500× into 50 m*M* HEPES–KOH, pH 7.5, 50 m*M* KCl, 5 m*M* ATP, 5 m*M* MgCl$_2$, 1% (w/v) polyethylene glycol 4000, 10 mg/ml soybean trypsin inhibitor, to deplete dithiothreitol (DTT). Diluted LSS was incubated with 1 m*M* NEM (from a freshly prepared 50 m*M* stock solution) for 15 min at 4°, then excess NEM quenched by incubation with 2 m*M* DTT for 2 min at 4°. In a control incubation, diluted LSS was mock treated using 1 m*M* NEM that had been preincubated with 2 m*M* DTT. ATP stability studies: LSS extracts were diluted 500× into 50 m*M* HEPES–KOH, pH 7.5, 50 m*M* KCl, 2 m*M* DTT, 5 m*M* MgCl$_2$, 1% (w/v) polyethylene glycol 4000, 10 mg/ml soybean trypsin inhibitor to deplete ATP. The depleted sample was incubated at 37° for 30 min, then returned to ice and ATP added to a final concentration of 1 m*M*. In a control incubation, the sample was mixed with ATP (to 1 m*M*) prior to 37° treatment. Fifty nanograms of the treated samples and controls were tested for NSF activity in duplicate, and the mean plotted. A background signal (see Fig. 2) of 685 cpm has been deducted from the data.

When fully resuspended, the sample is loaded at 0.5 ml/min onto a 20 × 3 cm column containing 60 ml Q Sepharose fast-flow (Pharmacia, Piscataway, NJ) previously equilibrated with Q200 buffer (in pilot trials we found that, under these conditions, ATP fails to bind to this resin at KCl concentrations greater than 160 m*M*). The column is washed with 2 vol of Q200 buffer at 0.5 ml/min, then bound NSF/Myc eluted by washing with 2 vol of Q400 (identical to Q200 buffer, but containing 400 m*M* KCl) at 1.0 ml/min. As well as yielding a substantial enrichment for NSF/Myc relative to other proteins (Table I), this rapid and simple column step is essential for the preparation of a sample sufficiently free of lipid and nucleic acids to permit anion-exchange chromatography using fast protein liquid chromatography (FPLC). The Q400-eluted material is mixed with an equal vol-

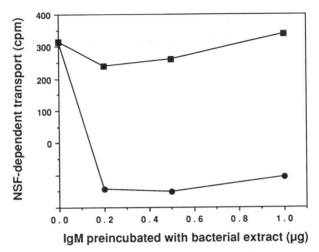

FIG. 4. NSF/Myc Activity in bacterial extract is inhibited by a monoclonal antibody raised against CHO cell NSF. Low-speed supernatant (56 ng) was added to an NSF-dependent transport assay cocktail such that the final KCl concentration was 65 mM (necessary to maintain the solubility of the 4A6 IgM; see Ref. 1). Increasing amounts of the 4A6 antibody (●) or a control IgM (■) were added and the mixture incubated for 15 min at 4°. Assays were in duplicate and the mean value plotted. A background (NSF/Myc-independent) signal of 656 cpm has been deducted.

TABLE I
PURIFICATION OF NSF/Myc[a]

Sample	Total protein (mg)	Total activity (units)	Specific activity (units/mg)	Enrichment	Yield (%)
100,000 g supernatant	213	2.8×10^{10}	1.3×10^8	1	100
10% polyethyleneglycol (PEG) precipitation	50.9	1.9×10^{10}	3.7×10^8	2.8	67.9
Fast Flow Q, 400 mM KCl batch elution	9.5	1.7×10^{10}	1.8×10^9	13.8	60.7
Mono Q pool	0.9	0.76×10^{10}	8.4×10^9	64.6	27.1

[a] Enrichment and yield are specified relative to the high-speed supernatant (HSS). Activity units for each sample are counts per minute of N-acetyl[^{3}H]glucosamine that became incorporated into vesicular stomatitis virus G protein as a result of addition of that sample to an NSF-dependent *in vitro* transport reaction. Counts per minute due to NSF-independent incorporation (387 cpm in these assays) were deducted from the data. All activity estimates were made after dilution of sample into the linear range for assay.

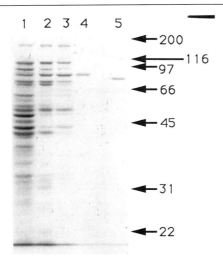

FIG. 5. Sodium dodecylsulfate gel electrophoresis of samples recovered at each stage of the NSF/Myc purification. The following amounts of total protein from each stage of the purification were subjected to SDS gel electrophoresis on a 7.5% (w/v) polyacrylamide gel, then visualized by staining with Coomassie Brilliant blue. Lane 1, 20 μg High-speed supernatant (HSS); lane 2, 7 μg 10% (w/v) PEG precipitate; lane 3, 5 μg Fast Flow Q 400 m*M* KCl elution; lane 4, 0.3 μg Mono Q FPLC pool; lane 5, 0.3 μg NSF purified from CHO cells by the method of Block *et al.*[1] Note the increased size of the NSF/Myc protein, which is 32 amino acids larger than NSF purified from CHO cells. Arrows indicate the migration of standard proteins of known molecular weight ($\times 10^{-3}$).

ume of Q0 buffer (containing no KCl) to dilute the KCl to a concentration of 200 m*M*, then concentrated using an Amicon (Danvers, MA) concentrator and YM30 membrane.

After concentration to 30 ml or less, the sample is loaded onto a 1-ml Mono Q FPLC column (preequilibrated in Q200 buffer) at 0.5 ml/min and the column washed with 5 vol of Q200 at 0.5 ml/min. Bound protein is eluted with a 20-ml linear gradient from 200 to 500 m*M* KCl, while maintaining other buffer conditions constant, and 0.5-ml fractions collected. The NSF activity elutes at 300 m*M* KCl, and 7.5% (w/v) SDS-PAGE shows the activity coincident with an 85K band that, following blotting, cross-reacts with the 9E10 antibody (data not shown). Conservative pooling of fractions at this stage is important to avoid contamination of NSF/Myc with a 50K protein that elutes from the column at a slightly higher KCl concentration. We normally recover two pools (which contain similar total amounts of NSF/Myc): the first is from the peak activity fractions and appears homogeneous when electrophoresed and visualized by Coomassie Brilliant blue staining; the second is derived from side fractions, and contains the 50K contaminant (present at a concentration similar to NSF/Myc). Although we use the pure fraction for most purposes we have encountered no problems using the less pure pool. Pooled fractions are dialyzed against 1000 vol of NSF buffer [50 mM HEPES–KOH, pH 7.5, 50 mM KCl, 0.5 mM ATP, 5 mM MgCl$_2$, 2 mM DTT, 10% (w/v) glycerol] at 4° for 2 hr using Spectrapor 2 dialysis tubing, then aliquots frozen in liquid nitrogen for storage at −70°. A standard 25-μl NSF-dependent *in vitro* transport assay is saturated by 20 ng of NSF/Myc purified in this manner (data not shown).

Of great advantage in the purification of NSF/Myc was the finding that this protein remains bound by Q Sepharose (and also diethylaminoethyl (DEAE)-cellulose; data not shown) at high salt concentrations (up to 300 mM KCl) at a pH of 7.0. In contrast, at pH 7.0 NSF purified from CHO cells does not bind DE-52 resin even at 100 mM KCl.[1] In subsequent studies, Tagaya et al.[10] have used the pTTQ18 expression vector to prepare NSF that does not contain the C-terminal epitope tag EQKLISEEDL. This protein still binds to Q Sepharose but can be eluted at much lower KCl concentrations than NSF/Myc. We believe it likely that the amino-terminal extension present in both forms of bacterially expressed NSF (and which contains five negatively charged amino acids; see Fig. 1) results in the recombinant protein having an increased affinity for anion-exchange resins. Much higher affinity then results from addition of the *myc* epitope tag, which contains four acidic residues, three of them in tandem adjacent to the most C-terminal amino acid of the protein. Use of this epitope may therefore be advantageous for purification of the recombinant protein, dependent on the context in which it is placed.

Acknowledgments

We thank T. Silhavy and members of the Silhavy laboratory for the use of their French press, and M. Wiedmann for helpful discussions.

[10] M. Tagaya, D. W. Wilson, M. Brunner, N. Arango, and J. E. Rothman, submitted for publication.

[30] Purification of Soluble *N*-Ethylmaleimide-Sensitive Fusion Attachment Proteins from Bovine Brain Microsomes

By Douglas O. Clary and James E. Rothman

Reconstitution of intra-Golgi vesicular transport *in vitro* has spurred the development of activity assays for required transport components. The first factor to be purified based on the transport assay was *N*-ethylmaleimide–sensitive fusion protein (NSF), isolated based on its ability to restore activity to an intra-Golgi transport assay inactivated with *N*-ethylmaleimide.[1] *N*-Ethylmaleimide-sensitive fusion protein was subsequently found to be a peripheral membrane protein active in the fusion stages of transport.[2] Reconstitution to transport competence of Golgi membrane fractions after extraction with high salt led to the discovery of at least two more peripheral membrane transport activities, termed fraction 1 and fraction 2 (Fr1 and Fr2).[3] The Fr2 activity was purified to apparent homogeneity, yielding a novel family of transport proteins now known as soluble NSF attachment proteins (SNAPs). Soluble NSF attachment protein activity had previously been characterized as a cytosolic protein cofactor required for NSF to rebind its integral membrane receptor.[4] The proteins purified from the Fr2 pool were found to be active in the intra-Golgi transport assay at the same stage as NSF (membrane fusion), and could facilitate the rebinding of NSF to its membrane receptor as well[5]; in fact, NSF was shown to bind to the Fr2 proteins directly, when tested in a solid-phase binding assay. Based on these observations, the Fr2 proteins were renamed SNAPs.[5]

There are at least three SNAP species found in bovine brain, each of which exhibits activity in the *in vitro* intra-Golgi transport assay and the NSF–membrane binding assay.[5] They are closely related in size, having apparent molecular masses of 35kDa (α-SNAP), 36kDa (β-SNAP), and 39kDa (γ-SNAP), and have similar chromatographic properties. Analysis of their primary structures by limited proteolysis[3] and by peptide and DNA sequencing (D. O. Clary and M. Brunner, unpublished observations, 1990)

[1] M. R. Block, B. S. Glick, C. A. Wilcox, F. T. Wieland, and J. E. Rothman, *Proc. Natl. Acad. Sci. USA* **85**, 7852 (1988).

[2] V. Malhotra, L. Orci, B. S. Glick, M. R. Block, and J. E. Rothman, *Cell* **54**, 221 (1988).

[3] D. O. Clary and J. E. Rothman, *J. Biol. Chem.* **265**, 10109 (1990).

[4] P. J. Weidman, P. Melançon, M. R. Block, and J. E. Rothman, *J. Cell Biol.* **108**, 1589 (1989).

[5] D. O. Clary, I. C. Griff, and J. E. Rothman, *Cell* **61**, 709 (1990).

indicates that each of the three proteins is distinct, although α-SNAP and β-SNAP are highly similar. At this point it is unclear whether the three SNAP species are homologs of each other that function independently, or whether they act together in some sort of a multisubunit complex. Soluble NSF attachment proteins have been detected in other tissues, although in lower abundance than in brain tissue (W. Whiteheart, personal communication, 1990). The yeast protein SEC17p has been identified tentatively as a homolog of α-SNAP,[5] and evidence has been presented that the yeast homolog of NSF, SEC18p, interacts with SEC17p *in vivo* and *in vitro*.[5,6]

Principle of Assay

Some of the factors required for transport (including the SNAPs) are found both in the cytosol and on the Golgi membranes used in the assay. These transport factors are likely to be peripheral membrane proteins that can cycle on and off the membranes during transport, and thus can be found in either a membrane-bound or soluble form, depending on the isolation conditions.[3] To assay for these activities, the Golgi membrane fractions are treated with 1 M KCl at 37°, to remove or inactivate peripheral transport factors that are associated with them. The extracted membranes (termed K-Golgi) are now dependent on the readdition of these peripheral transport activities to reconstitute their transport activity. N-Ethylmaleimide-sensitive fusion protein is a case in point, because while it is normally provided by the membrane fractions, it can be isolated in an active, cytosolic form. K-Golgi lack NSF activity, and NSF must be added to all K-Golgi transport reactions for reconstitution.[3]

The standard preparations of Chinese hamster ovary (CHO) cell cytosol and bovine brain cytosol can reconstitute transport with K-Golgi when supplemented with CHO NSF. This implies that any other peripheral transport proteins removed by the KCl treatment of the membranes can be supplied from soluble pools in these cytosol preparations. However, when yeast cytosol is tested under these conditions, it is unable to drive transport. Because yeast cytosol is able to reconstitute transport with untreated Golgi membrane fractions,[6,7] at least one peripheral transport activity found in yeast cytosol is missing or inactive under these conditions. In fact, analysis of the peripheral transport factors found in bovine brain cytosol revealed two complementing pools of activity, termed fraction 1 (Fr1) and fraction 2 (Fr2), both of which must be added to the K-Golgi/NSF/yeast cytosol

[6] C. A. Kaiser and R. Schekman, *Cell* **61**, 723 (1990).
[7] W. G. Dunphy, S. R. Pfeffer, D. O. Clary, B. W. Wattenberg, B. S. Glick, and J. E. Rothman, *Proc. Natl. Acad. Sci. USA* **83**, 1622 (1986).

assay to recover transport activity. Purification of the activity present in the Fr2 pool led to the isolation of the three SNAPs.[3]

The transport assay for SNAPs is therefore based on a classical biochemical fractionation. KCl-treated Golgi fractions are dependent on four soluble pools of transport activity: purified CHO NSF, yeast cytosol, Fr1 (from bovine brain), and SNAP (or Fr2) activity. By omitting SNAP from the mixture, the intra-Golgi transport assay becomes dependent on the readdition of SNAP, and the amount of transport that is reconstituted becomes an assay for SNAP. Here we present a purification scheme for three SNAP species from bovine brain, a rich source of SNAP activity.

Materials and Solutions

Materials. DEAE-Cellulose, protease inhibitors, soybean trypsin inhibitor, and Reactive Red 120 are obtained from Sigma Chemical Co. (St. Louis, MO). Dithiothreitol (DTT), ATP, and UTP are from Boehringer Mannheim (Indianapolis, IN). Sephacryl S-300, phenyl-Sepharose, Sepharose CL-6B, Mono Q HR 5/5, and Superose 12 (preparative grade) are obtained from Pharmacia LKB (Piscataway, NJ). The sodium dodecyl sulfate-polyacrylamide gel electrophoresis (SDS-PAGE) molecular weight markers (low range) and gel-filtration markers were from Bio-Rad (Richmond, CA). Red agarose was prepared as described using Reactive Red 120 and Sepharose CL-6B.[3,8]

Solutions. The solutions are coded as follows: (T), includes 25 mM Tris, pH 7.8, at 0°; (P), includes 25 mM potassium phosphate, pH 7.0; (#K), includes KCl with the number indicating the concentration (mM); (D), includes 1 mM DTT; (G), includes 10% (v/v) glycerol. Brain buffer contains 0.3 M sucrose, 25 mM Tris, pH 7.8. Homogenization buffer is brain buffer plus 1 mM DTT, and the protease inhibitors leupeptin (10 μg/ml), aprotinin (2 μg/ml), pepstatin (1 μM), and phenylmethylsulfonyl fluoride (1 mM). Extraction buffer is 25 mM Tris, pH 7.8, 1 M KCl, 1 mM DTT, and the protease inhibitors listed above. SNAP dilution buffer is 50 KTD plus 500 μg/ml soybean trypsin inhibitor and 5% (v/v) glycerol.

Assay for SNAP Activity

Preparation of Cytosols and Membranes. Wild-type CHO cell cytosol, donor Golgi fractions from vesicular stomatitis virus (VSV)-infected CHO 15B cells, and acceptor Golgi fractions prepared from wild-type CHO cells

[8] R. K. Scopes, *in* "Protein Purification," 2nd Ed., p. 146. Springer-Verlag, New York, 1987.

are prepared as has been described.[9] KCl-treated Golgi (K-Golgi) membranes are prepared by diluting 1 vol of either donor or acceptor Golgi membranes (in approximately 1 M sucrose, 10 mM Tris, pH 7.4) with 1 vol of water and 1 vol of 3 M KCl, for a final concentration of 1 M KCl. The mixture is incubated for 10 min at 37° and cooled on ice. The extracted membranes are loaded onto a sucrose step gradient containing 1 vol of 35% (w/w) sucrose and 3 vol 25% (w/w) sucrose, each in 10 mM Tris, pH 7.4. The gradients are centrifuged 30 min at 50,000 rpm [240,000 g, Beckman (Palo Alto, CA) SW50.1], at 4°. The membranes are recovered at the 25–35% interface in the original volume. Donor and acceptor Golgi fractions are treated separately, recovered, pooled, and frozen. Yeast cytosol is prepared as described[7] from the SEY2102.1 strain (*MAT*a *SUC2-Δ9 ade2-101 ura3-52 gal2 leu2-3,112 Δpep4::LEU2;* S. Emr, California Institute of Technology, Pasadena, CA). Brain cytosol is prepared as described,[10] except that the dialysis of the cytosol is against 50 KTD instead of 25 mM Tris, pH 8.0. The NSF is purified from CHO cell homogenates as described.[1] All membrane and cytosol fractions are frozen in aliquots in liquid nitrogen and stored at −80°.

Preparation of Fraction 1. Chromatography on Sephacryl S-300 is used to resolve the Fr1 and SNAP pools of activity found in bovine brain cytosol. Once the pools are separated, the Fr1 pool can be used in the K-Golgi transport assay to measure SNAP activity specifically.

Bovine brain cytosol (25 ml) is precipitated with ammonium sulfate at 60% saturation, pelleted, and resuspended in 6 ml 50 KTD, yielding approximately 35 mg/ml protein. The resuspended ammonium sulfate fraction is loaded onto a 350-ml (67 × 2.6 cm) S-300 column equilibrated in 50 KTD. The column is developed with 50 KTD at a flow rate of 88 ml/hr; 6-ml fractions are collected. The peak of Fr1 activity occurs just after the void volume of the column, while SNAP activity elutes in a peak with an apparent molecular mass of 35kDa (Fig. 1). A fraction near the void volume of the column is used in the standard SNAP activity assay (see below) to locate the peak of SNAP activity; the complementary experiment with a fraction containing SNAP activity can be performed to locate the peak of Fr1 (Fig. 1). The first column fractions that have significant protein and Fr1 activity and do not contain SNAP activity (typically five) are pooled and concentrated to 6 ml by ultrafiltration on a PM-30 membrane (Amicon, Danvers, MA). Fraction 1 is frozen in aliquots in liquid nitrogen and stored at −80°.

K-Golgi Transport Assay. The assay used to reconstitute cis to medial

[9] W. E. Balch, W. G. Dunphy, W. A. Braell, and J. E. Rothman, *Cell* **39**, 405 (1984).
[10] B. W. Wattenberg and J. E. Rothman, *J. Biol. Chem.* **261**, 2208 (1986).

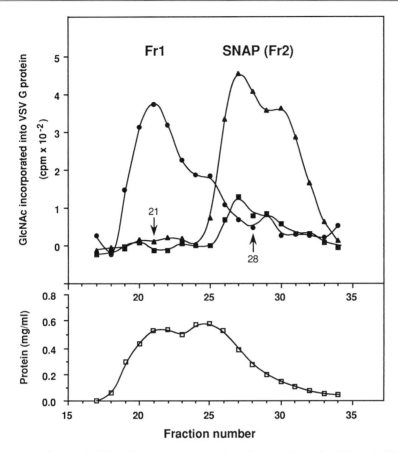

FIG. 1. Sephacryl S-300 chromatography resolves the complementing Fr1 and SNAP (Fr2) activities. A 20-ml Sephacryl S-300 column equilibrated in 50 KTD was loaded with 250 μl (3 mg) of bovine brain cytosol; 0.4-ml fractions were collected. Four microliters of each fraction was tested for activity alone (filled squares), mixed with fraction 21 (filled triangles), or with fraction 28 (filled circles) in incubations containing yeast cytosol, K-Golgi (donor and acceptor mixed), and CHO NSF under standard K-Golgi assay conditions. Fractions 21 and 28 are indicated with arrows. Transport was measured by the incorporation of tritiated GlcNAc into VSV G protein. Protein concentration (open squares) was determined with the Bio-Rad protein assay, using bovine serum albumin as a standard and a correction factor (actual concentration = 2.5 × measured concentration).

transport with KCl-treated Golgi membranes differs in some respects from the standard Golgi transport assay.[9] Typically, 5 μl (2–3 μg) of KCl-treated mixed donor and acceptor Golgi (in 1 M sucrose, 10 mM Tris, pH 7.4) is incubated with cytosol fractions (0–10 μl) to be tested for transport activity in the following final buffer conditions: 20 mM KCl, 25 mM N-2-hydroxyethylpiperazine-N'-2-ethanesulfonic acid (HEPES)–KOH, pH 7.0, 2.5 mM magnesium acetate, 1 mM dithiothreitol, 10 μM palmitoyl-CoA, 100 μM ATP, 250 μM UTP, 2.5 mM creatine phosphate, 11 IU/ml creatine phosphokinase. The volume of the reaction is 25 μl, and transport is measured by the addition of 0.3 μCi of UDP-N-acetyl-[6-^{3}H]-D glucosamine (10–30 Ci/mmol) to the reaction. Palmitoyl-CoA is added as reisolation of the membranes lowers their endogenous pool of acyl-CoA and partially inhibits their transport activity.[11] Incubations are performed at 30° for 120 min; KCl-treated Golgi have little activity at 37° regardless of the cytosol used. As all standard cytosols used in the transport reaction lack endogenous NSF activity due to its inactivation, and KCl treatment removes NSF activity from the membranes, all reactions are supplemented with saturating amounts (5 ng) of CHO NSF. After the incubation period, the assays are stopped with detergent buffer, and the VSV G protein is immunoprecipitated and counted exactly as described.[9]

SNAP Transport Assay. The K-Golgi transport assay is modified to measure specifically SNAP activity by the addition of yeast cytosol and a Fr1 pool. A typical reaction contains K-Golgi (mixed donor and acceptor, as above), 20 μg yeast cytosol, 5 ng CHO NSF, 23 μg Fr1 pool, and the sample to be tested for SNAP activity, with assay conditions as above. Concentrated and purified samples are diluted into SNAP dilution buffer before testing for activity.

Purification of SNAP Activity

Step 1: Preparation of Brain Membrane Extract. Fresh bovine brains are stored in brain buffer at 0° until use. The meninges are removed and the brains rinsed in brain buffer. The bulk of the white matter is removed to give a total weight of approximately 1 kg. All further steps in the purification are performed at 4°. The tissue is rinsed in homogenization buffer and minced. Two volumes of homogenization buffer is added to the tissue, and it is homogenized in a Waring blender for 20 sec. A crude membrane fraction is prepared by centrifuging the homogenate for 1 hr at 9000 rpm in a Sorvall (DuPont, Wilmington, DE) GS3 rotor. The supernatant is discarded, and the pellet rehomogenized as above, using 2 vol of homogenization buffer. Again, the homogenate is spun for 1 hr at 9000 rpm. This crude membrane preparation is now free from the majority of the cytosolic protein. The membranes are mixed with 2 vol of extraction buffer, homogenized 20 sec in the blender, and the membranes are re-

---

[11] B. S. Glick and J. E. Rothman, *Nature (London)* **326,** 309 (1987).

moved with a 1-hr, 9000-rpm spin, yielding an extract enriched in SNAP activity. The extract is clarified by centrifuging for 45 min at 45,000 rpm in a Ti 45 rotor (Beckman) and dialyzed overnight against 50 KTD buffer.

Step 2: Ammonium Sulfate Precipitation. The dialysate is centrifuged 9000 rpm for 1 hr to remove the precipitate. The clarified extract is typically 1700 ml with a concentration of 2 mg/ml. Solid ammonium sulfate is added to give a 30% saturated solution, and the extract is stirred for 30 min; the precipitate is removed with a 1-hr, 9000-rpm spin and discarded. The supernatant is brought to 50% ammonium sulfate saturation, and stirred for 30 min. The precipitate is collected as before; the supernatant is discarded. The pellets are resuspended in TD, in a volume that yields a conductivity below that of 100 mM KCl, usually around 500 ml.

Step 3: DEAE-Cellulose Chromatography. The protein solution is loaded at 2 ml/min onto a 5 × 15 cm DEAE-cellulose column equilibrated in 50 KTD. The column is washed with 50 ml of 50 KTD, and the SNAP activity eluted with a 750-ml 50–400 KTD gradient. The elution is done at 90 ml/hr, and 15-ml fractions are collected. Each fraction (0.1 μl) is assayed for SNAP activity; the peak of activity elutes at 210 mM KTD. The fractions containing significant activity (typically seven) are pooled, and the protein precipitated with the addition of solid ammonium sulfate to 60% saturation. The precipitate is pelleted at 15,000 rpm for 20 min, and resuspended in 15 ml of 100 KTDG. The solution is dialyzed twice against 1 liter 100 KTDG, for 45 min each.

Step 4: Red Agarose Chromatography. The protein pool from the DEAE-cellulose column is loaded on a 22.5 × 2.5 cm red agarose column equilibrated in 100 KTDG. The flow rate is 33 ml/hr, and 8-ml fractions are collected. After the sample is loaded, the column is washed with 25 ml of 100 KTDG, and then the eluant is changed to 200 KTDG. Fractions are collected well past the emergence of the salt bump, and 0.2 μl assayed for activity. The SNAP activity begins to elute about half-way into the flow-through peak of protein, and continues past the 200 mM KCl bump. Fractions containing SNAP activity (but avoiding the majority of the flow-through protein) are pooled, typically yielding 95 ml at 0.4 mg/ml.

Step 5: Phenyl-Sepharose Chromatography. The pool from the red agarose chromatography is loaded at a flow rate of 50 ml/hr onto a 15 × 1.5 cm column of phenyl-Sepharose equilibrated in 200 KTDG. The column is washed with 30 ml of 50 KTD plus 10% (v/v) ethylene glycol, and then developed with a 200-ml gradient of 10 to 75% ethylene glycol in 50 KTD. During this period, the flow rate is reduced to 14 ml/hr and 8-ml fractions are collected. After the gradient is finished, 30-μl aliquots of the fractions are analyzed by SDS-PAGE; 0.2 μl is assayed for activity. Activity is found in almost all fractions of the elution, often showing two peaks at

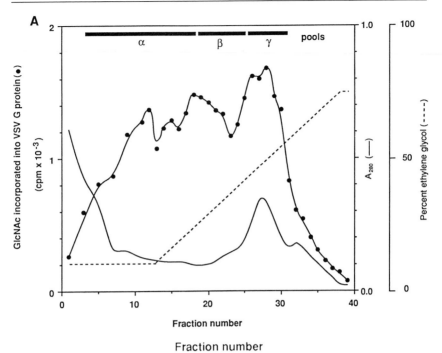

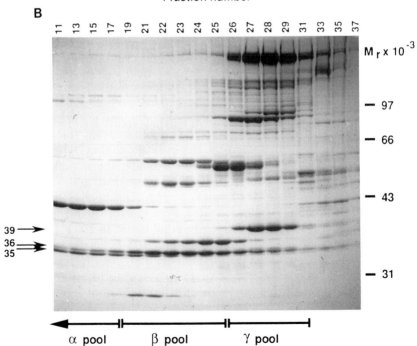

about 40 and 55% ethylene glycol (Fig. 2A). The 35K (α), 36K (β), and 39K (γ) SNAP proteins are partially resolved at this step, leading to a broad activity profile. The fractions are divided into three pools, according to their predominant SNAP species (Fig. 2B). Pool sizes for α-SNAP through γ-SNAP are typically 60, 30, and 20 ml.

Step 6: Mono Q (Phosphate) Chromatography. Each of the three SNAP pools is diluted with an equal volume of 50 KTD to reduce the ethylene glycol concentration. The Mono Q (phosphate) step is then applied to each pool in succession. A pool is loaded onto the Mono Q column, which has been equilibrated in 50 KPDG. The column is washed in 50 KPDG to remove the residual Tris, and the developed with a 25-ml gradient of 50–400 KPDG. Fractions of 0.5 ml are collected. The α-SNAP and β-SNAP coelute at approximately 250 mM KCl; γ-SNAP elutes at about 270 mM KCl, but the activity profile will usually show an additional peak at 250 mM KCl due to the presence of β-SNAP. Aliquots are analyzed by SDS-PAGE, and 0.01 μl is assayed for SNAP activity.

Step 7: Superose 12 Chromatography. For each pool, fractions from the Mono Q (phosphate) step containing significant amounts of SNAP activity (usually five) are pooled and loaded onto a 100-ml Superose 12 column (1.6 × 50 cm) equilibrated in 50 KTDG. The column flow rate is 0.25 ml/min, and 1.0-ml fractions are collected. All SNAPs elute from milliliters 56 to 61, (with an apparent molecular weight of 35K–40K), and these fractions are pooled.

Step 8: Mono Q (Tris) Chromatography. The Superose 12 pools are loaded onto the Mono Q column, previously equilibrated in 50 KTDG. The columns are developed with a 30-ml, 50–400 mM KTDG gradient. Fractions of 0.5 ml are collected: α- through γ-SNAP elute at approximately 250, 260, and 270 mM, respectively. Fractions are analyzed as for

FIG. 2. Chromatography of SNAPs on phenyl-Sepharose. The SNAP pool from the Reactive Red 120-Sepharose column was loaded onto a phenyl-Sepharose column equilibrated in 200 KTDG. The column was washed and eluted as described (see text). Four-milliliter fractions were collected starting at the eleventh milliliter of the wash. (A) The fractions were analyzed for SNAP activity (filled circles) and protein concentration ($A280$; solid line). Ethylene glycol concentration is shown as a dotted line. The black bars labeled α, β, and γ refer to pools used in succeeding purification steps (see text). (B) Sodium dodecyl sulfate gel electrophoresis of the protein population in the phenyl-Sepharose eluate fractions. Fraction numbers are shown at the top; 30 μl of each fraction was electrophoresed on a 10% (w/v) SDS-PAGE gel and stained with Coomassie blue R-250. The relative molecular weights of the marker proteins are shown at right: phosphorylase b (97,000), bovine serum albumin (66,000), ovalbumin (43,000), and carbonate dehydratase (31,000). Bars at the bottom refer to the α, β, and γ pools used in succeeding chromatographic steps, and to the predominant SNAP species found in each pool. The positions of the 35K, 36K, and 39K SNAP proteins are marked.

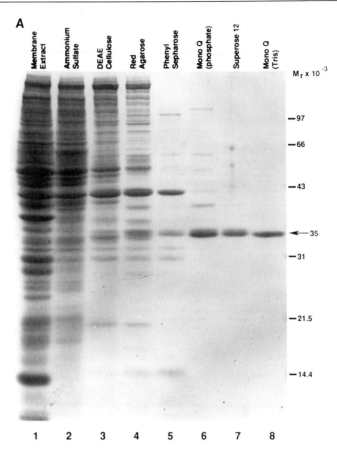

FIG. 3. Sodium dodecyl sulfate gel electrophoresis of α-SNAP purification pools, and the purified SNAP species. (A) The protein pools from an α-SNAP purification were analyzed on a 12.5% (w/v) SDS-PAGE gel in the following amounts and stained with Coomassie blue R-250: lane 1, membrane KCl extract, 94 μg; lane 2, 30–50% ammonium sulfate cut, 92 μg; lane 3, DEAE-cellulose pool, 26 μg; lane 4, red agarose, 15 μg; lane 5, phenyl-Sepharose pool, 4.4 μg; lane 6, Mono Q (phosphate) pool, 3.0 μg; lane 7, Superose 12 pool, 2.1 μg; lane 8, Mono Q (Tris) pool, 1.8 μg. (B) SDS-PAGE analysis of the three purified SNAP proteins. One microgram of each SNAP protein was subjected to electrophoresis on a 10% (w/v) SDS-PAGE gel and stained with Coomassie blue R-250. The molecular weight markers shown are those described in Fig. 2B, together with soybean trypsin inhibitor (21,500) and lysozyme (14,400). The arrows indicate the position of the 35K(α), 36K(β), and 39K(γ) SNAP proteins.

Mono Q (phosphate), above. The γ-SNAP should be pure after the Mono Q (Tris) chromatography (Fig. 3B). The α- and β-SNAPs are likely to partially contaminate each other at this stage. Therefore, fractions containing predominantly α-SNAP from both chromatography runs are

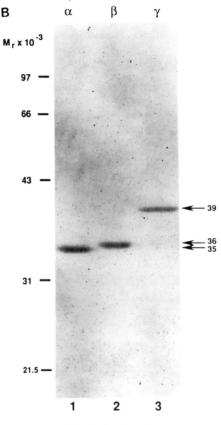

FIG. 3. *(continued)*

pooled, diluted twofold with TD, and rechromatographed. Conservative pooling from the second Mono Q (Tris) step yields α-SNAP that is >95% free from β-SNAP. The same procedure is followed for a preparation of β-SNAP that is free from α-SNAP (Fig. 3B). Fractions from the last Mono Q step for each protein are pooled, aliquoted, and frozen in liquid nitrogen.

Comments on the Purification. The SNAPs behave quite similarly through many chromatography steps. Their separation is best achieved on phenyl-Sepharose (Fig. 2) and Mono Q (Tris and phosphate).[3] Careful pooling after these chromatography steps will allow >95% pure preparations of each SNAP (Fig. 3B). The combination of Mono Q chromatography with both the phosphate and Tris buffers is critical, because β-SNAP

TABLE I
PURIFICATION OF α-SNAP

Fraction	Volume (ml)	Total protein (mg)	Total activity[a] (U)	Specific activity (U/mg)	Purification (fold)	Yield (%)
Membrane salt extract	1660	3100	5.18×10^6	1.67×10^3	(1)	(100)
30–50% $(NH_4)_2SO_4$	500	1320	5.71×10^6	4.34×10^3	2.6	110
DEAE-cellulose	136	144	2.53×10^6	1.76×10^4	10.5	49
Red agarose	95	36.1	1.96×10^6	5.43×10^4	32.5	38
Phenyl-Sepharose	62	6.77	6.00×10^5	8.87×10^4	53.1	12
Mono Q (phosphate)	2.5	0.938	4.47×10^5	4.76×10^5	285	8.6
Superose 12	6	0.642	4.08×10^5	6.36×10^5	381	7.9
Mono Q (Tris)	2	0.144	9.93×10^4	6.90×10^5	413	1.9

[a] The activity in each pool was determined by assaying dilutions of the pool in the SNAP transport assay such that the activity value fell into the initial, linear portion of the titration curve. Units (U) are defined as 1000 cpm of [³H]GlcNAc incorporated into VSV G protein. All activity determinations were done in the same experiment in duplicate.

can largely be removed from γ-SNAP using phosphate buffer (pH 7.0), and α-SNAP partially separates from β-SNAP using Tris buffer (pH 7.8).

Table I shows a summary of an α-SNAP purification. A total of 144 μg of protein was obtained, with a yield of approximately 2% of the starting activity (equivalent to approximately 4% when all three SNAPs are taken into account). The low yield is at least partially due to the strong tendency of SNAPs to stick to surfaces, especially late in the purification when little other protein remains. Sodium dodecyl sulfate gel electrophoretic analysis of the α-SNAP purification is shown in Fig. 3A. After the Superose 12 column, the major contaminant of α-SNAP is β-SNAP (Fig. 3A, lanes 7 and 8; Fig. 3B, lanes 1 and 2). For many purposes (including general use in the K-Golgi transport assay) a mixture of α- and β-SNAPs is suitable; therefore, several Mono Q chromatogaphy steps at the end of the purification can be eliminated, leading to a greater yield of protein. Wherever possible, dilute pure SNAP into SNAP dilution buffer (which contains 500 μg/ml soybean trypsin inhibitor) before use, as this will limit the loss of protein during the experiment.

Acknowledgment

This research was supported by a National Institutes of Health Grant (DK27044) to J.E.R.

[31] Purification of Coat Protomers

By M. Gerard Waters, Con J. M. Beckers, and James E. Rothman

Introduction

Transport through the Golgi apparatus occurs by vesicular transport,[1] that is, vesicles bud from one compartment (e.g., the cis cisternae), are specifically targeted to the next compartment (e.g., the medial cisternae), and subsequently fuse with the Golgi membrane. Although the vesicles possess a cytoplasmic protein coat, it is not composed of clathrin.[2-4] Golgi transport vesicle have been purified[4,5] and shown to contain a set of coat proteins, termed COPs (for coat proteins), which include α-COP (160K), β-COP (110K), γ-COP (98K), and δ-COP (61K). β-Coat protein exhibits limited but significant homology to β-adaptin,[6] a protein of similar size on clathrin-coated vesicles. We have purified a protein complex, which we term "coatomer" (short for coat protomer), that contains α-COP, β-COP, γ-COP, δ-COP, and proteins of 36K, 35K, and 20K.[7] Our working hypothesis is that the complex represents the assembly unit, or protomer, of the coat of Golgi COP-coated transport vesicles. We originally identified and purified coatomer because its chromatographic properties were similar to a transport factor required for cis- to medial-Golgi transport.[8] We found that β-COP resides exclusively in coatomer, and this has allowed us to monitor purification of coatomer by Western blotting with a monoclonal antibody against β-COP.[9] Although the original protocol used bovine brain as a starting material we have found that bovine liver is a superior source and thus purification from liver is described here.

[1] J. E. Rothman and L. Orci, *FASEB J.* **4**, 1460 (1990).
[2] G. Griffiths, S. Pfeiffer, K. Simons, and K. Matlin, *J. Cell Biol.* **101**, 949 (1985).
[3] L. Orci, B. S. Glick, and J. E. Rothman, *Cell* **46**, 171 (1986).
[4] V. Malhotra, T. Serafini, L. Orci, J. C. Shepherd, and J. E. Rothman, *Cell* **58**, 329 (1989).
[5] T. Serafini, G. Stenbeck, A. Brecht, F. Lottspeich, L. Orci, J. E. Rothman, and F. T. Wieland, *Nature (London)* **349**, 215 (1991).
[6] R. Duden, G. Griffiths, R. Frank, P. Allgood, and T. E. Kreis, *Cell* **64**, 649 (1991).
[7] M. G. Waters, T. Serafini, and J. E. Rothman, *Nature (London)* **349**, 248 (1991).
[8] M. G. Waters, D. O. Clary, and J. E. Rothman, *J. Cell Biol.* in press (1992).
[9] V. J. Allan and T. E. Kreis, *J. Cell Biol.* **103**, 2229 (1986).

Method

General Procedures

Protein concentrations are determined with the Bio-Rad (Richmond, CA) protein assay kit. All stated pH values are at room temperature. Conductivity is converted to salt concentration by using a standard curve of either 25 mM Tris-HCl, pH 7.4, 1 mM dithiothreitol (DTT), 10% (w/v) glycerol containing from 0 to 1 M KCl or 25 mM Tris-HCl, pH 7.4, 200 mM KCl, 1 mM DTT, 10% (w/v) glycerol containing from 0 to 500 mM potassium phosphate.

Detection of Coatomer by Immunoblotting

Aliquots of column fractions are run on sodium dodecyl sulfate (SDS)-10% (w/v) polyacrylamide minigels and blotted on to 0.45-μm nitrocellulose under standard conditions.[10,11] All manipulations are done at room temperature. The nitrocellulose is blocked in 5% (w/v) nonfat dried milk in phosphate-buffered saline (PBS) with 0.1% (w/v) Tween 20 (M/PBS/T) for 30 min, then incubated in a 1:200 dilution (in M/PBS/T) of tissue culture supernatant from M3A5 hybridoma cells[9] for 1 hr. The blot is washed four times, 5 min each, with M/PBS/T, incubated in a 1:2000 dilution (in M/PBS/T) of affinity-purified goat anti-mouse IgG (heavy and light chain) horseradish peroxidase conjugate (Bio-Rad) for 1 hr, washed as before, rinsed twice quickly with water, and developed with diaminobenzidine with nickel enhancement.[12]

Preparation of Bovine Liver Cytosol. Cow liver is obtained immediately after slaughtering, placed in ice-cold 25 mM Tris-HCl, pH 7.4, 320 mM sucrose, and kept on ice until use. Four hundred and fifty grams of tissue is placed in a glass Waring blender, which is then filled to the top (usually 800–810 ml) with 25 mM Tris-HCl, pH 8.0, 500 mM KCl, 250 mM sucrose, 2 mM ethylene glycol-bis(β-amino ethyl ether)N,N,N',N'-tetraacetic acid (EGTA), 1 mM DTT, 1 mM phenylmethylsulfonyl fluoride (PMSF), 0.5 mM 1:10 phenanthroline, 2 μM pepstatin A, 2 μg/ml aprotinin, and 0.5 μg/ml leupeptin [DTT and protease inhibitors are added just prior to use from the following stocks: 1 M DTT, 500 mM PMSF in dimethyl sulfoxide (DMSO), 500 mM 1:10 phenanthroline in ethanol, 2 mM pepstatin A in methanol, 10 mg/ml aprotinin in 100 mM Tris-HCl, pH 7.8, 2

[10] U. K. Laemmli, *Nature (London)* **227,** 680 (1970).
[11] H. Towbin, T. Staehlin, and J. Gordon, *Proc. Natl. Acad. Sci. USA* **76,** 4350 (1979).
[12] E. Harlow and D. Lane, "Antibodies: A Laboratory Manual." Cold Spring Harbor Lab., Cold Spring Harbor, New York, 1988.

mg/ml leupeptin]. The tissue is homogenized two times for 30 sec, separated by a 2-min cooling period.

The homogenate is centrifuged at 8500 rpm in a Sorvall (Norwalk, CT) GS3 rotor (7456 g_{av}, k factor 2989) for 1 hr at 4°. The supernatants are decanted, pooled, and centrifuged at 39,000 rpm in a Beckman (Palo Alto, CA) 55Ti rotor (119,000 g_{av}, k factor 177) for 2 hr at 4°. The supernatants are decanted, pooled, and dialyzed in Spectra/por 2 dialysis bags (Spectrum Medical Industries, Los Angeles, CA) (molecular weight cut-off 12,000–14,000, 6.4 ml/cm) against 30 liters of 25 mM Tris-HCl, pH 7.4, 50 mM KCl, 1 mM DTT at 4°. After at least 2 hr, dialysis is continued against fresh buffer for 4–12 hr. The dialyzed material is collected and clarified by centrifugation in a Sorvall GS3 rotor at 8500 rpm for 1 hr at 4°. The supernatants are pooled. This material is termed bovine liver cytosol (Fig. 1, lane 1).

Ammonium Sulfate Precipitation. The concentration of the bovine liver cytosol is adjusted to 10 mg/ml by dilution with fresh dialysis buffer. Ethylenediaminetetraacetic acid (EDTA) is added to 1 mM (from a 200 mM NaEDTA, pH 7, stock). The solution is transferred to a 2-liter beaker packed in ice and solid ammonium sulfate is added slowly with stirring to a final concentration of 40% saturation at 0° (add 0.229 g/ml). After dissolution of the salt, the solution is stirred for 30 min at 0° and then centrifuged in a Sorvall GS3 rotor at 8500 rpm for 1 hr at 4°. The supernatants are discarded and the pellets resuspended in 40 ml of 25 mM Tris-HCl, pH 7.4, 1 mM DTT, 10% glycerol with 10 strokes in a 40-ml Dounce (Wheaton, Millville, NJ) homogenizer with the B pestle and then diluted to 400 ml in the same buffer. The conductivity of the solution is measured and then enough resuspension buffer is added so as to bring the final conductivity to the equivalent of 25 mM Tris-HCl, pH 7.4, 100 mM KCl, 1 mM DTT, 10% glycerol. Insoluble material is removed by centrifugation in a Sorvall GS3 rotor at 8500 rpm for 10 min at 4°. The resuspended, clarified material is termed ammonium sulfate precipitate (Fig. 1, lane 2).

DEAE Chromatography. The clarified solution is loaded at 8 ml/min onto an approximately 700-ml DEAE-cellulose (Sigma, St. Louis, MO) column (5-cm i.d.) equilibrated in 25 mM Tris-HCl, pH 7.4, 100 mM KCl, 1 mM DTT, 10% glycerol. The column is washed with 300 ml of equilibration buffer and then eluted with a 0.3 mM/ml KCl gradient (i.e., a 1333-ml gradient from 25 mM Tris-HCl, pH 7.4, 100 mM KCl, 1 mM DTT, 10% glycerol to 25 mM Tris-HCl, pH 7.4, 500 mM KCl, 1 mM DTT, 10% glycerol).

The fractions containing β-COP, as judged by immunoblotting, are pooled and termed DEAE pool (Fig. 1, lane 3). β-Coat protein usually peaks from about 220 to 300 mM KCl.

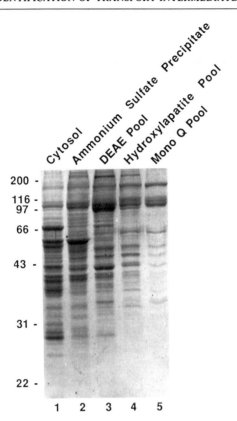

FIG. 1. Protein profiles of fractions through the coatomer purification. Twenty micrograms of bovine liver cytosol (lane 1), 20 μg of ammonium sulfate precipitate (lane 2), 20 μg of DEAE pool (lane 3), 10 μg of hydroxylapatite pool (lane 4), and 5 μg of Mono Q pool (lane 5) were separated on an SDS-10% (w/v) polyacrylamide minigel and visualized with Coomassie blue. Protein standards (Bio-Rad; $M_r \times 10^{-3}$) are indicated on the left-hand side.

Hydroxylapatite Chromatography. The KCl concentration of the DEAE pool is determined by measuring the conductivity, and then adjusting to 25 mM Tris-HCl, pH 7.4, 200 mM KCl, 1 mM DTT, 10% glycerol by addition of 25 mM Tris-HCl, pH 7.4, 1 mM DTT, 10% glycerol. Phosphate is added to 10 mM by addition of a 1/50 volume of 25 mM Tris-HCl, pH 7.4, 200 mM KCl, 1 mM DTT, 10% glycerol, 500 mM potassium phosphate.

The solution is loaded at 1.25 ml/min onto an approximately 140-ml hydroxylapatite (HA Ultrogel from IBF Biotechnics, Savage, MD) column (2.6-cm i.d.) equilibrated in 25 mM Tris-HCl, pH 7.4, 200 mM KCl,

1 mM DTT, 10% glycerol, 10 mM potassium phosphate. The column is washed with 70 ml of the same buffer and then eluted with a 0.5 mM/ml phosphate gradient in 25 mM Tris-HCl, pH 7.4, 200 mM KCl, 1 mM DTT, 10% glycerol.

The peak fractions containing β-COP immunoreactivity are pooled, usually from about 55 to 120 mM potassium phosphate. This material is called hydroxylapatite pool (Fig. 1, lane 4).

Mono Q Chromatography. The hydroxylapatite pool is diluted with 25 mM Tris-HCl, pH 7.4, 1 mM DTT, 10% glycerol to a conductivity equivalent to that of 25 mM Tris-HCl, pH 7.4, 150 mM KCl, 1 mM DTT, 10% glycerol and then centrifuged in a Sorvall GS3 rotor at 8500 rpm for 10 min at 4°. The supernatant is loaded at 3 ml/min onto an 8-ml Mono Q (Pharmacia LKB Biotechnology, Piscataway, NJ) column equilibrated in 25 mM Tris-HCl, pH 7.4, 150 mM KCl, 1 mM DTT, 10% glycerol. The protein is eluted with a 2.2 mM/ml KCl gradient in 25 mM Tris-HCl, pH 7.4, 1 mM DTT, 10% glycerol.

The peak fractions containing β-COP are pooled, usually from about 340 to 380 mM KCl. This material is termed Mono Q pool (Fig. 1, lane 5).

Purification by Gel Filtration. Coatomer can be purified directly from the Mono Q pool by gel filtration. The resulting complex is soluble at physiological salt but is usually slightly contaminated with other proteins, most notably one of 115K. This procedure works well with bovine liver because coatomer is highly enriched in the Mono Q pool. It is not recommended for bovine brain because the Mono Q pool is more heterogeneous.

Concentration on hydroxylapatite: The Mono Q pool is brought to 1 mM potassium phosphate by addition of a 1/1000 volume of 1 M potassium phosphate, pH 7.4. The solution is loaded onto a 1-ml hydroxylapatite (HTP; Bio-Rad) column (0.5-cm i.d.) at 0.2 ml/min and immediately eluted (i.e., no wash step) with 25 mM N-2-hydroxyethylpiperazine-N'-2-ethanesulfonic acid (HEPES)–KOH, pH 7.4, 200 mM KCl, 1 mM DTT, 10% glycerol, 200 mM potassium phosphate. Fractions (0.5 ml) of the eluate are collected and a protein assay performed. The peak fractions (usually two or three fractions) are pooled.

Superose 6 chromatography: The concentrated Mono Q pool is centrifuged for 10 min at 4° in a microfuge to remove any aggregates and the supernatant is sieved through onto a 100-ml preparative-grade Superose 6 (Pharmacia) column (1.6-cm i.d.) equilibrated in 25 mM HEPES–KOH, pH 7.4, 200 mM KCl, 10% glycerol at 0.3 ml/min. Fractions (2 ml) are collected and aliquots analyzed by electrophoresis and Coomassie blue staining (Fig. 2). The fractions containing coatomer, a complex of 160K, 110K, 98K, 61K, 36K, 35K, and 20K proteins (Fig. 2, fractions 16–20), are pooled. Coatomer gel filters with a Stokes radius of about 10 nm.

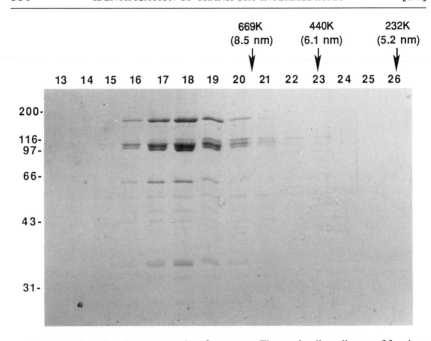

FIG. 2. Gel-filtration chromatography of coatomer. Three-microliter aliquots of fractions from a preparative-grade Superose 6 column were separated on an SDS-10% (w/v) polyacrylamide minigel and visualized with Coomassie blue. Protein standards for gel filtration (Pharmacia) were bovine thyroid thyroglobulin (669 K, 8.5-nm Stokes radius), horse spleen ferritin (440 K, 6.1-nm Stokes radius), and bovine liver catalase (232 K, 5.2-nm Stokes radius). SDS-PAGE protein standards (Bio-Rad; $\times 10^{-3}$) are indicated on the left-hand side.

Purification by Isoelectric Precipitation. Coatomer can also be purified by isoelectric precipitation followed by gel filtration.[7] This procedure has the advantage that the coatomer obtained is highly purified. However, the low-pH treatment changes the solubility of the complex such that it must be maintained in high (1 M) salt.

Isoelectric precipitation: The Mono Q pool is transferred to Spectra/por 2 dialysis bags and dialyzed extensively against 25 mM NaMES [2-(N-morpholino)ethanesulfonic acid, sodium salt], pH 5.8, 1 mM DTT, 10% glycerol at 4°. The solution is centrifuged in a Sorvall SS34 rotor at 10,000 rpm (7796 g_{av}, K factor 3007) for 15 min at 4°. The supernatant is aspirated and the pellet washed with 10 ml dialysis buffer (buffer is added, the tube vortexed, spun again as above, and the supernatant aspirated again). The pellet is resuspended in 1 ml of 25 mM Tris-HCl, pH 8.0, 1 M KCl, 1 mM DTT, 10% glycerol.

Superose 6 chromatography: The resuspended isoelectric precipitate

TABLE I
PURIFICATION OF COATOMER

Material	Total protein (mg)	Total coatomer (mg)	Yield (%)
Bovine liver cytosol	12,745	7.5	100
Ammonium sulfate precipitate	3,061	5.4	72
DEAE pool	237	6.2	83
Hydroxylapatite pool	57	4.1	55
Mono Q pool	8.9	3.6	48
Superose 6 pool	4.5	3.0	40

is centrifuged for 10 min in a microfuge to remove any insoluble material and the supernatant is sieved through onto a 100-ml preparative-grade Superose 6 (Pharmacia) column (1.6-cm i.d.) equilibrated in 25 mM Tris-HCl, pH 8.0, 1 M KCl, 1 mM DTT, 10% glycerol at 0.3 ml/min. Fractions (2 ml) are collected and aliquots analyzed by electrophoresis and Coomassie blue staining. The fractions containing coatomer, a complex of 160K, 110K, 98K, 61K, 36K, 35K, and 20K proteins, are pooled. Coatomer gel filters with a Stokes radius of about 10 nm.

Quantitation of the Purification. Quantitative immunoblotting is used to determine the yield of the coatomer during its purification (Table I). Sodium dodecyl sulfate-gel electrophoresis, Western blotting, and incubations with the primary (M3A5 supernatant) and secondary antibodies are carried out as described above. For detection, however, the ECL Western blotting detection system (Amersham, Arlington Heights, IL) is used. After incubation with the ECL reagent the blots are wrapped in Saran wrap and exposed to X-AR5 film (Kodak, Rochester, NY). The developed film is overlaid onto the blot, the regions containing β-COP are excised, and light production in each of these is determined in a Beckman LS6000 liquid scintillation system equipped with a single photon monitor.

Standard curves are obtained using coatomer purified using isoelectric precipitation and Superose 6 chromatography as described above and are prepared for each quantitation experiment. In general, the amount of light emitted is found to be linearly proportional to the amount of coatomer over a range of 10 to 400 ng.

Acknowledgments

The authors thank Thomas Kreis (European Molecular Biology Laboratory) for generously providing the anti-β-COP monoclonal antibody (M3A5)-producing hybridoma.

[32] Purification of Yeast Sec23 Protein by Complementation of Mutant Cell Lysates Deficient in Endoplasmic Reticulum-to-Golgi Transport

By Linda Hicke, Tohru Yoshihisa, and Randy W. Schekman

Temperature-sensitive mutations in *SEC* and *BET* genes have been used to genetically define many of the proteins required for transport from the endoplasmic reticulum (ER) to the Golgi in yeast.[1,2] To elicit the precise function of these proteins and of as yet unidentified factors, we wish to reconstitute ER-to-Golgi transport with purified components. As a first step toward this goal, ER-to-Golgi transport has been reconstituted in semiintact yeast cells and with microsomal membrane fractions. This transport requires cytosol, Ca^{2+}, and ATP.[3-5] More specifically, ER-to-Golgi transport *in vitro* has been shown to depend on the *SEC12, SEC18, SEC23, YPT1,* and *SAR1* gene products.[3-8] In this chapter we describe the purification to homogeneity of functional Sec23 protein using the ER-to-Golgi transport assay to monitor its activity.

The *SEC23* gene product is an 84K protein that is peripherally associated with intracellular membrane and can be converted into a soluble form by a number of treatments, including vigorous mechanical agitation during yeast cell lysis.[9] Sec23p appears to function in ER-to-Golgi transport during budding of a vesicular intermediate as transport vesicles do not form in *sec23* yeast *in vivo* or *in vitro*.[10] Transport in lysates prepared from *sec23* mutant cells is temperature sensitive and can be restored at the nonpermissive temperature by wild-type cytosol fractions that contain Sec23p.[3] This assay has been used to purify active Sec23p and an associated 105K protein from cells carrying a single chromosomal copy of

[1] P. Novick, C. Field, and R. Schekman, *Cell* **21**, 205 (1980).
[2] A. P. Newman and S. Ferro-Novick, *J. Cell Biol.* **105**, 1587 (1987).
[3] D. Baker, L. Hicke, M. Rexach, M. Schleyer, and R. Schekman, *Cell* **54**, 335 (1988).
[4] H. Ruohola, A. K. Kabcenell, and S. Ferro-Novick, *J. Cell Biol.* **107**, 1465 (1988).
[5] D. Baker, L. Wuestehube, R. Schekman, D. Botstein, and N. Segev, *Proc. Natl. Acad. Sci. USA* **87**, 355 (1989).
[6] R. A. Bacon, A. Salminen, H. Ruohola, P. Novick, and S. Ferro-Novick, *J. Cell Biol.* **109**, 1015 (1989).
[7] M. F. Rexach and R. W. Schekman, *J. Cell Biol.* **114**, 219 (1991).
[8] C. d'Enfert, L. J. Wuestehube, T. Lila, and R. Schekman, *J. Cell Biol.* **114**, 663 (1991).
[9] L. Hicke and R. Schekman, *EMBO J.* **8**, 1677 (1989).
[10] C. Kaiser and R. Schekman, *Cell* **61**, 723 (1990).
[11] L. Hicke, T. Yoshihisa, and R. Schekman, *Mol. Biol. Cell* **3**, 667 (1992).

SEC23 and from cells overproducing Sec23p from multiple plasmid-borne copies of the gene.[12]

Materials

Propagation of Yeast Strains

YPD growth medium:
> 2% (w/v) Bacto-peptone (Difco Laboratories, Inc., Detroit, MI), 1% (w/v) Yeast extract (Difco Laboratories), 2% (w/v) glucose

Casamino acids growth medium: 6.7 g/liter yeast nitrogen base without amino acids (Difco Laboratories), 1% (w/v) vitamin assay casamino acids (Difco Laboratories), 2% (w/v) glucose, 0.01% (w/v) adenine, 0.01% (w/v) methionine, 0.01% (w/v) histidine, 0.01% (w/v) tryptophan, 0.002% (w/v) uracil (for propagation of cells that do not carry a plasmid)

pCF23-2μm (multicopy) yeast plasmid carrying the wild-type *SEC23* gene

ER-to-Golgi Transport Assay

sec23 semiintact yeast cells (strain RSY281: *sec23-1 ura3-52 his4-619 MATα*)

sec23 yeast cytosol (RSY281)

Wild-type yeast cytosol (strain RSY255: *ura3-52 leu2-3, 112 MATα*)

[35]S-Labeled prepro-α-factor (translated and radiolabeled *in vitro*)

α-1,6-Mannose antiserum

Protein A-Sepharose CL-4B (Pharmacia, Piscataway, NJ)

Buffer 88:
> 0.15 *M* Potassium acetate, 0.25 *M* sorbitol, 50 m*M* magnesium acetate, 20 m*M* *N*-2-hydroxyethylpiperazine-*N'*-2-ethanesulfonic acid (HEPES)–KOH, pH 6.8

Purification of Sec23 Protein

Glass beads, 0.5 mm in diameter (BioSpec Products, Bartlesville, OK)

Bead-Beater (BioSpec Products)

DEAE Sepharose Fast Flow (Pharmacia)

S-Sepharose Fast Flow (Pharmacia)

Sephacryl S-300 HR (Pharmacia)

---

[12] D. Baker and R. Schekman, *Methods Cell Biol.* **31,** 127 (1989).

Protein Assay

Bio-Rad protein assay reagent (Bio-Rad Laboratories, Richmond, CA)
Protein standard–γ-globulin (Sigma Chemical Company, St. Louis, MO)

Assay for Endoplasmic Reticulum-to-Golgi Transport in Semiintact sec23 Yeast

Preparation of sec23 Membranes and sec23 Cytosol. To observe temperature-dependent ER-to-Golgi transport in *sec23* mutant lysates, both membranes and cytosol are prepared from *sec23* cells. Components are isolated from temperature-sensitive *sec23* yeast grown at the permissive temperature (24°), as preincubation of the cells at the nonpermissive temperature (37°) before lysis causes irreversible inactivation of both membranes and cytosol. The transport-competent membrane fraction is contributed by a semiintact yeast cell preparation in which cells are gently lysed by the slow freezing of yeast spheroplasts over liquid nitrogen vapors. As the peripheral association of Sec23p with intracellular membrane is not greatly disrupted by this lysis technique, a large amount of Sec23 protein remains associated with the membrane fraction used in this assay. We believe this is true for both the wild-type and mutant forms of Sec23p as the *sec23-1* mutation does not quantitatively alter the association of the protein with membrane (L. Hicke, unpublished observations, 1989). Wild-type and *sec23* cytosolic fractions are prepared by vigorously vortexing whole cells in buffer 88 with glass beads (described in detail in Ref. 12). Under these relatively harsh lysis conditions a large fraction of Sec23p is released into a soluble form and, on centrifugation of the lysate at 100,000 g for 60 min at 4°, is found in the cytosolic fraction. The protein concentration of this cytosol is typically 15–20 mg/ml.

Preparation of semiintact wild-type yeast has been described in detail elsewhere.[12] The protocol for preparing the membranes from *sec23* cells is similar except that all incubations are performed at 24°. *sec23* cells are grown in YPD medium at 24° until they reach an OD_{600} of 2–4. Cells are then harvested and converted to spheroplasts with lyticase at 24°. The extent of cell wall digestion is monitored by measuring the decrease in OD_{600} of a 1:100 dilution of the cell suspension in distilled H_2O, which should drop to 10–15% of its original value after 15 min of incubation. However, in many preparations (~50%) from *sec23* cells the OD_{600} decreases to ~25% of the initial value and then remains resistant to further spheroplast formation even on addition of more lyticase. This correlates with inability of the resulting membranes to translocate prepro-α-factor and therefore these cells are discarded. The reason for this behavior is not understood. Such variability is rare when preparing membranes from wild-type cells, nor is it related to the density at which cells were harvested or to variation in the conditions of spheroplast formation.

Spheroplasts are regenerated at 24° after removing the lyticase-containing buffer, washed, and transferred to a freezing buffer containing low osmotic support. Aliquots (200 μl) of cells in freezing buffer are transferred to 1.5-ml microcentrifuge tubes that are suspended ~ 10 cm above liquid nitrogen contained in an ice bucket. Spheroplasts are allowed to freeze slowly in the liquid nitrogen vapor during a 40-min period and then stored at − 80° for up to 2 months.

To generate semiintact cells for use in a transport assay, a 200-μl aliquot of spheroplasts is thawed quickly in a 24° water bath, washed 3 times in buffer 88 with brief centrifugation in a microcentrifuge (~ 10 sec), and resuspended in buffer 88 to a volume appropriate for 12–16 transport reactions.

Temperature-Sensitive Endoplasmic Reticulum-to-Golgi Transport in sec23 Mutant Lysates. The *in vitro* reaction that reproduces ER-to-Golgi transport with *sec23* components is shown schematically in Fig. 1. [^{35}S]Methionine-labeled prepro-α-factor is introduced into thawed, semiintact *sec23* cells in a posttranslational translocation reaction incubated for 15 min at 10°, a temperature that does not allow ER-to-Golgi transport.[12] Translocation of prepro-α-factor into *sec23* membranes is as efficient as translocation into wild-type membranes, indicating that there is not a nonspecific defect in the mutant membranes. After the translocation reaction is completed the membranes are washed and mixed with ATP and with cytosol prepared either from *sec23* cells or from *sec23* cells containing a centromere plasmid bearing the wild-type *SEC23* gene. In a typical experiment shown in Fig. 2, aliquots of each reaction were incubated at 15, 25, or 30° for 25–45 min. Reactions that contained *SEC23* gene product proceeded efficiently at all temperatures. The reactions supplemented with *sec23* cytosol were markedly temperature sensitive, as transport at 30° was reduced fivefold relative to transport at 15°. Higher temperatuees were not tested because the efficiency of the wild-type reaction decreased rapidly above 30°. *sec23* cytosol did not inhibit reactions containing the *SEC23* gene product. In addition, reactions containing membranes from wild-type cells incubated with *sec23* cytosol showed no defect in transport, probably because the membrane fraction contains a significant amount of wild-type Sec23 protein.

Complementation of sec23 Mutant Components with Wild-Type Sec23 Protein. To establish the *in vitro* reaction as a reliable assay for identifying Sec23p activity during purification of the protein, we have shown that

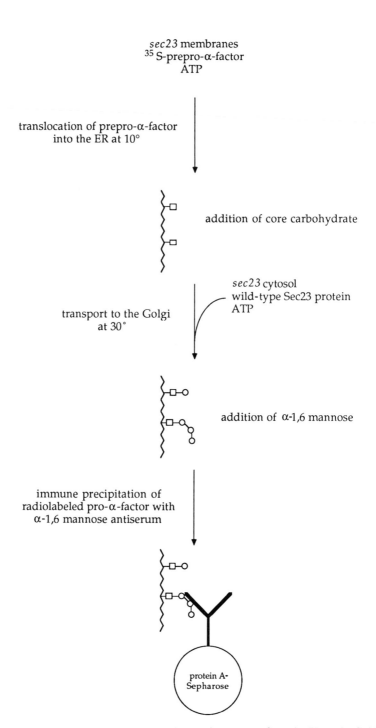

FIG. 1. *In vitro* assay for Sec23p activity in protein transport from the ER to the Golgi.

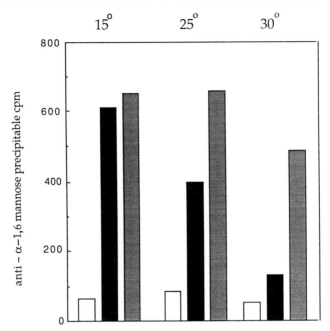

FIG. 2. *In vitro* transport with mutant *sec23* components is temperature sensitive. *sec23* membranes were incubated with ^{35}S-labeled prepro-α-factor under conditions that allow translocation but not transport. Washed membranes containing pro-α-factor were then mixed with buffer alone (open bars), with cytosol from *sec23* cells (black bars), or with cytosol from *sec23* cells carrying the wild-type *SEC23* gene on a plasmid (shaded bars). Aliquots of each mixture were incubated at 15, 25, or 30°, the reactions were stopped by the addition of SDS, and each sample was immune precipitated with α1,6-mannose antiserum. The amount of radioactive protein present in each immune precipitation was determined by scintillation counting and is represented in counts per minute (cpm) on the vertical axis of this histogram.

wild-type Sec23p, separated from the bulk of cytosolic protein, overcomes the temperature-sensitive defect in *sec23* cytosol. In a typical experiment shown in Fig. 3, cytosol was prepared from wild-type cells lysed by vigorous homogenization with glass beads.[10] A 1.5-ml aliquot of cytosol (31 mg protein) in buffer 88 was loaded onto a 75 ml (1.5 × 46 cm) Sephacryl S-300 gel-filtration column equilibrated in buffer 88. The column was eluted in the same buffer at ~ 15 ml/hr. Fractions (1.5 ml) were collected and total protein in each fraction was determined using the Bio-Rad protein assay. Column fractions were analyzed by immunoblot with Sec23p antiserum and separate aliquots were concentrated approximately 10-fold in a Centricon 30 microconcentrator (Amicon, Lexington, MA) and assayed for the ability to restore transport at 30° in lysates that

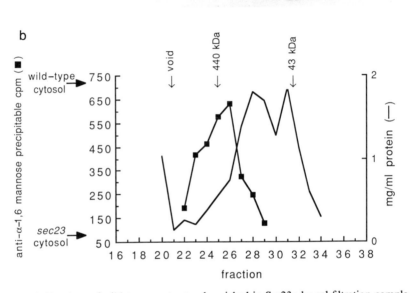

FIG. 3. Fractions of wild-type yeast cytosol enriched in Sec23p by gel filtration complement *sec23* mutant lysates in an *in vitro* assay for ER-to-Golgi transport. Cytosol was
prepared from wild-type yeast and loaded onto a Sephacryl S-300 gel-filtration column.
Fractions were collected, assayed for total protein, and immunoblotted with Sec23p antiserum. Eight fractions that included the peak of Sec23 protein were each concentrated
approximately 10-fold and tested for the ability to restore ER-to-Golgi transport in *sec23*
lysates at 30°. (a) Immunoblot of fractions 20–32 with Sec23p antiserum. (b) Sec23p activity
and total protein content of fractions. Sec23p activity in fractions 21–29 was measured by the
addition of each fraction to an *in vitro* transport reaction containing *sec23* membranes and
sec23 cytosol. Transport in the presence of each fraction is measured by the amount of
anti-α-1,6-mannose-precipitable counts per minute generated in the reaction. The horizontal
arrows indicate the levels of transport that occurred in *sec23* membranes in the presence of
sec23 cytosol alone or in *sec23* membranes mixed with wild-type cytosol. Fractions have no
transport activity above background if *sec23* cytosol is omitted from the reaction mixture.
Vertical arrows denote the elution of standards used to calibrate the Sephacryl S-300 column.
The void volume was determined by the elution of blue dextran.

contained *sec23* membranes and *sec23* cytosol. Figure 3a shows that fraction 26 contained the peak of Sec23p immunoreactive material. The same fraction contained the peak of Sec23p complementing activity (Fig. 3b). The peak was one fraction displaced from a marker protein, ferritin (440 kDa), suggesting Sec23p was part of a large complex or oligomer.

Complementation activity required both *sec23* cytosol and an aliquot of the active column fractions. No activity was detected with column fractions mixed alone with *sec23* membranes. Thus, wild-type Sec23p is the soluble molecule responsible for restoring transport, at the nonpermissive temperature, to a reaction containing *sec23* membranes and *sec23* cytosol.

Purification of Functional Sec23 Protein

Assay for Sec23p Activity and Definition of Activity Units. Sec23p activity is determined with the following assay (see Fig. 1). Purification sample (1–25 μl) is added to 100–200 μg *sec23* cytosol on ice. *sec23* membranes carrying translocated, ^{35}S-labeled pro-α-factor are added to the cytosol–Sec23p mixture and the samples transferred to 30° for 45 min. Reactions are stopped by the addition of sodium dodecyl sulfate (SDS) and transport is quantified by determining the amount of radiolabeled protein precipitated with α-1,6-mannose antiserum. Reactions containing *sec23* cytosol with no added Sec23p and reactions containing cytosol prepared from wild-type cells are carried out at 30° in every set of assays. A measurement of the background in each assay is obtained by preparing a complete reaction on ice and adding SDS to 1% (w/v) at the beginning of the incubation period.

Glycerol concentrations above 1% (v/v) in the *in vitro* reactions inhibit transport, therefore glycerol is not included in buffers during the purification. However, pure samples of Sec23p are routinely stored in 10% (v/v) glycerol at −85° until further use.

One unit of Sec23p activity is defined as 10% of the difference in counts per minute precipitable with α-1,6 mannose antiserum between reactions conducted at 30° containing only mutant Sec23p and those containing a saturating amount of functional, wild-type Sec23p.

Purification Scheme. Sec23p exhibits unusual binding to ion-exchange materials that facilitated its purification. The protein binds both a cation-exchange matrix, S-Sepharose, in 50 m*M* potassium acetate at neutral pH, and the weak anion-exchange matrix DEAE-Sepharose, in 50 m*M* potassium acetate at neutral pH. These ion-exchange and gel-filtration materials are employed in the purification scheme developed and shown in Fig. 4.

Purification of Sec23p from Cells Overproducing the Protein. Cytosol

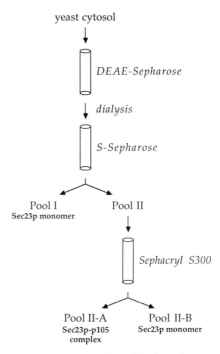

FIG. 4. Sec23 protein purification scheme.

prepared from yeast bearing the *SEC23* gene on a multicopy plasmid contains significantly more Sec23p activity than an equal amount of cytosol prepared from wild-type cells, therefore a soluble fraction of Sec23p-overproducing yeast is used as the starting source for purification. Sec23p purified from these cells exists in two active forms, as a monomer and as a large multimer with a 105K polypeptide (p105).

In a typical preparation, 12 liters of a wild-type yeast strain transformed with the multicopy *SEC23* plasmid, pCF23, was grown to mid-logarithmic phase (OD_{600} = 4–6) in casamino acids medium lacking uracil in a 16-liter SF116 fermenter (New Brunswick, Edison, NJ). Cells were harvested in a Sharples centrifuge (Pennwalt, Warminster, PA) and purification was initiated by lysing the yeast cells by vigorous homogenization with glass beads. Cells (35–45 g wet weight) were washed once in cold distilled H_2O and once in cold buffer 88, then resuspended in 250 ml buffer 88, 0.5 mM phenylmethylsulfonyl fluoride (PMSF), 1 mM dithiothreitol (DTT) and transferred to a 360-ml bead beater (Bio-Spec Products, Bartlesville, OK) half-filled with acid-washed glass beads (0.5 mm in diameter). Cells were lysed in the bead beater at 4° with six to eight 1-min bursts. Lysis efficiency was examined with a light microscope and was always greater than 50%. The lysate was removed from the glass beads with a pipette, the beads were rinsed with an equal volume of buffer 88, and the rinse combined with the lysate. Unbroken cells and rapidly sedimenting membranes were removed by centrifugation at 4000 g in a GSA rotor (Sorvall/DuPont, Wilmington, DE) for 10 min at 4°, and the supernatant fraction was further centrifuged at 100,000 g in a Beckman (Palo Alto, CA) Ti45 rotor for 1 hr at 4° to remove all remaining membranes. Cytosol was split into three equal volumes and either used immediately or quick-frozen in liquid N_2 and stored at $-85°$ until needed.

Purification was initiated with one-third the volume of cytosol obtained from the preparation described above (typically 0.5–1 g cytosolic protein). Cytosol was adjusted to 0.5 M potassium acetate with 5 M potassium acetate and the conductivity measured. Cytosol in 0.5 M potassium acetate was applied at 55 ml/hr to a 45-ml DEAE-Sepharose column (2.5 × 9.5 cm) equilibrated in buffer 88, 0.5 M potassium acetate. The column was then washed with 1.5 column volumes buffer 88, 0.5 M potassium acetate and eluted with a 180-ml linear gradient of 0.5–0.75 M potassium acetate in buffer 88. Fractions (9 ml each) were assayed for Sec23p activity, total protein, and conductivity. In assays for Sec23p activity the high concentration of salt in samples eluted from the DEAE-Sepharose medium was compensated for by the addition of buffer 88 without potassium acetate so that the final concentration of potassium acetate was 0.15 M in each reaction. Recovery of Sec23p activity from this column was typically 50–75%.

Fractions containing the peak of Sec23p activity were pooled and dialyzed overnight at 4° against buffer 88, 50 mM potassium acetate until the conductivity reached that of the buffer. Dialysis at this step not only reduced the salt concentration, but also resulted in a twofold increase in activity, perhaps due to the removal of a low molecular weight inhibitor of the transport reaction. After dialysis of the protein into buffer containing 50 mM potassium acetate, the sample was applied to a 5-ml S-Sepharose Fast Flow (1.0 × 6.0 cm) column equilibrated in the same buffer. Five-milliliter fractions were collected during the loading of the column. The column was washed with 6 ml buffer 88, 50 mM potassium acetate and protein was released with a 25-ml linear gradient of 50 mM–0.5 M potassium acetate in buffer 88. One-milliliter fractions were collected during the wash and elution steps. All flow rates were ~100 ml/hr to minimize the time Sec23p was associated with the column medium. Activity, total protein, and Sec23p antigen cross-reactivity were determined in all fractions. Sec23p activity eluted in two peaks that were pooled separately. The separation of these peaks was reproducible in several preparations, although occasionally the first peak appeared as a shoulder of the second. In a number of preparations, the loss of activity on S-Sepharose was significant. High flow rates reduced binding of Sec23p activity to S-Sepharose; however, slower flow rates that allowed complete binding of Sec23p to the column caused unacceptable loss (> 90%) of activity. Sec23p was stable at the salt concentrations used in this step. It is possible that another protein required for full Sec23p activity is separated from Sec23p during this chromatograph step; however, addition of S-Sepharose flow-through fractions to eluted fractions containing Sec23p does not stimulate the activity. Alternatively, loss of activity on the cation-exchange medium may be due to a disruptive interaction between the charged beads and the Sec23 protein.

The first peak of activity (pool I) collected from S-Sepharose contained ~ 11% of the initial soluble Sec23p activity (Table I) and consisted of a single 84K polypeptide (Fig. 5, lane 4) that cross-reacted with Sec23p antiserum.[11] In contrast, the second peak (pool II) contained Sec23p and a mixture of several other proteins (Fig. 5, lane 5). Pool I and pool II were each independently fractionated by gel filtration on a 350-ml Sephacryl S-300 column (2.5 × 70 cm) equilibrated in buffer 88. Fractions of 5.5 ml were collected at a flow rate of 50 ml/hr. Fractions collected from Sephacryl S-300 were assayed for Sec23p activity, and aliquots were evaluated for protein content by SDS-polyacrylamide gel electrophoresis (PAGE) developed with silver stain.

TABLE I
PURIFICATION OF Sec23p ACTIVITY

Fraction	Total protein (mg)	Activity (units/ml)	Total activity (units)	Specific activity (units/mg)	Activity yield (%)	Purification (fold)
Cytosol	1,000	2,710 ± 390	87,800	88	100	1
DEAE-Sepharose before dialysis	37.5	878 ± 13	55,300	1,480	63.0	17
DEAE-Sepharose after dialysis	33.8	1,380 ± 140	103,500	3,060	118	35
S-Sepharose: pool I	0.53	1,730 ± 160	9,510	17,800	10.8	202
S-Sepharose: pool II	3.3	2,250 ± 330	17,300	5,260	19.7	60
Sephacryl S-300: pool II-A	1.4	283	9,340	6,740	10.6	76
Sephacryl S-300: pool II-B	1.3	97	3,200	2,490	3.6	28

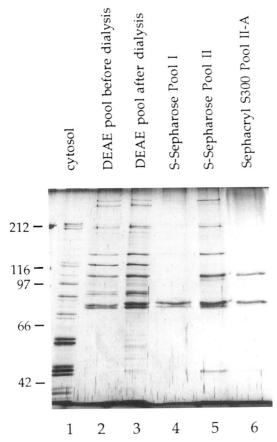

Fig. 5. Protein composition of fractions generated during Sec23 protein purification. Samples of protein from each step of a Sec23p preparation from yeast overproducing Sec23p were diluted into Laemmli sample buffer and electrophoresed on a 7.5% (w/v) polyacrylamide gel. Approximately 1.5 mg total protein was loaded in each lane. Lane 1, total cytosolic protein; lane 2, Sec23 activity pooled from DEAE-Sepharose, before dialysis; lane 3, DEAE-Sepharose pool, after dialysis; lane 4, S-Sepharose pool I; lane 5, S-Sepharose pool II; lane 6, Sephacryl S-300 pool II-A (Sec23p activity that eluted as a protein of 260K on gel filtration of pool II). Markers in left-hand side; M_r ($\times 10^{-3}$).

The transport activity recovered from gel filtration of pool I on Sephacryl S-300 eluted as expected for an ~70K protein and coincided with a single 84K polypeptide observed on a 7.5% SDS gel stained with silver. Staining of the same fractions electrophoresed on a 12.5% polyacrylamide gel indicated they contained no low molecular weight components that may migrate at the dye front of a 7.5% gel.[11] This behavior was consistent with that expected for monomeric Sec23p. In contrast, the Sec23p activity contained in pool II separated into two peaks on gel filtration, one that migrated as a 200K–300K species (pool II-A) and another, containing less activity, that behaved as a small protein of 50K–100K (pool II-B).[11]

Although a large fraction of purified, overexpressed Sec23p activity existed as a monomer, a significant amount behaved as a much larger form. To determine whether the large form of Sec23p was a homomultimer or whether Sec23p was associated with other polypeptides we examined the protein content of pools II-A and II-B by silver staining. Coincident with the two peaks of transport activity, two peaks of an 84K polypeptide that cross-reacted with Sec23p antiserum were observed in gel-filtered fractions of pool II. The fractions that correspond to activity collected in pool II-B contained only Sec23p that migrated as a monomer (50K–100K). The Sec23p in fractions containing pool II-A activity behaved as a 260K protein on the column and copurified precisely with a 105K polypeptide present in stoichiometric amounts (Fig. 5, lane 6). Although the amount of silver-stained 84K protein in pool II-A fractions was clearly less than that in pool II-B fractions, pool II-A contained approximately threefold more transport activity (Table I).

The purification data presented in Table I indicates that 10% of the Sec23p activity from yeast overproducing Sec23p can be obtained in two chromatographic steps as pure Sec23p monomer with high specific activity (pool I). A further 10% of the activity was recovered after a further gel-filtration step as Sec23p in association with a 105K protein (pool II-A). Although a substantial amount of additional monomeric Sec23p was resolved from the Sec23p multimer on the gel-filtration column, it had low specific activity. The dilution of the protein to the low concentrations that occurred during this step may have caused a loss of activity of one or both forms of Sec23p. Alternatively, there may have been heterogeneity in the monomeric Sec23p protein that was reflected in its affinity for S-Sepharose. The protein that bound more tightly to S-Sepharose and was eluted in the second peak in higher salt may have been inherently less active.

The silver-stained gel presented in Fig. 5 shows the major proteins present at each step of the purification. Approximately equivalent amounts of total protein were loaded in each lane. The 84K band that is Sec23p is not observed in this staining of crude cytosol (Fig. 5, lane 1), yet it is already present as one of the major components after one purification step (Fig. 5, lanes 2 and 3). Pure Sec23p monomer is obtained after additional fractionation over S-Sepharose (pool I; Fig. 5, lane 4), in addition to Sec23p–p105 complex contaminated with several other polypeptides (pool II; Fig. 5, lane 5). Gel filtration of pool II results in pure Sec23p complex containing only the Sec23p and p105 polypeptides. The polypeptide that runs with slightly higher electrophoretic mobility than Sec23p in these samples is thought to be a proteolytic breakdown product of Sec23p as it cross-reacts with Sec23p antiserum and its presence is variable from preparation to preparation.

Purification of Sec23p from Cells Containing One SEC23 Gene. A modified purification procedure has been developed to optimize the isolation of Sec23p–p105 complex from untransformed cells. *pep4,* a strain carrying a mutation in the vacuolar protease A structural gene, is used to minimize proteolysis during the purification. In a typical preparation the strain was propagated in 12 liters of YPD at 30° to $OD_{600} = 5-7$, harvested, and washed as described above. The cell pellet (40–70 g wet weight) was resuspended in 175 ml buffer 88, 1 mM DTT, 0.5 mM PMSF, 0.5 μg/ml leupeptin, and 0.7 μg/ml pepstatin A, lysed in a bead beater (Bio-Spec), and the cytosolic and membrane fractions separated. To recover Sec23p still associated with membranes, the pellet fraction was mixed with 100 ml buffer 88, 0.75 M potassium acetate, 1 mM DTT, 0.5 mM PMSF, 0.5 μg/ml leupeptin, 0.7 μg/ml pepstatin A and resuspended with three strokes in a Potter–Elvehjem homogenizer. Another 100 ml of the same buffer was added, the suspension stirred for 30 min at 4°, and centrifuged at 100,000 g for 60 min at 4°. The supernatant fraction from this salt extraction was combined with the cytosol fraction and the entire mixture was adjusted to 0.5 M potassium acetate and used for purification.

Isolation of Sec23p from untransformed cells was performed on an approximately threefold larger scale than from the overproducer strain. Cytosolic protein (8.2 g) was loaded onto a DEAE-Sepharose column (200 ml bed volume, 5 × 10 cm) and bound proteins were eluted with a 900-ml 0.5–0.85 M potassium acetate linear gradient in buffer 88. After dialysis, protein was applied to a 16-ml S-Sepharose column (1.5 × 9.0 cm) and eluted with a 90-ml 50–0.5 mM potassium acetate gradient in buffer 8. The single peak of Sec23p activity was pooled and concentrated by adsorbtion to and elution from a 1.5-ml bed volume DEAE-Sepharose column equilibrated in buffer 88. The column was washed with 6 vol of buffer 88 and eluted with buffer 88 containing 0.75 M potassium acetate. Fractions containing proteins eluted in 0.75 M potassium acetate were pooled and subjected to gel filtration on 150 ml Sephacryl S-300 (1.5 × 84 cm). A single peak of activity that behaved as a protein of ~260K eluted from this column and contained two polypeptides, Sec23p and p105.[11] Approximately 2–3 mg Sec23p complex was obtained from this preparation.

Acknowledgments

L.H. was supported by a predoctoral fellowship from the National Science Foundation, and T.Y. by the Howard Hughes Medical Research Foundation. The work was supported by grants from the National Institutes of Health and the Howard Hughes Medical Research Foundation.

[33] Purification of Sec4 Protein from *Saccharomyces Cerevisiae* and *Escherichia coli*

By Peter Novick, Michelle D. Garrett, Patrick Brennwald, and Alisa K. Kabcenell

Introduction

Genetic analysis of protein export in the yeast *Saccharomyces cerevisiae* has identified a large number of genes whose protein products play essential roles in the many steps of transport of proteins from their site of synthesis on the endoplasmic reticulum to their ultimate release at the cell surface. In some cases, analysis of the sequence of the gene has revealed important clues to the possible mechanism of function of the gene product. One of the first such examples is the *SEC4* gene, which is required at the final stage of transport to the cell surface.[1] The sequence of *SEC4* revealed that the encoded protein is a member of the *ras* superfamily, defined by the presence of conserved domains that together constitute a high-affinity guanine nucleotide-binding site.[2,3] While all of the members of this superfamily share at least 30% sequence identify with each other, a subfamily, known commonly as the Rab proteins, share more extensive sequence identity with Sec4p.[3] Various lines of evidence suggest that members of the Rab protein family play roles analogous to Sec4p, regulating the many different vesicular transport events in eukaryotic cells.[4]

Sec4p undergoes a cycle of GTP binding and hydrolysis that may be coupled to a cycle of subcellular localization in which Sec4p first attaches to post-Golgi vesicles that go on to fuse with the plasma membrane. Sec4p then recycles through a soluble pool to reassociate with a new round of vesicles. Membrane attachment of Sec4p, Ypt1p, and the many Rab ho-

[1] P. Novick, C. Field, and R. Schekman, *Cell* **21**, 205 (1980).
[2] A. Salminen and P. Novick, *Cell* **49**, 527 (1987).
[3] A. Valencia, P. Chardin, A. Wittinghofer, and C. Sander, *Biochemistry* **30**, 4637 (1991).
[4] W. E. Balch, *Trends Biochem. Sci.* **15**, 473 (1990).

mologs requires prior modification of carboxy-terminal cysteine moieties by the addition of isoprenyl lipids. One approach toward understanding the role of Sec4p and its many homologs is to pursue the accessory proteins that function in conjunction with them. These include proteins that stimulate the hydrolysis of bound GTP,[5,6] proteins that stimulate the release of bound GDP allowing exchange for GTP, proteins that modify the carboxy terminus to facilitate membrane association,[7] and proteins that allow dissociation from membranes for recycling.[8] To assay these accessory proteins it is useful to have available large amounts of pure protein. We present procedures here for producing and purifying large amounts of native Sec4p from either yeast or bacteria. The purification of Sec4p from yeast has been previously described[9] and the purification of Sec4 from bacteria has been adapted from the yeast protocol. Because bacteria lack prenylation enzymes, bacterially produced Sec4p is a convenient substrate for modification studies.

Expression of Sec4p in *Saccharomyces cerevisiae* and Preparation of a Cell Lysate

To produce large amounts of Sec4 protein in yeast, the gene is expressed under the control of the *GAL1* promoter. This promoter is strongly transcribed in the presence of galactose, but in the absence of the inducer expression is greatly reduced.[10] This is important because high levels of overexpression of Sec4p are detrimental to yeast and will ultimately arrest growth. The *SEC4* gene was placed under control of the *GAL1* promoter by first trimming the upstream noncoding region of *SEC4* to only 8 base pairs (bp) and then ligating it into the *Bam*HI site of the expression vector pNB187. This construction has been described in detail.[11] Because this vector is maintained at only single copy number in yeast, the *GAL1-SEC4* construction was subcloned into YEp13, a multicopy yeast vector containing the *LEU2* gene as a selectable marker[12] to further increase the level of expression. The construction of this plasmid, designated pNB283 (see Fig. 1A) has been described in detail.[9] pNB283 was introduced into the yeast strain NY603 (*MAT*a *leu2-3, 112 ura3-52 GAL⁺ pep4::URA3*) by alkali

[5] J. Becker, T. Tan, H. Trepte, and D. Gallwitz, *Eur. Mol. Biol. J.* **10,** 785 (1991).

[6] E. S. Burnstein, K. Linko-Stentz, Z. Lu, and I. Macara, *J. Biol. Chem.* **266,** 2689 (1991).

[7] G. Rossi, Y. Jiang, A. Newman, and S. Ferro-Novick, *Nature (London)* **351,** 158 (1991).

[8] S. Araki, A. Kikuchi, Y. Hata, M. Isomura, and Y. Takai, *J. Biol. Chem.* **265,** 13007 (1990).

[9] A. K. Kabcenell, B. Goud, J. Northup, and P. Novick, *J. Biol. Chem.* **265,** 9366 (1990).

[10] J. C. Schneider and L. Guarente, this series, Vol. 194, p. 373.

[11] B. Goud, A. Salminen, N. C. Walworth, and P. Novick, *Cell* **53,** 753 (1988).

[12] J. R. Broach, J. N. Stathern, and J. B. Hicks, *Gene* **8,** 121 (1979).

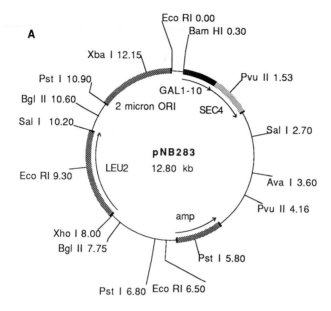

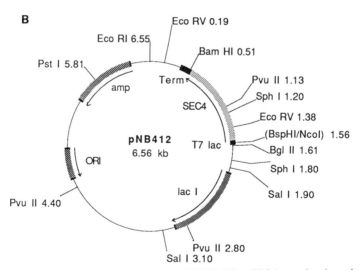

FIG. 1. (A) Yeast Sec4 expression vector pNB238. The *SEC4* gene has been inserted behind the *GAL1* promoter on a vector maintained at high copy number in yeast. (B) Bacterial expression vector pNB412. The *SEC4* gene has been inserted between the T7 promotor and terminator (Term).

cation treatment[13] and transformants were selected at 25° on minimal medium containing 0.7% (w/v) yeast nitrogen base without amino acids (Difco, Detroit, MI) and supplemented with 2% (w/v) glucose. A recipient strain harboring a disruption of the *PEP4* gene was utilized as *PEP4* encodes the vacuolar hydrolase protease A, which has been implicated in the activation of several other vacuolar zymogens.[14,15] Elimination of this enzyme, therefore, reduces the endogenous protease activity in a yeast lysate. Each time the resulting strain, NY671, is grown from a frozen stock it should be screened for the maintenance of the Pep4⁻ phenotype by assaying for the absence of carboxypeptidase Y activity.[16]

A stationary culture is prepared by inoculating 25 ml of minimal medium containing 2.5% (w/v) raffinose [from a 10% (w/v) stock] with a single colony of NY671 and incubating for 4 days at 25° with aeration. Raffinose will not induce the *GAL1* promoter, nor will it interfere with rapid induction following subsequent transfer to galactose medium. This stationary culture is then used to inoculate 2 liters of minimal medium supplemented with raffinose and growth is continued at 25° with aeration for 21 hr. The A_{600} of this exponential culture should be between 0.6 and 2.0. To induce expression of the episomal *SEC4* gene, 1600 A_{600} units of the cell suspension is pelleted in the GH-3.7 rotor of a Beckman (Palo Alto, CA) GPR centrifuge by spinning for 10 min at 3000 rpm at room temperature, and the cells are resuspended in 4 liters of minimal medium containing 2% (w/v) galactose [from a 20% (w/v) stock], and incubated at 25° with shaking. After 24 hr of incubation, the cells, at an A_{600} of 1.3–1.5, are harvested as above and washed once with deionized water. All subsequent steps are performed at 4°.

The cell pellet is resuspended in 20 ml of lysis buffer [0.3 M sorbitol, 20 mM Tris-HCl, pH 8.0, 10 mM NaCl, 5 mM MgCl$_2$, 1 mM dithiothreitol (DTT), 1 mM phenylmethylsulfonyl fluoride (PMSF), and 1 μg/ml each of antipain, leupeptin, aprotinin, chymostatin, and pepstatin A] at 0–4°. The cell suspension is added to the prechilled 85-ml chamber of a Bead-beater (BioSpec Products, Bartlesville, OK) half-filled with 0.5-mm glass beads. Additional lysis buffer is added to fill the chamber completely and thereby exclude all air. The chamber is sealed, surrounded by an ice water bath, and the cells are homogenized for a total of 3 min in six 30-sec

[13] H. Ito, Y. Fukada, K. Murata, and A. Kimura, *J. Bacteriol.* **153**, 163 (1983).
[14] G. Ammer, C. P. Hunter, J. H. Rothman, G. C. Saari, L. A. Valls, and T. H. Stevens, *Mol. Cell. Biol.* **6**, 2409 (1986).
[15] C. A. Woolford, L. B. Daniels, F. J. Park, E. W. Jones, J. N. Van Arsdell, and M. A. Innis, *Mol. Cell. Biol.* **6**, 2500 (1986).
[16] E. W. Jones, this series, Vol. 194, p. 428.

intervals. Between each 30-sec interval the chamber should be allowed to cool for 3 min. For a detailed discussion on the use of the BioSpec Bead-beater see Jazwinski.[17] The resulting lysate is recovered from the chamber, centrifuged at 3500 rpm for 10 min to remove unbroken cells and debris, and the supernatant spun in a Beckman Ti70 rotor at 34,000 rpm (90,000 g) for 60 min to pellet membranes. A layer of lipid forms at the top of the tubes and should be discarded. The high-speed supernatant should contain approximately 400 mg protein in a volume of 30–40 ml and can be used immediately or rapidly frozen in liquid nitrogen and stored at −80°.

Expression of Sec4 Protein in *Escherichia coli* and Preparation of a Cell Lysate

To express native Sec4 protein in *Escherichia coli,* we use the T7 expression system.[18] By this system the coding sequence of interest is inserted behind a T7 RNA polymerase promoter and then introduced into a bacterial strain containing a T7 RNA polymerase gene under control of the inducible *lacUV5* promoter. Production of T7 RNA polymerase in this strain is induced with isopropylthio-β-D-galactoside (IPTG), which in turn yields high-level expression of the foreign gene. We have made use of the polymerase chain reaction (PCR) to engineer a restriction site such that we could ligate the *SEC4* coding sequence in the optimal position behind the T7 promoter of the expression vector pET11d. This vector contains the gene *10* ribosome-binding site with an *Nco*I site at the initiating AUG and a strong transcriptional terminator following the *Bam*HI site. Two oligonucleotides have been employed for amplifying *SEC4:* one incorporates a *Bsp*HI site at the initiating AUG and the second incorporates a *Bam*HI site downstream of the *SEC4* termination codon. A *Bsp*HI site was chosen in primer I because digestion with this enzyme yields an overhang that is compatible with the *Nco*I site in pET11d and retains a serine as the second residue. The resulting construction is then used to transform BL21(DE3) cells that contain the IPTG-inducible T7 RNA polymerase gene. The transformants are plated on ZB (10 g Bacto-tryptone, 5 g NaCl, 15 g agar per liter) plates containing 100 μg/ml of ampicillin (Amp) and grown overnight at 37°.

To test the transformants for expression of Sec4p, multiple independent colonies are transferred from the ZB-Amp plate into 5 ml of ZB supplemented with 100 μg/ml ampicillin (Boehringer Mannheim, India-

---

[17] S. M. Jazwinski, this series, Vol. 182, p. 154.
[18] F. W. Studier, A. H. Rosenberg, J. J. Dunn, and J. W. Dubendorff, this series, Vol. 185, p. 60.

napolis, IN) and grown overnight at 37° until freshly saturated (less than 12 hr). Frozen stocks are prepared and stored at −80° so that all subsequent cultures can be started directly from them. Fifty microliters of each overnight culture is then inoculated into 5 ml of ZB-Amp and grown at 37° for 3–4 hr until the A_{600} is between 0.4 and 0.6. IPTG is added to each tube to a final concentration of 0.4 mM and cultures incubated at 37° for 2 hr with good aeration. One milliliter of each culture is harvested by centrifugation for 30 sec at top speed in a microfuge and the pellet is resuspended in 150 μl of sodium dodecyl sulfate (SDS) sample buffer and boiled for 5 min. These samples are analyzed for expression of Sec4 protein by SDS-polyacrylamide gel electrophoresis (PAGE) on a 15% (w/v) SDS-polyacrylamide gel, which is then stained with Coomassie Brilliant blue. All four transformants that we analyzed showed some expression of a protein of the predicted size of Sec4p, but the levels varied quite dramatically. The identity of the overexpressed protein can be confirmed by immunoblots using anti-Sec4p sera.

High-level expression of proteins in bacteria can lead to formation of an insoluble aggregate. Therefore the transformants showing the best expression of Sec4p are tested for solubility of the protein at varying temperatures and times of induction. For each transformant, two 50-ml cultures are prepared by inoculating 100 ml of ZB-Amp with 1 ml of an overnight culture and growing one 50-ml aliquot at 30° and the other at 37°, until the A_{600} of the cells is between 0.4 and 0.6. Induction is then started by the addition of IPTG (0.4 mM, final concentration) and growth is continued at the same temperature with the removal of 5-ml aliquots from each culture 1, 2, and 3 hr after the start of induction. These samples are harvested by centrifugation in the GH-3.7 rotor of a Beckman GPR centrifuge at 3000 rpm at 4°. The pellets are resuspended in ice-cold lysis buffer (20 mM Tris-HCl, pH 8.0, 100 mM NaCl, 5 mM MgCl$_2$, 1 mM DTT, 1 mM PMSF, and 1 μg/ml each of antipain, leupeptin, aprotinin, chymostatin, and pepstatin A) and transferred to 1.5-ml microfuge tubes. The samples are then sonicated on ice for 30 sec, in three 10-sec intervals with cooling on ice for 1 min between each 10-sec sonication (we used a Sonifier cell disruptor (Branson, Danbury, CT) model W185 on setting 5.5). The resulting sonicate is centrifuged at top speed in a microfuge and the pellet resuspended in a volume of lysis buffer equal to that of the sonicate supernatant. Equal volumes of sonicate supernatant and pellet fractions can then be analysed by SDS-PAGE and Coomassie Brilliant blue staining of the gel.

We found that expression of Sec4p in the sonicate supernatant (i.e., soluble Sec4p) was different for the two transformants that we tested and also varied with the temperature and time of induction. Consequently, for the transformant expressing the greatest amount of soluble Sec4p (strain

NRB412d), we found that the optimal conditions for expression of this protein were growth of the cells at 30° and induction with IPTG for 1 hr at 30°.

To produce a larger scale induction of Sec4p, 5 ml of ZB medium containing 0.1 mg/ml ampicillin is inoculated from a stock of the strain NRB412d stored at −80° and grown overnight at 37° with aeration. This is then used to inoculate 500 ml of ZB broth containing 0.1 mg/ml ampicillin and growth is continued at 30° with aeration until the A_{600} is between 0.4 and 0.6. To induce expression of the *SEC4* gene, 0.4 mM IPTG (final concentration) is added to the culture and growth is continued for 1 hr at 30° with aeration. Following this induction the cells are harvested by centrifugation for 10 min at 5000 rpm at 4° (we used a Beckman J2.21 centrifuge with a JA-10 rotor). The cell pellet can be used immediately or rapidly frozen and stored at −80°.

The pelleted cells are resuspended in 7 ml of ice-cold lysis buffer and the cell suspension is sonicated on ice for 6 min, in three 2-min intervals with cooling on ice for 1 min between each sonication. The resulting sonicate is then centrifuged at 11,000 rpm for 15 min at 4° (we used a Beckman J2.21 centrifuge with a JA-20 rotor) to remove unbroken cells and debris. The supernatant should contain approximately 60 mg of protein and should be used immediately.

Guanine Nucleotide-Binding Assay

GTP-binding activity is used to monitor the Sec4 protein throughout its purification. While in principle this assay does not uniquely detect Sec4 because there are many other GTP-binding proteins in both *S. cerevisiae* and *E. coli* lysates, they are, in fact, quite minor in abundance with respect to Sec4p due to the high level of expression achieved. This is reflected by the observation that the level of soluble GTP-binding activity increases over 40-fold on Sec4p induction in yeast and over 150-fold on Sec4p induction in *E. coli*. Guanine nucleotide binding is assayed by filtration.[19] Briefly, 5 μl of column fractions is diluted into 20 μl of buffer A [50 mM sodium 4-(2-hydroxyethyl)-1-piperazineethanesulfonic acid (HEPES), 5 mM MgCl$_2$, pH 8.0, 200 mM NaCl, 1 mM EDTA, 1 mM DTT, and 0.1% (w/v) Lubrol 12A9]. The binding reaction is initiated with the addition of 25 μl of buffer A containing 4 μM GTPγS (Boehringer Mannheim, Indianapolis, IN) and [^{35}S]GTPγS (New England Nuclear, Boston, MA) at a specific activity of approximately 5000 cpm/pmol. Following a 60-min incubation at 30°, the reaction mixtures are diluted with 4 ml of ice-cold

[19] J. K. Northup, M. D. Smigel, and A. G. Gilman, *J. Biol. Chem.* **257**, 11416 (1982).

20 mM Tris-HCl, pH 8.0, 100 mM NaCl, 25 mM MgCl$_2$, and rapidly filtered through 25-mm type HA filters (Millipore, Bedford, MA). A sampling manifold and vaccuum pressure pump (Millipore) are convenient in this regard. Wild-type Sec4p isolated from either yeast or bacteria is predominantly bound to GDP. Because the off-rate of GDP from Sec4p is only about 4 min in the presence of 5 mM MgCl$_2$ at 30°, no preincubation under conditions of micromolar MgCl$_2$ is necessary to increase the rate of nucleotide exchange, as it is with other members of the *ras* superfamily. The filters are washed five times with 2 ml of cold buffer, dried under a heat lamp, and counted in 10 ml of Ecoscint scintillation fluid (National Diagnostics, Manville, NJ). The specific radioactivity of the nucleotide is determined by diluting the radioactive mixture 1 : 80 and spotting 20 μl (1 pmol) onto each of three filters, which are then dried and counted as above.

Purification of Sec4p from Yeast

Gel filtration is an effective initial purification step because Sec4p is a small protein that, when overproduced, accumulates in a soluble, monomeric form. Approximately 350 mg of protein derived from a high-speed supernatant fraction of NY671 cells in a volume of 30 ml is applied to an 800-ml Sephacryl S-100 column (Pharmacia, Piscataway, NJ; 5 × 44 cm). The column is eluted with TMD buffer (20 mM Tris-HCl, pH 8.0, 5 mM MgCl$_2$, 1 mM DTT) supplemented with 100 mM NaCl, and 9-ml fractions are collected. It is convenient, but not essential, to follow the A_{280} with an in-line UV detector and chart recorder. The peak of [^{35}S]GTPγS-binding activity elutes as a symmetrical peak after the bulk of other cell proteins (Fig. 2). Relative to the yeast high-speed supernatant, this purification step gives approximately 20-fold purification with 82% recovery of activity in a volume of 38 ml (see Table I).

Sec4p binds only weakly to DEAE-Sephacel (under our buffer conditions). Thus the second stage of the purification involves a step in which Sec4p fails to bind to this ion-exchange resin while a number of contaminating proteins are retained. The pooled Sephacryl S-100 fractions are combined and then diluted twofold with an equal volume of TMD buffer to a final concentration of 50 mM NaCl. It is then loaded onto a 50-ml DEAE-Sephacel column (2.5 × 11.5 cm; Pharmacia) equilibrated in TMD buffer with 50 mM NaCl. The column is washed with this buffer, and 4.4-ml fractions are collected. The flow-through and wash fractions containing protein as measured by absorbance at 280 nm are combined. It is not necessary to assay for GTPγS binding during this step.

In the final stage of purification the NaCl concentration is reduced to

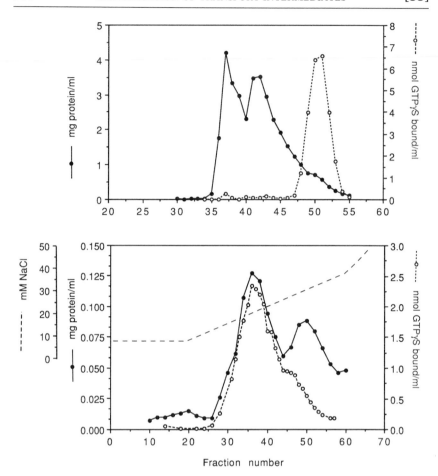

FIG. 2. *Top:* Elution profile of Sec4 activity (nmol GTPγS bound/ml) in a yeast lysate from a Sephacryl S-100 column. *Bottom:* Elution profile of Sec4 activity (nmol GTPγS bound/ml) from a DEAE-Sephacel column on elution with a gradient of NaCl.

the level at which Sec4p is retained on a DEAE-Sephacel column and it is then eluted with an NaCl gradient. To lower the NaCl concentration without diluting the protein to the extent that Sec4p is denatured the pool is first concentrated to 5 ml by pressure filtration in a stirred cell containing an Amicon (Danvers, MA) PM10 membrane and then diluted fivefold with TMD buffer for a final NaCl concentration of 10 mM. The pool is applied to a 30-ml DEAE-Sephacel column (2.5 × 7 cm) equilibrated in TMD buffer with 10 mM NaCl, and the adsorbed protein is eluted with a 150-ml linear gradient of 10–50 mM NaCl in TMD buffer. Fractions of

TABLE I
PURIFICATION OF Sec4 FROM YEAST AND BACTERIA

| Source and fraction | GTPγS binding | | | | | |
	Total protein (mg)	Total volume (ml)	Total activity (nmol)	Specific activity (nmol/mg)	Yield (%)	Purification (-fold)
Yeast						
Soluble lysate	350.0	40.0	232	0.7	100	1
S-100	14.8	38.0	191	12.9	82	20
DEAE flow-through	7.9	95.0	159	20.2	69	31
DEAE pool	2.1	2.8	65	30.9	28	47
Bacteria						
Soluble lysate	60	6.0	21.8	0.36	100	1
S-100	2.9	8.0	24.9	8.6	114	24
DEAE flow-through	1.6	1.5	19.5	12.1	89	34

2.8 ml are collected and alternate fractions are assayed for GTPγS-binding activity. The peak of GTPγS-binding activity elutes at approximately 20 mM NaCl and is pooled and concentrated approximately 10-fold by pressure filtration in a stirred cell. Prior to storage at −80°, the purified Sec4 protein should be diluted one-fourth with 6.4 mM dipalmitoylphosphatidylcholine (Sigma, St. Louis, MO), 32 mM 3-[3-cholamidopropyldimethylammonio]-1-propane sulfonate (CHAPS; Calbiochem, San Diego, CA), and 50 mM NaCl in TMD buffer and then rapidly frozen in small aliquots in sialyzed glass tubes. The addition of detergent and phospholipid appears to stabilize Sec4p against repeated freeze-thaw cycles. The final preparation contains only a single species as visualized by SDS-PAGE and silver staining. The overall purification is typically 47-fold with a yield of approximately 2 mg.

Purification of Sec4p from *Escherichia coli*

This procedure is performed as described above for purification of Sec4p from yeast except that column sizes are scaled down because the lysate is prepared from only 500 ml of cells and the final ion-exchange column is omitted. The 6 ml of supernatant (from centrifugation of the sonicate at 13,800 g) containing approximately 60 mg of protein is applied to a 170-ml Sephacryl S-100 column (1.5 cm × 96 cm; Pharmacia). The column is eluted with TMD buffer supplemented with 100 mM NaCl, 2.2-ml fractions are collected and the peak of GTPγS-binding activity pooled. This step gives approximately 24-fold purification with excellent recovery of GTPγS-binding activity in a volume of 8 ml (see Table I). The pooled fractions are then diluted with an equal volume of TMD buffer to give a final concentration of 50 mM NaCl. This is loaded onto a 12-ml DEAE-Sephacel column (2.5 × 2.5 cm) equilibrated with TMD buffer containing 50 mM NaCl. The column is washed with the same buffer until the absorbance at 280 nm returns to baseline. The peak of GTPγS-binding activity is pooled and concentrated approximately 18-fold to 1.5 ml by pressure filtration in a stirred cell containing an Amicon PM10 membrane. The protein is then rapidly frozen and stored at −80°. The overall purification is approximately 34-fold (see Table I), with a yield of about 500 μg of active protein and a purity of approximately 80–90% as analyzed by SDS-PAGE and Coomassie Brilliant blue staining of the gel. The GDP off-rate, GTP on-rate, and intrinsic GTP hydrolysis rate of the bacterially produced Sec4p are quite similar to those found for Sec4p isolated from yeast.

Acknowledgments

This work was supported by Grants GM35370 and CA46218 to P.N. from the National Institutes of Health. M.G. was supported by the Lucille P. Markey Charitable Trust, P.B. was supported by the Damon Runyon Walter Winchell Cancer Research Fund, and A.K.K. was supported by the Jane Coffin Childs Memorial Fund for Medical Research and by a Swebilius Cancer Research Award.

[34] Preparation of Recombinant ADP-Ribosylation Factor

By PAUL A. RANDAZZO, OFRA WEISS, and RICHARD A. KAHN

Introduction

ADP-ribosylation factor (ARF) proteins were originally identified and purified based on an *in vitro* activity as the protein cofactor required for efficient ADP-ribosylation of G$_s$ by cholera toxin[1,2] (for recent reviews see Kahn[3,4]). Subsequently, ARF was shown to be a 21-kDa GTP-binding

[1] L. S. Schleifer, R. A. Kahn, E. Hanski, J. K. Northup, P. C. Sternweis, and A. G. Gilman, *J. Biol. Chem.* **257,** 20 (1982).

[2] R. A. Kahn and A. G. Gilman, *J. Biol. Chem.* **259,** 6228 (1984).

[3] R. A. Kahn, in "G Proteins" (L. Birnbaumer and R. Iyengar, eds.), p. 201. Academic Press, Orlando, Florida, 1990.

[4] R. A. Kahn, this series, Vol. 195, p. 233.

protein that is active only in the GTP-bound form.[5] ADP-ribosylation factor activity and/or immunoreactivity have been found in every eukaryotic cell tested but is absent from *Escherichia coli*.[3] Multiple genes for ARF have been found in every cell type examined, including two homologous genes for ARF in the yeast *Saccharomyces cerevisiae*,[6] two identified in bovine cDNA libraries,[7,8] and three found in human cDNA libraries.[9-11] A large number of ARF and ARF-like cDNAs have been identified[12] and have led to the establishment of ARF and related proteins as the fourth subfamily of the RAS superfamily of low molecular weight GTP-binding proteins. Hallmarks for this subfamily of proteins include myristoylation at the amino-terminal glycine instead of carboxy-terminal processing (common to the other three subfamilies in the RAS superfamily) and specific protein sequence motifs, e.g., D(V/I)GGQ instead of DTAGQ, as found in the other three subfamilies at the second consensus GTP-binding domain.[3] ADP-ribosylation factor is also highly functionally conserved,[11] and can be distinguished from other GTP-binding proteins, including the ARF-like proteins,[13] by two criteria that define a *bona fide* ARF from a structurally related protein. The ARF proteins have activity in the ARF assay, as cofactor for cholera toxin-dependent ADP-ribosylation,[11] and expression of an ARF protein can rescue the lethal double-mutant *arf1⁻arf2⁻* in yeast.[11] Examples of ARF-related genes include *arl*[13] from *Drosophila melanogaster* and *SAR1*[14] and *CIN4*[15] found in *S. cerevisiae*. Each of these gene products are in the ARF subfamily of proteins but lack ARF activities, as defined above. We will limit further discussion to activities found in association with these functionally defined ARF proteins.

ADP-ribosylation factor has been implicated as a critical component in the protein secretory machinery in yeast (*S. cerevisiae*) and mammalian cells. Disruption of the *ARF1* gene in yeast causes a defect in N-glycosyla-

[5] R. A. Kahn and A. G. Gilman, *J. Biol. Chem.* **261**, 7906 (1986).

[6] T. Stearns, R. A. Kahn, D. Botstein, and M. A. Hoyt, *Mol. Cell. Biol.* **10**, 6690 (1990).

[7] J. Sewell and R. A. Kahn, *Proc. Natl. Acad. Sci. USA* **85**, 4620 (1988).

[8] S. R. Price, M. Nightingale, S. C. Tsai, K. C. Williamson, R. Adamik, H. C. Chen, J. Moss, and M. Vaughan, *Proc. Natl. Acad. Sci. USA* **85**, 5488 (1988).

[9] Z. Peng, I. Calver, J. Clark, L. Helman, R. A. Kahn, and H. Kung, *BioFactors* **2**, 45 (1989).

[10] D. A. Bobak, M. S. Nightingale, J. J. Murtagh, S. R. Price, J. Moss, and M. Vaughan, *Proc. Natl. Acad. Sci. USA* **86**, 6101 (1989).

[11] R. A. Kahn, F. G. Kern, J. Clark, E. P. Gelman, and C. Rulka, *J. Biol. Chem.* **266**, 2606 (1991).

[12] J. Clark and R. A. Kahn, unpublished observations, 1990.

[13] J. W. Tamkun, R. A. Kahn, M. Kissinger, B. J. Brizuela, C. Rulka, M. P. Scott, and J. A. Kennison, *Proc. Natl. Acad. Sci. USA* **88**, 3120 (1991).

[14] A. Nakano and M. Muramatsu, *J. Cell Biol.* **109**, 2677 (1989).

[15] T. Stearns, R. A. Kahn, M. A. Hoyt, and D. Botstein, in preparation.

tion, similar to that seen for *ypt1* mutants, as evidenced by the formation of incompletely glycosylated invertase.[16] Furthermore, the *arf1⁻* mutant shows synthetic lethality with *ypt1-1*, *sec21-1*, and *bet2-1*, each of which has been shown to cause defects in protein secretion.[16] In NIH 3T3,[16] Chinese hamster ovary (CHO),[17] and normal rat kidney (NRK) cells,[18] immunocytochemical techniques have allowed the localization of ARF to Golgi membranes and Golgi-derived vesicles. In fact, ARF has been shown to be an abundant protein on non-clathrin-coated vesicles.[17] The binding of ARF proteins and β-COP to Golgi membranes has been found to be rapidly and specifically blocked by addition of brefeldin A, both *in vivo* and *in vitro*.[18] This has been taken as support for a model for ARF action involving cycling between soluble and particulate pools, similar to the model proposed for SEC4.[19] It has proved difficult to demonstrate definitively a requirement for ARF proteins in a specific step of the protein secretory pathway as ARF is a ubiquitous and abundant protein, present in both cytosol and membrane preparations. None of the antibodies currently available have neutralizing properties or the ability to immunoprecipitate the native proteins. The discovery of potent and specific peptide inhibitors of ARF activities has allowed the demonstration of an absolute requirement for ARF in *in vitro* secretion assays[20,21] and clearly shown a role for ARF proteins in secretion. Peptides derived from the N terminus of either mARF1p or mARF4p inhibited ARF association with a Golgi membrane fraction from CHO cells,[22] blocked vesicle budding from a similar membrane preparation,[20] and blocked ER-to-Golgi[21] and intra-Golgi transport,[20] measured *in vitro* using CHO cellular fractions. Although a role for ARF in protein secretion appears established, much more work is needed to clarify the roles of GTP binding and hydrolysis by ARF as a regulatory event in secretion, the specific step which requires ARF (e.g., budding, fusion, etc.), and the site of action of brefeldin A, and to answer many more questions of interest to both workers in the protein secretion field as

---

[16] T. Stearns, M. C. Willingham, D. Botstein, and R. A. Kahn, *Proc. Natl. Acad. Sci. USA* **87**, 1238 (1990).

[17] T. Serafini, L. Orci, M. Amherdt, M. Brunner, R. A. Kahn, and J. E. Rothman, *Cell* **67**, 239 (1991).

[18] J. Donaldson, R. A. Kahn, J. Lippincott-Schwartz, and R. D. Klausner, *Science* **254**, 1197 (1991).

[19] N. C. Walworth, B. Goud, A. Kastan-Kabcenell, and P. J. Novick, *EMBO J.* **8**, 1685 (1989).

[20] R. A. Kahn, P. Randazzo, T. Serafini, O. Weiss, C. Rulka, J. Clark, M. Amherdt, P. Roller, L. Orci, J. E. Rothman, *J. Biol. Chem.* **267**, 13039 (1992).

[21] W. Balch, R. A. Kahn, and R. Schwaninger, *J. Biol. Chem.* **267**, 13053 (1992).

[22] J. G. Donaldson, D. Cassel, R. A. Kahn, and R. D. Klausner, *Proc. Natl. Acad. Sci. USA* in press (1992).

well as those elucidating mechanisms for members of the RAS superfamily of proteins.

In all tissues thus far examined, ARF is purified as a closely spaced doublet containing an unknown number of gene products.[2,9] The use of a single gene product is desirable for studies aimed at elucidating the role(s) of individual proteins in a specific biochemical reaction, for example examining the role of ARF in secretion. Thus, rather than using ARFs purified from mammalian or yeast cells we have made use of the expression system of Studier and Moffatt[23] to overproduce specific *ARF* gene products for use in biochemical studies of protein function. The recombinant protein is indistinguishable from that purified from tissues in terms of nucleotide-binding kinetics and activity in cholera toxin-catalyzed ADP-ribosylation of G_s.[24] The only known covalent modification to ARF proteins is N-terminal myristoylation,[3] which is essential for function *in vivo* but has no effect on activity in the ARF assay. While bacteria are incapable of protein N-myristoylation, it has been shown that coinduction of both the protein of interest and the yeast myristoyl-CoA: protein *N*-myristoyltransferase (NMT) results in the formation of properly myristoylated proteins in bacteria.[25] As we begin to reconstitute recombinant ARF proteins into more crude systems that allow studies of function it is clear that having both myristoylated and nonmyristoylated forms of the protein is useful. The methods used in obtaining both nonmyristoylated and myristoylated ARF proteins are described below.

Methods

The methods described below allow the expression and purification of ARF proteins in bacteria. Typically, ARF can constitute as much as 10% of total bacterial protein, which results in a yield of 1–5 mg of purified ARF from 1 liter of cells. ADP-ribosylation factor can be monitored during purification and quantitated using a radioligand binding assay described by Kahn and Gilman,[5] a modification of the method described by Northup *et al.*[26] The best estimate of final purity is obtained from densitometric scanning of stained sodium dodecyl sulfate (SDS) gels or quantitation of protein-associated GDP.[5]

[23] F. W. Studier and B. A. Moffatt, *J. Mol. Biol.* **189**, 113 (1986).
[24] O. Weiss, J. Holden, C. Rulka, and R. A. Kahn, *J. Biol. Chem.* **264**, 21066 (1989).
[25] R. J. Duronio, E. Jackson-Machelski, R. O. Heuckeroth, P. O. Olins, C. S. Devine, W. Yonemoto, L. W. Slice, S. S. Taylor, and J. I. Gordon, *Proc. Natl. Acad. Sci. USA* **87**, 1506 (1990).
[26] J. K. Northup, M. D. Smigel, and A. G. Gilman, *J. Biol. Chem.* **257**, 11416 (1982).

Protein Expression

High-level expression of several ARF and ARF-related proteins has been achieved by using the T7 polymerase/promoter system described by Studier *et al.*[23,27] The construction of the ARF expression plasmid, pOW12, has been described previously.[24] Briefly, this entails insertion of the coding region of the gene into the pET3C vector[23,27] such that the initiating methionine is part of an *Nde*I restriction site that is located at the optimal distance from a T7 promoter to direct maximal expression. BL21(DE3) cells[27] contain the T7 polymerase structural gene under control of the *lacUV5* promoter and is thus induced by isopropyl-β-D-thiogalactopyranoside (IPTG). BL21(DE3) cells transformed with pOW12 are grown in LB medium containing 100 μg/ml ampicillin at 37°. When the $A_{600} \approx 1.0$, IPTG is added to a final concentration of 1 mM. Ninety minutes later, the cells are harvested by centrifugation at 4000 g for 15 min at 4° and may be stored at −80°.

Purification

Purification of the recombinant protein is accomplished in two steps, batch elution from DEAE-Sephacel and a gel-filtration column. This protocol routinely results in a product that is at least 85% pure and has been used successfully for recombinant human ARF1p[24] and ARF4p,[11] *drosophila* ARL1p,[13] and wild-type and eight point mutants of yeast ARF1p.[28]

1. The cell pellet from a 1-liter culture is resuspended in 20 ml of 50 mM Tris-HCl, pH 8.0, 40 mM ethylenediaminetetraacetic acid (EDTA), 25% (w/v) sucrose, 1 mg/ml lysozyme and incubated at room temperature for 15 min. The cells are lysed by the addition of 8 ml of 0.2% (w/v) Triton X-100, 50 mM Tris-HCl, pH 8.0, 100 mM MgCl$_2$. After 10 min at 4°, the suspension is clarified by centrifugation at 100,000-g for 60 min at 4°.

2. The supernatant is loaded onto a 50-ml DEAE-Sephacel column, previously equilibrated in three column volumes of 20 mM Tris-HCl, pH 7.4, 1 mM EDTA, 1 mM dithiothreitol, 50 mM NaCl. The flow through and a wash of one column volume are collected and concentrated to 4 ml by ultrafiltration using an Amicon (Danvers, MA) YM10 membrane.

3. The concentrate is applied to an Ultrogel AcA 54 column (200 ml, 40 × 2.5 cm; IBF Biotechnics, Columbia, MD) that has been equilibrated with 20 mM N-2-hydroxyethylpiperazine-N'-2-ethanesulfonic acid (HEPES), pH 7.4, 1 mM EDTA, 100 mM NaCl, 1 mM dithiothreitol

[27] A. H. Rosenberg, B. N. Lade, D. Chui, S. W. Lin, J. J. Dunn, and F. W. Studier, *Gene* **56**, 125 (1987).
[28] R. A. Kahn, J. Clark, and C. Rulka, unpublished observations, 1990.

(DTT), and 2 mM MgCl$_2$. The column is developed in the same buffer at a flow rate of 20 ml/hr and 2.5-ml fractions are collected. The Coomassie blue-stained polyacrylamide gels of protein from each step of the purification are shown in Fig 1. The fractions containing pure ARF are pooled and concentrated by ultrafiltration to 1 mg/ml and stored at −80°.

GTP-Binding Assay

ADP-ribosylation factor is easily detected by nucleotide binding and this is the method used in our laboratory for monitoring the elution of recombinant ARF from chromatographic columns during purification. As previously described,[5] nucleotide exchange on ARF requires phospholipid and detergent and is most efficient at low concentrations of Mg^{2+}. Hence, the binding cocktail contains 25 mM HEPES, pH 7.4–8.0, with

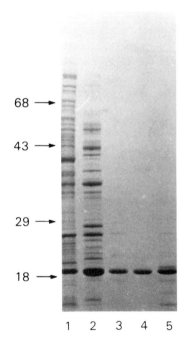

FIG. 1. Purification of mARF1p from bacteria. Coomassie blue stained-SDS polyacrylamide gel fractionation of proteins from each step of the purification of mARF1p from BL21 (DE3) cells transfected with pOW12. Lane 1, total cell extract from cells induced with IPTG; lane 2, pooled DEAE-Sephacel fractions; lane 3, pooled Ultrogel AcA 54 fractions (purified recombinant mARF1p); lane 4, pooled Ultrogel AcA 54 fractions from bacteria transfected with expression vectors for mARF1 and N-myristoyltransferase; lane 5, ARF purified from bovine brain as previously described.[4] Molecular weight markers ($\times 10^{-3}$) on left-hand side.

100 mM NaCl, 1 mM EDTA, 0.5 mM MgCl$_2$, 1 mM DTT, 3 mM L-α-dimyristoylphosphatidylcholine (DMPC), 0.1% (w/v) sodium cholate, and 1 μM [α-^{32}P]GDP (10,000 cpm/pmol), [γ-^{32}P]GTP (10,000 cpm/pmol), or [^{35}S]GTPγS (10,000 cpm/pmol).[4,5] The DMPC is freshly prepared by sonication of a 30 mM solution in 20 mM HEPES, 1 mM EDTA, and 2 mM MgCl$_2$ immediately prior to performing the binding assay. The turbid suspension is sonicated until it becomes translucent and then maintained at 30° until diluted into the binding cocktail. The protein is incubated in the binding cocktail for 1 hr at 30° in a total volume of 50 μl and exchange is stopped by the addition of ice-cold TNMD buffer: 2 ml of Tris, pH 7.4, 100 mM NaCl, 10 mM MgCl$_2$, and 1 mM DTT. The bound and free nucleotides are separated by filtration through 25-mm BA85 nitrocellulose filters (Schleicher & Schuell, Keene, NH). The filters are washed six times with 2 ml ice-cold TNMD. ADP-ribosylation factor–nucleotide complex is retained and can be quantified by scintillation spectroscopy. Binding typically increases linearly for the first 30 min and reaches a plateau within 1 to 3 hr. Binding stoichiometries tend to be poor, especially for triphosphates and triphosphate analogs, as the protein contains 1 mol bound GDP/mol protein. As GDP has at least a 100-fold greater affinity for the protein than the triphosphates the stoichiometry of GTP binding is no more than 0.1. Hence, ARF is not easily quantified by this assay.

Quantitation

Typically, the ARF prepared by the protocol above is 85% pure (Fig. 1) which can be assessed by Coomassie blue staining of protein fractionated on polyacrylamide gels. As active ARF is purified as a 1 : 1 complex with GDP,[5] the concentration of active ARF can also be assessed by determining the amount of nucleotide released after heat denaturation of a given amount of the purified protein.[5,24]

Comments

1. The time of induction can be varied. With some proteins, accumulation continues beyond 90 min and yields are improved with longer (3–4 hr) inductions.

2. Batch elution from DEAE-Sephacel is the most variable step. Elution with higher concentrations of NaCl increases the yield of ARF but generally results in a less pure preparation after gel filtration. A significant fraction of ARF (as much as 50%) is also found in the 100,000 g pellet. This has been purified in the presence of 7 M urea and results in a nucleotide-free preparation of ARF that is much less stable[24] than ARF purified by the protocol presented in this chapter.

3. This approach has provided our laboratory with purified recombinant ARF proteins from several sources. It has also been modified to obtain myristoylated recombinant proteins. In this case, the bacteria are cotransfected with the plasmid containing the ARF coding region as well as pBB131,[25] a plasmid containing the yeast N-myristoyltransferase gene *NMT1* under the same T7 promoter,[25] and a marker for kanamycin resistance to allow selection for both plasmids. The BL21(DE3) cells are then grown in LB medium containing 50 μg/ml kanamycin and 100 μg/ml ampicillin. On induction with IPTG we also add 200 μM myristic acid [125 mM stock in 50% (v/v) methanol is freshly prepared]. Purification, GTP-binding determination, and quantitation are all performed as for the nonmyristoylated protein. Using this approach, approximately 10–60% of the ARF is myristoylated as determined by sequencing of the amino terminus of the purified proteins.[28] The myristoylated protein cannot be separated from the unmodified protein using ion-exchange, gel-filtration, or hydrophobic interaction chromatography, including phenyl-Sepharose or heptylamine-Sepharose.[28] As is true for the nonmyristoylated recombinant protein, the myristoylated recombinant protein is indistinguishable from ARF purified from bovine brain in terms of nucleotide binding kinetics and activity in the cholera toxin-catalyzed reaction.[28]

[35] Ypt Proteins in Yeast

By PETER WAGNER, LUDGER HENGST, and DIETER GALLWITZ

Introduction

The *ras* superfamily of genes encodes structurally related, guanine nucleotide-binding proteins of similar size (around 200 amino acid residues) and biochemical properties.[1-3] From work with yeast mutants, it is evident that members of the Ypt subfamily of proteins,[4] including Ypt1

[1] H. R. Bourne, D. A. Sanders, and F. McCormick, *Nature (London)* **348**, 125 (1990).
[2] H. R. Bourne, D. A. Sanders, and F. McCormick, *Nature (London)* **349**, 117 (1991).
[3] A. Hall, *Science* **249**, 635 (1990).
[4] D. Gallwitz, H. Haubruck, C. Molenaar, R. Prange, M. Puzicha, H. D. Schmitt, C. Vorgias, and P. Wagner, *in* "The Guanine–Nucleotide Binding Proteins" (L. Bosch, B. Kraal, and A. Parmeggiani, eds.), NATO ASI Ser. A, Vol. 165, p. 257. Plenum, New York, 1989.

protein (Ypt1p)[5-7] and Sec4p,[8,9] fulfill essential functions in the vesicular transport of the secretory pathway. The localization of mammalian Ypt/Rab proteins to different compartments of the exocytic and endocytic pathway,[10-14] and the essential role of GTP hydrolysis for *in vitro* protein transport[15-18] suggest a similar function of small GTPases in mammalian cells.

This chapter deals mainly with Ypt1p[19] and Ypt6p[20] from the budding yeast *Saccharomyces cerevisiae,* two of the six presently known members of the Ypt subfamily in this unicellular eukaryote. Ypt1p is required for cell viability.[21,22] Among other phenotypic alterations, yeast cells depleted of Ypt1p[6] or conditional lethal *ypt1* mutants at the nonpermissive temperature[5,6,23] accumulate endoplasmic reticulum (ER), 50-nm vesicles, and the ER forms of core-glycosylated proteins bound to reach the plasma membrane or the vacuole, suggesting a role of this GTPase in protein transport from the ER to the Golgi apparatus.[24] This is supported by the findings that anti-Ypt1p antibodies inhibit ER-to-Golgi transport *in vitro*[7] and that several suppressors of the loss of *YPT1* gene function act, like Ypt1p, at an early stage of the secretory pathway.[25,26] The *YPT6* gene encodes a GTPase that is highly homologous to the mammalian Rab6p.[20]

[5] N. Segev, J. Mulholland, and D. Botstein, *Cell* **52,** 915 (1988).

[6] H. D. Schmitt, M. Puzicha, and D. Gallwitz, *Cell* **53,** 635 (1988).

[7] D. Baker, L. Wuestehube, R. Scheckman, D. Botstein, and N. Segev, *Proc. Natl. Acad. Sci. USA* **87,** 355 (1990).

[8] A. Salminen, and P. J. Novick, *Cell* **49,** 527 (1987).

[9] N. C. Walworth, B. Goud, A. K. Kabcenell, and P. J. Novick, *EMBO J.* **8,** 1685 (1989).

[10] P. Chavrier, R. G. Parton, H. P. Hauri, K. Simons, and M. Zerial, *Cell* **62,** 317 (1990).

[11] G. Fischer von Mollard, G. A. Mignery, M. Baumert, M. S. Perin, T. J. Hanson, P. M. Burger, R. Jahn, and T. C. Südhof, *Proc. Natl. Acad. Sci. USA* **87,** 1988 (1990).

[12] S. Araki, A. Kikuchi, Y. Hata, M. Isomura, and Y. Takai, *J. Biol. Chem.* **265,** 13007 (1990).

[13] B. Goud, A. Zahraoui, A. Tavitian, and J. Saraste, *Nature (London)* **345,** 553 (1990).

[14] J.-P. Gorvel, P. Chavrier, M. Zerial, and J. Gruenberg, *Cell* **64,** 915 (1991).

[15] P. Melançon, B. S. Glick, V. Malhotra, P. J. Weidman, T. Serafini, M. L. Gleason, L. Orci, and J. E. Rothmann, *Cell* **51,** 1053 (1987).

[16] Y. Goda and S. R. Pfeffer, *Cell* **55,** 309 (1988).

[17] C. J. M. Beckers and W. E. Balch, *J. Cell Biol.* **108,** 1245 (1989).

[18] S. A. Tooze, U. Weiss, and W. B. Huttner, *Nature (London)* **347,** 207 (1990).

[19] D. Gallwitz, C. Donath, and C. Sander, *Nature (London)* **306,** 704 (1983).

[20] L. Hengst and D. Gallwitz, unpublished results, 1991.

[21] H. D. Schmitt, P. Wagner, E. Pfaff, and D. Gallwitz, *Cell* **47,** 401 (1986).

[22] N. Segev and D. Botstein, *Mol. Cell. Biol.* **7,** 2367 (1987).

[23] J. Becker, T. J. Tan, H.-H. Trepte, and D. Gallwitz, *EMBO J.* **10,** 785 (1991).

[24] H. R. Bourne, *Cell* **53,** 669 (1988).

[25] C. Dascher, R. Ossig, D. Gallwitz, and H. D. Schmitt, *Mol. Cell. Biol.* **11,** 872 (1991).

[26] R. Ossig, C. Dascher, H.-H. Trepte, H. D. Schmitt, and D. Gallwitz, *Mol. Cell. Biol.* **11,** 2980 (1991).

YPT6 null mutuants[20] share many phenotypic alterations with a yeast mutant, *vps11 (end1*[27]*)*, that is defective in vacuolar protein sorting and endocytosis.

A detailed functional analysis of Ypt1p and other related proteins, their precise mode of action, their biochemical properties, and their interaction with cellular membranes and other proteins requires that purified GTP-binding proteins and monospecific antibodies be available. We describe here the methods that we have found to be useful for the bacterial production and the purification of Ypt proteins, and for the generation of specific antibodies to these proteins. We also describe procedures for measuring nucleotide binding and the intrinsic GTPase activity of Ypt1p.

The Sec4 protein, which also belongs to the Ypt subfamily of proteins in *S. cerevisiae,* is being dealt with in [33] of this volume.

Bacterial Expression of Yeast Ypt Proteins

General Considerations

For the biochemical and structural characterization of Ypt proteins and for the generation of antibodies to these proteins, large quantities of purified protein are required. The methods of choice are to express the genes or cDNAs encoding the GTP-binding proteins either in bacteria using appropriate expression vectors or in insect cells using the baculovirus expression system. Overproduction of Ypt1p in *S. cerevisiae* is possible, but the amount of protein per cell seems to be restricted due to a complex regulation of *YPT1* gene expression.[21] For functional analyses of Ypt proteins in *in vitro* transport systems or for analyzing the interaction of Ypt proteins with other cellular proteins, it might be necessary to use the GTP-binding proteins produced in eukaryotic cells as the *Escherichia coli*-expressed proteins lack the C-terminal lipid moieties that are required for efficient membrane association of Ypt proteins.[9,28,29] However, bacterially produced Ypt1p has been successfully employed for determining its nucleotide-binding properties and intrinsic GTPase activity[30] as well as for analyzing the acceleration of its low intrinsic GTPase activity by a seemingly Ypt-specific GTPase-activating protein, yptGAP, from porcine liver[23] and yeast.[31]

[27] V. Dulić and H. Riezmann, *EMBO J.* **8,** 1349 (1989).
[28] C. M. T. Molenaar, R. Prange, and D. Gallwitz, *EMBO J.* **7,** 971 (1988).
[29] G. Rossi, Y. Jiang, A. P. Newman, and S. Ferro-Novick, *Nature (London)* **351,** 158 (1991).
[30] P. Wagner, C. M. T. Molenaar, A. J. G. Rauh, R. Brökel, H. D. Schmitt, and D. Gallwitz, *EMBO J.* **6,** 2373 (1987).
[31] T. J. Tan, P. Vollmer, and D. Gallwitz, *FEBS Lett.* **291,** 322 (1991).

Bacterial Expression Vector

For the production in *E. coli* of Ypt1p, Ypt6p, and other GTP-binding proteins, a pBR322-derived expression vector with an unique *Nde*I restriction site providing the ATG translation initiation codon was designed.[30] The *Nde*I site was generated by mutating 2 bp next to the ATG of the *lacZ* gene in the bacteriophage M13mp8.[32,33] A 352-bp *Ava*II/*Hind*III fragment of the modified bacteriophage containing the regulatory elements of the *lac* operon and the multicloning region downstream from P*lac* was blunt-end ligated to the 2.3-kb *Pvu*II/*Eco*RI fragment of pBR322 from which the *Nde*I restriction site had been deleted. The resulting plasmid, pLN, has lost a functional *rop* gene that negatively regulates plasmid copy number in *E. coli*.[34] Through ligation of the filled-in *Hind*III site of the M13mp8 polylinker region with the *Pvu*II site of pBR322, the last 13 codons and the 3'-flanking region of the *rop* gene are fused to the multicloning site, providing a transcription termination signal for genes inserted into the multicloning site of the expression vector (Fig. 1).

An *Nde*I restriction site was also created by oligonucleotide-directed mutagenesis at the ATG initiation codon of the *YPT1* and the *YPT6* genes, allowing the insertion of the protein-coding region of the wild-type genes or their mutated versions as *Nde*I/*Bam*HI fragments into pLN and avoiding the production of hybrid protein. Expression of the foreign genes inserted into pLN is under direct control of the regulatory sequences of the *lac* operon. To prevent basal levels of foreign gene expression, *E. coli* strains JM101,[35] TG1,[36] or RB791[37] are the hosts of choice. These strains carry a mutated allele of the *lac* operon, *lacI*^{q}, which leads to an overproduction of repressor molecules ensuring the reprimed state of the additional operator copies of the expression vector in the cell.[33] After adding isopropyl-β-D-thiogalactopyranoside (IPTG), a potent synthetic inductor of the *lac* operon, a dramatic increase of Ypt1p and Ypt6p was observed, reaching a maximal level after 3–4 hr of induction. The bacterially produced proteins are readily soluble and amount to about 1–3% of the total cellular protein.

Preparation of Bacterial Protein Extracts

Two slightly different methods are used in our laboratory for preparing crude protein extracts from bacteria producing yeast GTP-binding pro-

[32] J. Messing and J. Vieira, *Gene* **19**, 263 (1982).
[33] C. Yanisch-Perron, J. Vieira, and J. Messing, *Gene* **33**, 103 (1985).
[34] T. Som and J.-I. Tomizawa, *Proc. Natl. Acad. Sci. USA* **80**, 3232 (1983).
[35] J. Messing, *Recomb. DNA Tech. Bull.* **2** (2), 43 (1979).
[36] T. J. Gibson, Ph.D. Thesis, Cambridge Univ.
[37] R. Brent and M. Ptashne, *Proc. Natl. Acad. Sci USA* **78**, 4204 (1981).

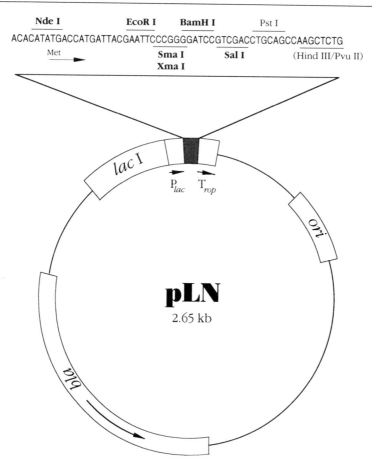

Nde I   EcoR I   BamH I   Pst I

ACACATATGACCATGATTACGAATTCCCGGGGATCCGTCGACCTGCAGCCAAGCTCTG

Met   Sma I   Sal I   (Hind III/Pvu II)

Xma I

P_{lac} T_{rop}

lacI

ori

pLN

2.65 kb

bla

FIG. 1. Bacterial expression vector pLN. The replication origin *(ori)*, the ampicillin resistance gene *(bla)*, and the termination region of the *rop* gene (T_{ROP}) are derived from plasmid pBR322. The regulatory region of the *lac* operon *(lacI)*, including the *lac* promoter (P_{lac}), and the multicloning site (dark box) are from phage M13mp8. The newly created *Nde*I restriction site provides the ATG translation initiation codon and is used with one of the 3′-located singular restriction sites (bold letters) for cDNA insertion.

teins. One involves the use of lysozyme to digest the bacterial cell wall, the other rests on sonication in the presence of a nonionic detergent.

Method 1

1. Grow 10 liters of *E. coli* harboring the recombinant pLN expression vector in L broth (5 g/liter Bacto-yeast extract, 10 g/liter Bacto-tryptone,

10 g/liter NaCl, 5 ml/liter 1 N NaOH) containing 50 μg/ml ampicillin at 37° to early log phase (OD_{595}: 0.3–0.5).

2. Induce Ypt1p or Ypt6p production by adding IPTG to a final concentration of 0.5 mM and continue bacterial growth for 3–4 hr with good aeration.

3. Harvest cells by centrifugation at 10,000 g for 10 min at 4° (yield: 20–25 g wet weight) and wash twice with 100 ml each of ice-cold washing buffer (50 mM Tris-HCl, pH 8.0, 100 mM NaCl). Cells can be stored frozen at −80°.

4. Suspend cells by vortexing after adding 12 ml of 50 mM Tris-HCl, pH 8.0, 10% (w/v) sucrose, 3 ml lysozyme (5 mg/ml), and 3 ml 500 mM ethylenediaminetetraacetic acid (EDTA), and incubate for 30 min at 37°. Add 30 ml of a detergent lysis mix [50 mM Tris-HCl, pH 8.0, 0.1% (v/v) Triton X-100, 62.5 mM EDTA] and continue incubation for further 45 min at 37°.

5. Sonicate cell suspension at 50 W (Sonifire B15; Branson, Danbury, CT) for a total of 3 min, allowing the suspension to cool on ice every 30 sec.

6. Clarify the lysate from cell debris and unbroken cells by centrifugation at 12,000 g for 20 min at 4°. The supernatant is used for chromatographic purification of the GTP-binding protein.

Method 2

Steps 1–3 are identical to those described under method 1.

4. Resuspend cells by vortexing in lysis buffer [20 mM Tris-HCL, pH 7.0, 10 mM MgCl$_2$, 1 mM NaN$_3$, 1 mM phenylmethylsulfonyl fluoride (PMSF), 1.5% (v/v) Triton X-100] at 3 ml buffer/g cells (wet weight).

5. Sonicate cells for 15 min on ice at a rate of 50% pulse time at 75 W using a Branson Sonifire B15 with a macrotip.

6. Remove cell debris and unbroken cells at 25,000 g for 10 min at 4°. Centrifuge the supernatant for an additional 30 min at 100,000 g and 4°.

Purification of Bacterial Expressed Ypt Proteins

Using a three-step procedure, we have successfully purified to near homogeneity a large variety of bacterially produced small GTP-binding proteins. The procedure includes ion-exchange chromatography on DEAE-Sephacel (Pharmacia, Freiburg, Germany) or DEAE-Sepharose CL-6B (Pharmacia), ammonium sulfate precipitation, and gel filtration on Sephadex G-150 or Superdex 200 (Pharmacia). We prefer DEAE-Sepharose CL-6B and Superdex 200, and prepacked columns have given the best results. The pH optimum for the binding of different GTPases to the ion exchanger, and the $(NH_4)_2SO_4$ concentration for precipitating different GTPases, vary and must be determined for each protein in a pilot experiment. To follow the purification, a simple GTP dot-blot analysis for monitoring different column fractions is recommended. The final purification of the GTP-binding protein can be assessed by sodium dodecyl sulfate-polyacrylamide gel electrophoresis (SDS-PAGE) on a 12.5% (w/v) acrylamide gel. An example for the final purification step of Ypt6p is given in Fig. 2.

DEAE-Sepharose Chromatography

The chromatographic purification of proteins is performed in a cold room at 4°.

1. Soluble bacterial proteins from 25 g of cells (~80 ml) are carefully mixed with an equal volume of ice-cold buffer A (20 mM Tris-HCl, pH 8.5 for Ypt1p, pH 7.0 for Ypt6p, 10 mM MgCl$_2$, 1 mM NaN$_3$) and applied to a 500 ml DEAE-Sepharose CL-6B column equilibrated with buffer A at a flow rate of 2 ml/min.

2. The column is washed with at least two column volumes of buffer A.

3. Bound proteins are eluted with a 2-liter 50–300 mM NaCl gradient in buffer A.

4. Twenty to 100 μl of the column fractions is spotted onto a nitrocellulose filter to identify the Ypt protein-containing fractions by a GTP-binding assay (see GTP Dot-Blot, below).

5. Relevant fractions are pooled for $(NH_4)_2SO_4$ precipitation.

Ammonium Sulfate Precipitation

1. To the solution containing the GTP-binding protein, slowly add a saturated $(NH_4)_2SO_4$ solution while stirring to give a final concentration of 50% saturation (Ypt6p) or 55% saturation (Ypt1p) and leave on ice for at least 1 hr.

2. Pellet precipitated proteins by centrifugation at 12,000 g for 15 min at 4°. Discard the pellet.

3. Slowly add solid $(NH_4)_2SO_4$ to the supernatant while stirring to give a final concentration of 60% (Ypt6p) or 70% saturation (Ypt1p) and leave on ice for at least 1 hr.

4. Pellet precipitated proteins by centrifugation at 100,000 g for 30 min at 4°.

5. Solubilize precipitated proteins in 5 ml of buffer A and dialyze

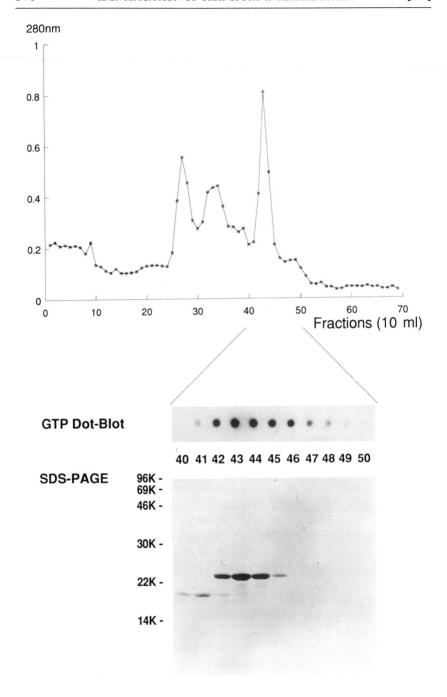

overnight in the cold room against buffer A containing 200 mM NaCl. Change the dialysis buffer several times.

6. Centrifuge the dialyzed protein solution at 12,000 g for 10 min before loading onto the Superdex 200 column.

Gel Filtration on Superdex 200

1. Apply protein solution to a prepacked 320-ml Superdex 200 column (Pharmacia) equilibrated with buffer A containing 200 mM NaCl.

2. Separate proteins at a flow rate of 0.5 ml/min and collect 10-ml fractions. Monitor the optical density at 280 nm.

3. Identify fractions containing Ypt1p, Ypt6p, or other GTPases to be purified by GTP dot-blot analysis.

4. Check relevant fractions by SDS-PAGE for the degree of purity of GTP-binding protein. On the average, 20–25 mg of Ypt1p or Ypt6p with a purity exceeding 90% is obtained. The protein is kept on ice for further analysis or shock-frozen in liquid nitrogen and stored at $-80°$.

GTP Dot-Blot

The GTP dot-blot is an easy and rapid procedure to identify chromatographic fractions containing the GTP-binding protein being purified.

1. Apply 20–100 μl of different column fractions to a nitrocellulose filter using a Bio-Rad (Richmond, CA) dot-blot chamber.

2. Rinse the filter immediately with binding buffer [50 mM NaH$_2$PO$_4$, pH 7.5, 10 μM MgCl$_2$, 2 mM dithiothreitol (DTT), 4 μM ATP, 0.3% (v/v) Tween 20] for 10 min at room temperature. Repeat the filter wash under identical conditions.

3. Incubate the nitrocellulose filter in binding buffer containing 1 μCi/ml of [α-^{32}P]GTP (3000 Ci/mmol; Amersham, Braunschweig, Germany) for 2 hr at room temperature.

4. Wash the filter six times for 2–5 min each in binding buffer. Dry the filter and expose to X-ray film [Kodak (Rochester, NY) X-Omat AR]. The filter can be developed after 10–60 min of exposure.

FIG. 2. Purification of bacterially produced *S. cerevisiae* Ypt6p by Superdex 200 chromatography following DEAE-Sepharose CL-6B chromatography and (NH$_4$)$_2$SO$_4$ fractionation. Starting material was total soluble protein from 25 g of *E. coli* RB791 cells expressing Ypt6p. Fractions (10 ml) were collected from which 50-μl aliquots were spotted onto a nitrocellulose filter for measuring GTP binding with [α-^{32}P]GTP (GTP dot-blot). Aliquots (25 μl) were subjected to SDS-PAGE [7.5% (w/v) stacking gel, 12.5% (w/v) separation gel, 18 × 14 cm] and electrophoresed for 3.5 hr at 150 V (constant). Proteins were stained with Coomassie Brilliant blue. Mobilities of molecular mass markers are indicated at left.

Nucleotide Binding of Ypt Proteins

General Considerations

Ypt proteins from the budding yeast *S. cerevisiae* and the evolutionary very distant fission yeast *Schizosaccharomyces pombe* bind guanine nucleotides specifically.[30,38-40] GDP/GTP binding of wild-type and mutant proteins can be assessed in two different ways, either with proteins electrophoretically transferred to nitrocellulose filters (GTP blot) or with purified proteins in solution.

Ypt proteins, which are purified to homogeneity, partially purified, or contained in total protein extracts from yeast or bacteria, can be separated by SDS-PAGE,[41] transferred to nitrocellulose filters, and probed for nucleotide binding.[30,42,43] Boiling of protein samples in the presence of SDS leads to complete denaturation of proteins and the release of protein-bound nucleotides. Proper refolding of small GTPases demands the removal of SDS, which is mainly achieved during the electrophoretic transfer from the gel to the filter and by treating the filter-bound proteins in an appropriate buffer. Radiolabeled GTP or GDP is then used in the binding reaction. Although this is a convenient procedure for a qualitative assessment of nucleotide binding, certain restrictions must be considered. First, the method seems to work well only with nucleotide-binding proteins of small size and high binding affinity. Second, different thermodynamics of refolding of wild-type and mutant proteins might be envisaged, and this does not allow direct comparison of their nucleotide-binding capacity.

As small GTPases are usually isolated in a complex with GDP,[44] accurate measurements of nucleotide-binding constants are possible only with nucleotide-free proteins.[45] Although nucleotide-free proteins are thermally less stable than proteins complexed with nucleotides, the association and dissociation rate constants have been determined for several Ras and Ras-related proteins,[46-48] including Ypt1p from yeast.[47]

[38] B. Goud, A. Salminen, N. C. Walworth, and P. J. Novick, *Cell* **53**, 753 (1988).

[39] L. Hengst, T. Lehmeier, and D. Gallwitz, *EMBO J.* **9**, 1949 (1990).

[40] H. Haubruck, U. Engelke, P. Mertins and D. Gallwitz, *EMBO J.* **9**, 1957 (1990).

[41] U.K. Laemmli, *Nature (London)* **227**, 680 (1970).

[42] J. P. McGrath, J. D. Capon, D. V. Goeddel, and A. D. Levinson, *Nature (London)* **310**, 644 (1984).

[43] E. G. Lapetina and B. R. Reep, *Proc. Natl. Acad. Sci. USA* **84**, 2261 (1987).

[44] M. Poe, E. M. Scolnick, and R. B. Steitz, *J. Biol. Chem.* **260**, 3906 (1985).

[45] J. Feuerstein, H. R. Kalbitzer, J. John, R. S. Goody, and A. Wittinghofer, *Eur. J. Biochem.* **162**, 49 (1987).

[46] J. Feuerstein, R. S. Goody, and A. Wittinghofer, *J. Biol. Chem.* **262**, 8455 (1987).

[47] P. Wagner, Ph.D. Thesis, Univ. of Marburg.

[48] M. Frech, I. Schlichting, A. Wittinghofer, and P. Chardin, *J. Biol. Chem.* **265**, 6353 (1990).

GTP Blot

In our laboratory, we use a method that has been modified from previously published procedures[42,43] and suggested to us by Zerial.[49]

Protocol

1. Separate proteins by SDS-PAGE, using a 12.5% (w/v) acrylamide gel.
2. Soak the gel for 30 min in 1 liter of 50 mM Tris-HCl, pH 7.5, 20% (v/v) glycerol.
3. Transfer proteins electrophoretically onto a nitrocellulose filter overnight at a constant current of 150 mA in transfer buffer (10 mM NaHCO$_3$/3 mM Na$_2$CO$_3$, pH 9.8).
4. After transfer, immediately rinse the blot twice for 10 min each in binding buffer [50 mM NaH$_2$PO$_4$, pH 7.5, 10 μM MgCl$_2$, 2 mM DTT, 0.3% (v/v) Tween 20, 4 μM ATP].
5. Incubate the nitrocellulose filter in binding buffer containing 1 μCi/ml [α-^{32}P]GTP (3000 Ci/mmol; Amersham) for 2 hr.
6. Wash the filter six times for 2 to 5 min each in binding buffer.
7. Dry the filter and expose to X-ray film.

Preparation of Nucleotide-Free Ypt1p

A high-performance liquid chromatography (HPLC) method developed by Feuerstein *et al.*[46] to study the kinetics of nucleotide–p21^{ras} interaction has been successfully used to prepare Mg^{2+}- and nucleotide-free yeast Ypt1p for affinity measurements.[47] This method is based on hydrophobic interaction chromatography and exploits the significantly increased nucleotide exchange rate in the absence of Mg^{2+}.[45,50]

Protocol

1. Fifty microliters of purified Ypt1p (5 mg/ml) is mixed with 500 μl 2 M (NH$_4$)$_2$SO$_4$, 10 mM EDTA, and incubated for 5 min at room temperature.
2. The mixture is injected into an HPLC system using a TSK-phenyl-5PW, 7.5 × 1 cm column (Beckman, Palo Alto, CA), applying a flow rate of 0.5 ml/min.
3. The column is washed with buffer A (20 mM Tris-HCl, pH 8.0,

---

[49] M. Zerial, personal communication, 1991.
[50] A. Hall and A. J. Self, *J. Biol. Chem.* **261**, 10963 (1986).

5 mM EDTA, 5 mM DTT) containing 1 M $(NH_4)_2SO_4$. Separated nucleotides and other impurities appear in the flow-through.

4. Nucleotide-free Ypt1p is eluted by applying a linear decreasing 1 M–10 mM $(NH_4)_2SO_4$ gradient in buffer A for 10 min.

5. Relevant fractions of 1 ml are collected and kept on ice.

6. Nucleotide-free Ypt1p is immediately desalted at 4° on a 1 × 10 cm column of Sephadex G-25 superfine equilibrated with 50 mM Tris-HCl, pH 8.0, 0.5 mM DTT and can be concentrated in Centricon 10 tubes (Amicon, Danvers, MA) at 4°.

More than 90% of the Ypt1p retains its nucleotide-binding capacity for several hours, provided the protein is kept on ice.

Kinetic Measurements

Nucleotide- and Mg^{2+}-free Ypt1p is used to determine the rate constants for association (k_A) and dissociation (k_D) of the protein and guanine nucleotides. Assuming a kinetic mechanism of nucleotide binding given by

$$Ypt1p + GDP \underset{k_D}{\overset{k_A}{\rightleftharpoons}} Ypt1p \cdot GDP \tag{1}$$

the association follows second-order and the dissociation first-order kinetics. If the GDP concentration is much higher than that of Ypt1p, conditions for pseudo-first-order kinetics can be applied for the association. In this case, the half-time is given by

$$t_{\frac{1}{2}} = 1/k_A \text{ (GDP)} \tag{2}$$

whereas for the dissociation the relation

$$t_{\frac{1}{2}} = \ln 2/k_D \tag{3}$$

is valid. The rate constant of binding (K) is calculated from the ratio of association and dissociation rate constants:

$$K = k_A/k_D \tag{4}$$

With an association rate constant of 8.2×10^5 M^{-1} sec^{-1} and a dissociation rate constant of 3.76×10^{-4} sec^{-1} at 30°, the binding affinity constant of Ypt1p was determined to be about 2.2×10^9 M^{-1},[47] which is in the same range as the binding constant for p21$^{ras} \cdot$ GDP.[46]

Determination of Association Rate Constant of Ypt1p and GDP

1. Incubate 3 nM nucleotide-free Ypt1p and 12 nM [8-^{3}H]GDP (10–15 Ci/mmol; Amersham) at the desired temperature in 15 ml of binding buffer (50 mM Tris-HCl, pH 8.0, 0.5 mM DTT, 10 mM MgCl$_2$).
2. Every 30 sec (reaction is complete within 30 min at 30°), withdraw 1-ml aliquots from the incubation mixture and immediately pass through nitrocellulose filters (BA85, 0.45 μm; Schleicher & Schuell, Dassel, Germany).
3. Wash immediately twice with 5 ml each of ice-cold binding buffer.
4. Dry filters for 10 min at 80° and transfer into scintillation vials. Dissolve filters in 5 ml of Filter-Solv (Beckman) by shaking vigorously for 30 min.
5. Determine radioactivity in a scintillation counter and plot radioactivity versus time.
6. Calculate association rate constant according to

$$k_A = 1/(t_{\frac{1}{2}} \text{ GDP}) \tag{5}$$

Determination of Dissociation Rate Constant of Ypt1p · GDP

1. Incubate 3 nM nucleotide-free Ypt1p and 12 nM [8-^{3}H]GDP at the desired temperature in 15 ml of binding buffer for 30 min.
2. Add unlabeled GDP to give a final concentration of 1 mM.
3. Every 10 min (the reaction is complete within about 3 hr at 30°), withdraw 500-μl aliquots from the incubation mixture and immediately pass through nitrocellulose filters.
4. Follow steps 3–5 of the previous protocol and calculate dissociation rate constant according to

$$k_D = \ln 2/t_{\frac{1}{2}} \tag{6}$$

Determination of Intrinsic GTPase Activity

Small GTP-binding proteins belonging to the Ras superfamily possess a slow intrinsic GTPase activity. At 37°, values of 2.8×10^{-2} min^{-1} for H-Ras p21,[51] 7.2×10^{-2} min^{-1} for Ralp,[48] 4×10^{-3} min^{-1} for Rap1Ap,[52] and $5 \times 10^{-3} - 5 \times 10^{-2}$ min^{-1} for different Rab proteins[53] have been obtained. A GTPase reaction rate constant of 6×10^{-3} min^{-1} has been measured for yeast Ypt1p at 30°.[30,47]

[51] J. John, I. Schlichting, E. Schiltz, P. Rösch, and A. Wittinghofer, *J. Biol. Chem.* **263**, 11792 (1988).
[52] M. Frech, J. John, V. Pizon, P. Chardin, A. Tavitian, R. Clark, F. McCormick, and A. Wittinghofer, *Science* **249**, 169 (1990).
[53] A. Zahraoui, N. Touchot, P. Chardin, and A. Tavitian, *J. Biol. Chem.* **264**, 12394 (1989).

The intrinsic GTPase activity of Ras and Ras-related proteins can be accelerated several hundredfold following interaction with specific GTPase-activating proteins (GAP).[2,3,54] GTPase-activating protein activities with specificity for small GTPases involved in intracellular transport, including yeast Ypt1p[31] and its mammalian counterpart Ypt1p/Rab1p,[23] have also been reported.

To measure the GTPase activity of Ypt proteins, the protein-bound nucleotide must be exchanged with either $[\gamma\text{-}^{32}\text{P}]$GTP or $[\alpha\text{-}^{32}\text{P}]$GTP in the absence of Mg^{2+}, and after adding $MgCl_2$, the time-dependent production of either $^{32}P_i$ or $[\alpha\text{-}^{32}\text{P}]$GDP is followed and used to calculate the hydrolysis rate constant. Alternatively, $[\gamma\text{-}^{32}\text{P}]$GTP remaining in the protein–nucleotide complex after different periods of incubation can be measured after filtration of the incubation mixture through nitrocellulose filters.

Protocol

1. Five hundred microliters of 0.5 μM purified Ypt1p, 50 mM Tris-HCl, pH 8.0, 1 mM DTT, 1 mM EDTA, 1 mM NaN$_3$, 2.5 μM unlabeled GTP, and 0.05 μM $[\gamma\text{-}^{32}\text{P}]$GTP (5000 Ci/mmol; Amersham) is incubated at 30° for 15 min to exchange protein-bound GDP with GTP.

2. The incubation mixture is immediately passed at 4° over a Sephadex G-25 column (NAP-5; Pharmacia) previously equilibrated with 10 ml of GTPase buffer (50 mM Tris-HCl, pH 8.0, 0.5 mM DTT, 1 mM ATP) to remove excess nucleotides. Fractions of about 100 μl are collected.

3. Fractions containing the protein–GTP complex are pooled and after adjusting the complex to 0.2 μM in 500 μl with GTPase buffer, GTP hydrolysis is started at 30° by the addition of $MgCl_2$ to a final concentration of 2.5 mM.

4. At different time points, duplicate samples of 25 μl are withdrawn from the incubation mixture and immediately filtered through nitrocellulose filters (BA85, 0.45 μm; Schleicher & Schuell).

5. Filters are washed twice with 5 ml each of ice-cold wash buffer (20 mM Tris-HCl, pH 8.0, 5 mM $MgCl_2$, 100 mM KCl, 10 mM NH$_4$Cl, 1 mM 2-mercaptoethanol) and dried at 80° for 10 min.

6. Filters are transferred to scintillation vials and dissolved with 5 ml of Filter-Solv (Beckman) by agitation for 30 min at 4°.

7. Radioactivity is counted and the data are plotted as the percentage of total Ypt1p·GTP complex remaining at different times of incubation.

[54] M. Trahey and F. McCormick, *Science* 230, 542 (1987).

Generation of Specific Anti-Ypt Protein Antibodies

Background

Antibodies specific for a given Ypt protein are invaluable tools for characterizing the protein in question (functional domains), for analyzing its intracellular localization (immunofluorescence, immunoblot analysis after cellular fractionation), and for studying its function (precipitation of complexes with other proteins, inhibition of defined *in vitro* transport reactions, etc.). Polyclonal anti-Ypt1p antibodies raised against bacterial fusion proteins[21,22] have been used for indirect immunofluorescence,[22] for studying membrane association of Ypt1p,[28,29] and for investigating the role of this protein in ER-to-Golgi vesicle transport *in vitro*.[7,55]

To date, 12 different members belonging to the Ras superfamily of proteins in *S. cerevisiae* are known. Because all of these proteins share several regions of highly conserved primary sequence, it is not a trivial task to obtain antibodies that do not cross-react with related antigens. However, the rather variable sequence of the C-terminal 40 amino acid residues in Ras and Ras-related proteins allows the generation of specific antibodies against synthetic C-terminal peptides. This approach has been used to differentiate several mammalian Ypt/Rab proteins and to localize them to different compartments of the exo- and endocytic pathway by immunoelectron microscopy.[10] The usefulness of this approach is also documented by our observation that a monoclonal anti-peptide antibody generated against the C-terminal 14 amino acids of the mouse Ypt1p/Rab1p(IQSTPVKQSGGGCC)[56] reacts specifically with the mouse protein but not with its *S. cerevisiae* homolog Ypt1p. The degree of primary structure identity between the mouse Ypt1p/Rab1p and the *S. cerevisiae* Ypt1p exceeds 70%,[57] but the C terminus of the yeast protein is LKGQSLTNTGGGCC.[19] Polyclonal antibodies against the latter peptide have been raised in rabbits and they do not cross-react with mouse Ypt1p/Rab1p.[56]

Another region of small GTPases to which anti-peptide antibodies with specificity for different members of the Ras superfamily might be obtained is the so-called effector domain[40,58] and its surrounding sequence. In p21,[H-*ras*] part of the effector domain and several amino acid residues

[55] N. Segev, *Science* **252**, 1553 (1991).
[56] M. Puzicha and D. Gallwitz, unpublished results, 1989.
[57] H. Haubruck, C. Disela, P. Wagner, and D. Gallwitz, *EMBO J.* **6**, 4049 (1987).
[58] J. S. Sigal, J. B. Gibbs, J. S. D'Alonzo, G. L. Temeles, B. S. Wolanski, S. H. Socher, and E. M. Scolnick, *Proc. Natl. Acad. Sci. USA* **83**, 952 (1986).

preceding it (amino acids 26–36) form a loop (L2)[59] that is the site of interaction with GAP.[60] This loop region differs in different superfamily members.[61] We have found that polyclonal anti-peptide antibodies raised against the corresponding loop region of *S. pombe* Ryh1p[39] did react with the fission yeast protein but not with Ypt6p of *S. cerevisiae*,[20] although the sequence to which the antibody was raised differs between the two proteins in only 3 out of 14 amino acids.

It is not the purpose of this chapter to provide details about raising, purifying, and using antibodies, and the reader is referred to other excellent reviews.[62,63] We instead describe a few principle methods that are used in our laboratory to generate and characterize antibodies against Ypt proteins and synthetic peptides.

Generation of Polyclonal Anti-Ypt1p Antibodies

At least two rabbits should be immunized, and 20 ml of blood should be collected to obtain preimmune serum for control purposes.

Immunization

1. Mix 500 μl of phosphate-buffered saline (PBS; 10 mM sodium phosphate buffer, pH 7.2, 150 mM NaCl) containing 1 mg of bacterially produced, purified Ypt1p with 500 μl of Freund's complete adjuvant and form a viscous emulsion.

2. Inject the emulsion subcutaneously at 10 different sites of the back of the rabbit.

3. At 2- to 3-week intervals, inject the same amount of protein emulsified in Freund's incomplete adjuvant following the procedure as described in step 2.

4. Collect 5–10 ml of blood from the marginal ear vein at 10–14 days after boosting the animal and test the serum for the presence of anti-Ypt1p antibodies. A sufficiently high titer of antibody is usually reached after the third boost.

5. Bleed the animal 2 weeks after the final boost and proceed to IgG purification.

[59] E. F. Pai, W. Kabsch, U. Krengel, K. C. Holmes, J. John, and A. Wittinghofer, *Nature (London)* **341**, 209 (1989).
[60] F. McCormick, *Cell* **56**, 5 (1989).
[61] A. Valencia, P. Chardin, A. Wittinghofer, and C. Sander, *Biochemistry* **30**, 4637 (1991).
[62] E. Harlow and D. Lane, "Antibodies: A Laboratory Manual." Cold Spring Harbor Lab., Cold Spring Harbor, New York, 1988.
[63] J. R. Pringle, A. E. M. Adams, D. G. Drubin, and B. K. Haarer, this series, Vol. 194, p. 565.

Isolation of IgG Fraction

1. Allow blood to clot for 1 hr at room temperature in 30-ml Corex tubes. Carefully remove and store the serum at 4° overnight. All further steps are performed at 4°.

2. Centrifuge for 20 min at 10,000 g and carefully remove the serum from the precipitate.

3. While stirring, slowly add 25 g of solid $(NH_4)_2SO_4$ to 100 ml serum.

4. Centrifuge for 20 min at 25,000 g, resuspend the precipitate in 25 ml of 1.75 M $(NH_4)_2SO_4$, and centrifuge again under the same conditions.

5. Dissolve the pellet in 100 ml of distilled water and dialyze against PBS for at least 12 hr with several buffer changes. The IgG fraction can be further purified by chromatography on DEAE- or protein A-Sepharose (optional).

Affinity Purification of Anti-Ypt1p Antibodies

The antibodies should be purified by affinity chromatography on an antigen-bound matrix. We use either CNBr-activated Sepharose 4B (Pharmacia) or AminoLink (Pierce, Rockford, IL) to immobilize purified Ypt1p.

Protocol

1. Fifty milligrams of bacterially produced, purified Ypt1p is coupled to 15 ml CNBr-activated Sepharose 4B according to the recommendations of the manufacturer.

2. Transfer the slurry to a glass column and wash with 500 ml PBS.

3. Slowly recirculate 50 ml of IgG (step 5 of the previous section) over the affinity column at 4° overnight.

4. Wash the column with 500 ml PBS.

5. Elute Ypt1p-specific antibodies with 100 mM glycine-HCl, pH 2.8. Collect 1-ml fractions and monitor the elution of protein at A_{280}.

6. Immediately neutralize by adding 50 μl of 1 M Tris-HCl, pH 9.5, to each fraction.

7. Pool fractions containing protein and dialyze against PBS containing 0.02% (w/v) NaN$_3$. Determine the protein concentration and store in aliquots (about 1 mg/ml) at $-80°$.

Calculation of Antibody Titer

1. Spot 1-μl aliquots of antigen [0.5 mg/ml in TBS (20 mM Tris-HCl, pH 7.4, 150 mM NaCl)] onto several nitrocellulose filters (1 × 1 cm) and let dry at room temperature.

2. Transfer one filter into each well of a cell culture dish (Multidish; Nunc, Roskilde, Denmark) and saturate unbound protein binding sites with 0.5 ml of TBS-BSA [TBS containing 2% (w/v) bovine serum albumin] for each filter. Incubate at room temperature for 1 hr.

3. Remove the solutions with a pipette and replace with 0.4 ml of TBS–bovine serum albumin (BSA) containing 1/50–1/30,000 dilutions of affinity-purified antibody or serum. Slowly shake the culture dish for 1 hr at room temperature.

4. Wash the filters five times for 3–5 min each with 0.4 ml of TBS.

5. To each filter, add 0.5 ml of horseradish peroxidase-linked donkey anti-rabbit IgG (Amersham) at a dilution of 33 μl/100 ml TBS–BSA and incubate for 1 hr.

6. Wash the filters as described under step 4.

7. Add 0.5 ml of a solution freshly prepared by mixing 6 mg of 4-chloro-1-naphthol (Bio-Rad, Richmond, CA) dissolved in 2 ml of ice-cold methanol, 10 ml TBS, and 6 μl 30% (v/v) H_2O_2 (Merck, Darmstadt, Germany). Incubate for 3–5 min and monitor the antigen–antibody reaction by color development.

Coupling of Synthetic Peptides to Keyhole Limpet Hemocyanin

Because of the importance of synthetic peptides for the generation of antibodies specific for different small GTPases, we describe two methods for coupling peptides to a carrier protein. We routinely use keyhole limpet hemocyanin (KLH) as carrier, which can be obtained in solubilized form in 50% glycerol (Calbiochem, San Diego, CA). Depending on the amino acid composition, the peptides are coupled to the carrier protein either with 1-ethyl-3-(3-dimethylaminopropyl) carbodiimide hydrochloride (EDC)[64] or with glutaraldehyde.[65,66] We use peptides 15 amino acid residues in length, but antibodies to 10-amino acid peptides have also been obtained.[47] Several kits for coupling synthetic peptides to carrier proteins are commercially available (Pierce).

EDC Coupling

1. Prepare a solution of 15 mg KLH (60 mg/ml in 50% glycerol; Calbiochem) in 600 μl of distilled water (pH adjusted to pH 3 with HCl).

2. Add 18 mg of EDC to the KLH solution and keep on ice for 15 min.

3. Dissolve 5 mg of peptide in 300 μl of 5% (w/v) (NH$_4$)HCO$_3$ and add 700 μl distilled water.

[64] M. Reichlin, this series, Vol. 70, p. 159.
[65] S. Bauminger and M. Wilchek, this series, Vol. 70, p. 151.
[66] T. E. Kreis, EMBO J. **5**, 931 (1986).

4. Mix the EDC-treated KLH (step 2) and peptide solution (step 3), and incubate with agitation for 3 hr at room temperature.

5. Dialyze overnight against several changes of PBS.

6. For immunization of rabbits, use 200 μg of coupled peptide and follow the procedure as described above.

Glutaraldehyde Coupling

1. Prepare a 2 ml solution of peptide and KLH (final concentrations, 3 and 7 mg/ml, respectively) in PBS. Perform the following steps in a fume hood.

2. Add 3.2 μl of a 12.5% (v/v) aqueous glutaraldehyde solution [stock solution 25%, electron microscopy (EM) grade I, Sigma, St. Louis, MO) store at $-80°$].

3. Stir for 5 min at room temperature on a magnetic stirrer.

4. Repeat steps 2 and 3 five times.

5. Follow steps 5 and 6 as described under EDC Coupling (above).

The protocols for immunization, preparation of serum, partial purification of IgG fraction, affinity purification, and calculation of antibody titer are the same as described above.

Acknowledgments

We are thankful to members of the Gallwitz laboratory for discussion and to K. Larson-Becker for secretarial assistance. This work was supported in part by grants to D.G. from the Deutsche Forschungsgemeinschaft and the Bundesministerium für Forschung und Technologie.

[36] Rab Proteins and Gene Family in Animals

By ARMAND TAVITIAN and AHMED ZAHRAOUI

Introduction

The acronym "rab" was first used in 1987 when four additional members of the *ras* gene family were isolated from a rat brain cDNA library by means of synthetic oligonucleotide probes.[1] It was readily observed that these mammalian *rab* genes constituted a distinct branch of the *ras* superfamily, more closely related to two genes discovered in *Saccharomyces*

---

[1] N. Touchot, P. Chardin, and A. Tavitian, *Proc. Natl. Acad. Sci. USA* **84,** 8210 (1987).

cerevisiae: the *YPT* gene found serendipitously in 1983 by Gallwitz *et al.*[2] and the then newly characterized *SEC4* gene involved in yeast secretion.[3]

rab genes have been the focus of intense research and, indeed, this branch of research has grown. Additional members have been characterized by several groups: the YPT cDNA of mouse, capable of complementing that of yeast, confirmed that *rab1 (rab1A)* was the homolog of the yeast *YPT* gene.[4] G proteins purified from bovine brain permitted the design of oligonucleotide that revealed additional members related to *rab3* (denoted *smg25A, smg25B,* and *smg25C*).[5] The search for human counterparts of the rat *rab* cDNAs in human cDNA libraries revealed the same four genes *(rab1, rab2, rab3, rab4)* and additionally *rab3B, rab5,* and *rab6*.[6] BRL-*ras*, isolated from a rat liver cell line cDNA library,[7] is referred to as *rab7*. Another clone, very close to *rab1A*, was isolated from a rat brain cDNA library and denoted *rab1B*.[8] More recently additional genes, *rab8, rab9, rab10,* and *rab11*, were characterized in canine cells.[9] The fission yeast *Schizosaccharomyces pombe* has three genes, *YPT1, YPT2,* and *Ryh,* for which the human counterpart is *rab6*.[10]

At present, there are some 16 members, characterized in mammals, that pertain to the *rab* branch of the family. This figure may represent only a fraction of the total number of the existing Rab proteins, even though different approaches have often led to the rediscovery of previously known members. One may foretell that there are, perhaps, as many as 30 different genes pertaining to the *rab* family.

The small G proteins of the Rab branch have molecular weights of 22K–27K. They are well conserved throughout the phylogeny and especially between mammalian species. For instance, the percentage of identity between the human Rab1A and the YPT protein is 75%. The rat and the human Rab1A proteins are identical. There is only one conservative change between the human and rat Rab3A proteins. The four domains of the Ras proteins that were shown to be involved in the binding of GTP/GDP and which correspond to some seven or eight amino acid residues around positions 15, 60, 115, and 145 are highly conserved. Some varia-

[2] D. Gallwitz, C. Donath, and C. Sander, *Nature (London)* **306,** 704 (1983).

[3] A. Salminen and P. J. Novick, *Cell* **49,** 527 (1987).

[4] H. Haubruck, R. Prange, C. Vorgias, and D. Gallwitz, *EMBO J.* **8,** 1427 (1989).

[5] Y. Matsui, A. Kikuchi, J. Kondo, T. Hishida, Y. Teranishi, and Y. Takai, *J. Biol. Chem.* **263,** 11071 (1988).

[6] A. Zahraoui, N. Touchot, P. Chardin, and A. Tavitian, *J. Biol. Chem.* **264,** 12394 (1989).

[7] C. Bucci, R. Frunzio, L. Chiarotti, A. L. Brown, M. M. Rechier, and C. B. Bruni, *Nucleic Acids Res.* **16,** 9979 (1988).

[8] E. Vielh, N. Touchot, A. Zaharoui, and A. Tavitian, *Nucelic Acids Res.* **17,** 1770 (1989).

[9] P. Chavrier, R. G. Parton, H. H. Hauri, K. Simons, and M. Zerial, *Cell* **62,** 317 (1990).

[10] L. Hengst, T. Lehmeier, and D. Gallwitz, *EMBO J.* **9,** 1949 (1990).

tions are observed, however, in region 10–17, where the glycines in position 12 and sometimes 13 of Ras are replaced by other conservative (or even nonconservative for Rab6) amino acids.

Two additional sequences corresponding to residues 35–40 (known as the effector region in Ras) and 51–85 are highly conserved among all the Rab proteins. In particular, region 35–40 points to different effectors from the Ras. There are probably different effectors for the Rab proteins; it is remarkable that the YPT2 protein, whose sequence in the effector domain is identical to the corresponding SEC4 sequence, can complement the SEC4 mutations and cannot complement the YPT1 mutations.

Isolation of *Rab* cDNAs

There are several well-defined procedures for the isolation of a cDNA insert coding for a protein of interest. In recent years different techniques have been developed to isolate *Rab* cDNAs encoding 23K–27K Ras-like small GTP-binding proteins.[1,4,9] The preferred technique utilizes cDNA libraries made in λgt10 cloning vector. Large number of plaques can be screened with different probes. In our laboratory, an improved oligonucleotide strategy has been applied to detect a number of *Rab* sequences in cDNA libraries. This strategy was based on the use of degenerate oligonucleotide probes coding for the DTAGQE amino acid sequence in positions 57–62 strictly conserved (except in Rap/D-Ras3 proteins) in Ras and Ras-related proteins identified so far. Moreover, this sequence is not found in the heterotrimeric G proteins and in other nucleotide-binding proteins. The degenerate oligonucleotide was used to screen a λgt10 rat brain cDNA library (for details see Ref. 1). Four cDNAs encoding four Rab proteins were isolated. The same method, utilizing a degenerate oligonucleotide based on the conserved sequence WDTAGQE in Rab, Rho, yeast YPT1, and SEC4 proteins, had led to the identification of new Rab proteins.[11] Other techniques, such as the screening of cDNA libraries at low stringency with known *Rab* cDNA probes, also led to the isolation of additional Rab proteins.[6]

Construction of Rab/pET-3c Expression Vector

Methods for expressing large amount of protein from cloned cDNA coding region introduced in *Escherichia coli* have been described and have proved valuable for the biochemical and functional analysis of proteins.

---

[11] P. Chavrier, M. Vingron, C. Sander, K. Simons, and M. Zerial, *Mol. Cell. Biol.* **10,** 6578 (1990).

Because most of the Rab proteins are not expressed efficiently in ptac vector,[12] we used the expression vector pET-3c. This system utilizes a bacteriophage T7 RNA polymerase/promoter system.[13] The pET-3c vector allows cloning of target DNAs at sites where they will be selectively and actively transcribed by T7 RNA polymerase *in vitro* and in *E. coli* cells. Transcription is controlled by the strong $\phi 10$ promoter for T7 RNA polymerase. pET-3c is a derivative of pBR322 that carries sequentially the bacteriophage T7 gene $\phi 10$ promoter, the ribosome-binding site, a translation start (ATG), and a *Bam*HI cloning site. The nucleotide sequence upstream from the gene *10* initiation codon contains a *Nde*I site that includes the ATG. Thus, the *Nde*I site is unique and can be useful for joining coding regions to the ATG initiator codon without making any change in the coding sequence. pET-3c was used for production of intact, native Rab proteins. In fact, any ATG start codon could be joined at the *Nde*I site (Fig. 1).

In Vitro Direct Mutagenesis

Site-directed mutagenesis to introduce an *Nde*I site upstream from the coding region is performed on single-stranded M13 templates with oligonucleotides synthesized on an automatic DNA synthesizer.

Step 1: Kinasing Oligonucleotide for Mutagenesis

Mix in an Eppendorf tube:

Oligonucleotide primer (100 pmol), 10 μl
10× kinase buffer (500 mM Tris-HC1, pH 8.0, 100 mM MgCl$_2$), 2 μl
Dithiothreitol (DTT; 100 mM), 1 μl
ATP (10 mM), 2 μl
Distilled water, 4 μl
Polynucleotide kinase, 5 units

Incubate for 30 min at 37°, 10 min at 70°, and then store at $-20°$.

Step 2: Annealing the Template, the Mutagenic Kinased Oligonucleotide, and the Universal M13 Sequencing Primer

Because the universal primer lies at the 5′ side of the mutagenic primer, it does not need to be kinased.

[12] J. Tucker, G. Sczakiel, J. Feuerstein, J. John, R. S. Goody, and A. Wittinghofer, *EMBO J.* **5**, 1351 (1986).
[13] A. H. Rosenberg, B. N. Lade, D. S. Chui, S. W. Lin, J. J. Dunn and F. W. Studier, *Gene* **56**, 125 (1987).

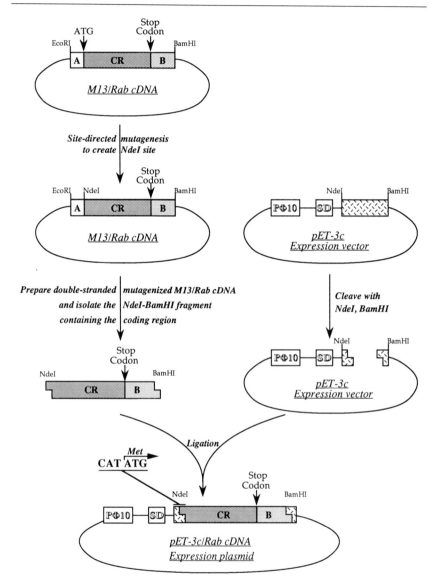

FIG. 1. Scheme for the construction of *rab*/pET-3c expression vector. pET-3c expression vector carries the bacteriophage T7 φ*10* promoter (Pφ*10*), a Shine–Delgarno (SD) sequence, and 11 amino acids of the T7 bacteriophage major capsid protein inserted between *Nde*I and *Bam*HI restriction sites. The *Nde*I site (CATATG) is located at the translation site and used to construct the *rab*/pET-3c plasmid. The Rab cDNA in M13 is represented by a box; A, noncoding sequence; CR, coding region; B, 3′-noncoding sequence.

Mix in an Eppendorf tube:

Kinased mutagenic oligonucleotide (10 pmol), 2 μl
M13 universal primer (10 pmol), 1 μl
M13 template (1 μg), 1 μl
10× TM buffer (100 mM Tris-HCl, pH 8.0, 100 mM MgCl$_2$, 1 μl
Distilled water, 5 μl

Heat to 80° and let cool to room temperature.

Step 3: Extension–Ligation

Add the following compounds to the tube:

10× TM buffer 1 μl
dNTPs (5 mM), 1 μl
ATP (5 mM), 1 μl
DTT (100 mM), 1 μl
Distilled water, 4 μl

Add 10 units of T4 DNA ligase and 2 units of DNA polymerase Klenow fragment (cloned) (Amersham, Arlington Heights, IL). Then incubate overnight at 14°.

Ligation mixture (0.5 to 1.0 μl) is used to transform *E. coli* strain JM105. M13 plaques are transferred to nitrocellulose filters and hybridized to ^{32}P-labeled oligonucleotide. Hybridization is carried out at a temperature corresponding to $T_m = 3GC + 2AT$, in 5 SSPE, 5× Denhardt's solution, 0.1% (v/v) sodium dodecyl sulfate (SDS) for 4 hr. Filters are washed in 2× SSC (1× SSC: 0.15 M NaCl, 0.015 M sodium citrate), 0.1% (w/v) SDS for 15 min at room temperature, and 20 min at $T_m + 2°$. Single-stranded DNA is prepared from positive plaques and sequenced to confirm the mutagenesis.

Expression of Rab Protein in *Escherichia coli*

1. After the introduction of the *Nde*I site upstream the *rab* coding sequence, prepare the corresponding double-stranded M13 vector, and cut with *Nde*I–*Bam*HI.

2. Clone the *Nde*I–*Bam*HI insert containing the complete *rab* coding sequence, including the ATG codon, into the bacteriophage T7 promoter expression plasmid pET-3c.

3. Transform *E. coli* strain BL21(DE3) *lysS* and select for ampicillin-resistant (50 μg/ml) and chloramphenicol-resistant (12.5 μg/ml) transformants. This strain is a lysogen bearing the bacteriophage T7 polymerase gene under the control of the *lacUV5* promoter. It also bears the plasmid

p*lysS* carrying the bacteriophage T7 lysozyme gene. T7 lysozyme is a specific inhibitor of T7 RNA polymerase. The presence of p*lysS* in a host that carries the inducible gene for T7 RNA polymerase increases the tolerance for toxic target plasmids.[14]

4. Identify the correct recombinant plasmid either by minipreparation, analysis or by hydridization with the appropriate probe.

5. Inoculate LB medium containing 100 and 12.5 μg/ml of ampicillin and chloramphenicol, respectively, and incubate overnight at 37° to obtain a saturated culture.

6. Dilute the saturated culture 1 : 50 in LB medium containing 50 μg/ml ampicillin and incubate for 1.5 hr at 37°. Remove 1 ml of the culture. Induce the remaining culture by adding isopropylthio-β-D-galactoside (IPTG) to a final concentration of O.4 m*M*. Continue the incubation of both cultures at 37° for 1 hr.

7. Centrifuge in an Eppendorf microfuge at 12,000 *g* for 3 min at 4°. Resuspend each pellet in 100 μl of SDS gel-loading buffer. Extracts from induced and uninduced cultures are boiled 5 min and 15 μl is subjected to a 15% (w/v) SDS-polyacrylamide gel. Clones expressing the Rab proteins are identified by Coomassie blue staining and GTP binding (see below). Amounts of produced Rab proteins are estimated by SDS-polyacrylamide gel electrophoresis (SDS-PAGE). They might attain as much as 5–10% of the total *E. coli* proteins.

Purification of Rab Proteins

Cell Culture

1. Inoculate 250 ml of an overnight culture of *E. coli* DE3 (BL21)p*lysS* carrying the *rab*/pET-3c recombinant plasmid in LB medium (with 50 μg of ampicillin/ml) in 5 liters of the same medium. Incubate for 1.5 hr at 37°.

2. Induce the production of the Rab proteins with 0.4 m*M* IPTG for 5 hr at 37°. The cells are harvested by centrifugation and washed once with buffer A (see below) plus 100 m*M* NaC1 to give ~ 30 g of cell paste, which is stored at −70°.

Isolation Procedure

The solubilization and purification procedures of Rab proteins are carried out according to the method previously described.[12] All the purification steps are carried out at 4°.

---

[14] B. A. Moffatt and F. W. Studier, *Cell* **49**, 221 (1987).

1. Frozen bacteria (30 g) are resuspended in 150 ml of buffer A [50 mM Tris-HCl, pH 7.8, 1 mM sodium azide, 0.5 mM DTT, 0.1 mM phenylmethylsulfonyl fluoride (PMSF)].

2. Add 300 μl of 0.5 M ethylenediaminetetraacetic acid (EDTA) and 60 mg of lysozyme. Homogenize the bacterial suspension. Incubate at 4° for 30 min. Add 850 μl of 5% (v/v) sodium deoxycholate and 20 mg of DNase. After a further 30-min incubation at 4°, the suspension (less viscous) is centrifuged at 25,000 rpm for 30 min in an SW28 rotor (Beckman, Palo Alto, CA).

3. QAE-Sepharose column: The supernatant (160 ml) is applied to a column of QAE-Sepharose fast flow (2.5 × 25 cm) (Pharmacia-LKB Biotechnology, Inc., Piscataway, NJ) equilibrated with 5 vol of buffer B (buffer A plus 10 mM MgCl$_2$). The column is washed with the same buffer until the OD$_{280\,nm}$ ≈ 0.100 and eluted with an 800-ml linear gradient from 0 to 0.4 M NaCl in buffer B. The flow rate is 25 ml/hr. Fractions of 5 ml are collected. Fractions containing Rab proteins (as detected by SDS-PAGE and GTP binding) are pooled.

4. Ammonium sulfate precipitation: Precipitate the pooled fraction to 60% saturation with ammonium sulfate. After 30 min, the precipitate is recovered by centrifugation, dissolved in 3 ml buffer B, and clarified again by centrifugation.

5. Load the dissolved precipitate on two consecutive columns of AcA 54 (IBF, France) (2.5 × 200 cm) equilibrated with buffer C (buffer B + 0.1 mM GDP and 200 mM NaCl). The column is eluted with buffer C at a flow rate of 20 ml/hr. Fractions (2.5 ml) are collected. The peak of GTP-binding activity is located and fractions containing Rab proteins relatively free from contaminants are pooled and brought to 70% saturation with ammonium sulfate. The precipitate is resuspended in 3 ml of buffer B and dialyzed against the same buffer at 4° overnight. Human Rab proteins with a purity exceeding 90–95% are obtained. Protein concentration is determined by the method of Bradford[15] using bovine serum albumin for calibration. [The remaining impurities can be removed by an additional purification step on a Mono Q column (Pharmacia-LKB Biotechnology, Inc.)

GTP-Binding Assay

Guanine nucleotide-binding proteins are a class of enzymes that perform diverse functions by the same general principles, namely a conformational change between an active GTP state and an inactive GDP state, which is triggered by the intrinsic GTPase activity.

---

[15] M. M. Bradford, *Anal. Biochem.* **72**, 248 (1976).

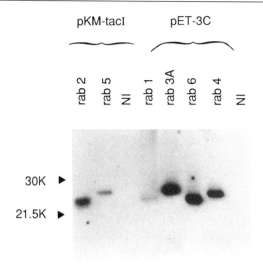

FIG. 2. GTP binding of human Rab proteins. Total proteins from induced and noninduced *E. coli* carrying *rab* expression vectors as indicated in the figure were solubilized by boiling for 5 min in SDS sample buffer, separated on SDS-PAGE, and transferred to nitrocellulose. The filter was probed with $[\alpha^{32}\text{P}]$GTP. NI, Noninduced bacteria transformed with pKM-*tac*I *(rab5)* or with pET-3c *(rab4)*.

The nucleotide-binding capacity of Rab protein can be analyzed by Western blot. This method has been shown to allow the renaturation of some GTP-binding proteins.[16] Under the conditions described below, the ability of Rab proteins to bind GTP after transfer onto nitrocellulose filters is highly variable (Fig. 2). No quantitative conclusions on the respective amounts of different proteins can be drawn by this method.

1. Rab proteins are separated on 15% (w/v) polyacrylamide-SDS gels. Gels are incubated for 30 min at 4° in phosphate-buffered saline (PBS: 50 mM Na$_2$HPO$_4$/KH$_2$PO$_4$, pH 7.2, O.12 M NaC1). The proteins are then electrophoretically transferred to nitrocellulose paper (BA85, 0.45 μm; Schleicher & Schuell, Keene, NH) in 25 mM Tris/0.15 mM glycine and 20% methanol as transfer buffer.

2. The transfer filter is washed for 1 hr in buffer A [10 mM Tris-HC1, pH 7.8, 0.15 M NaC1, 5% (w/v) bovine serum albumin (BSA), and 0.05% (v/v) Tween 20].

3. Incubate the filter for 30 min at room temperature in 10 ml of binding buffer [20 mM Tris-HC1, pH 7.8, 10 mM MgCl$_2$, 2 mM DTT,

[16] E. Lapetina and B. Reep, *Proc. Natl. Acad. Sci. USA* **84**, 2261 (1987).

0.1% (v/v) Nonidet P-40, and 0.3% (w/v) BSA] containing 30 μCi of [α-^{32}P]GTP (3000 Ci/mmol; Amersham).

4. Wash the filter briefly (twice, 3 min each) in 100 ml of buffer [10 mM Tris-HC1, pH 7.8, 0.15 M NaC1, 0.5% (v/v) BSA, 0.5% (v/v) Triton X-100, and 0.2% (v/v) SDS]. The filter is dried and autoradiographed (Fig. 2).

The same procedure could be used to analyze the binding of other nucleotides and to study the competition of guanine nucleotide binding with that of other nucleotides, in which case a 10-fold excess of nonlabeled nucleotide is present in the incubation solution. In our laboratory, we have performed the competition for GTP binding as follows: 1 μg of purified Rab protein was incubated for 10 min in 50 mM Tris buffer, pH 7.5, containing 1 mM DTT, 2 mM EDTA, 1 μM [α-^{32}P]GTP (10 Ci/mmol) in the presence or absence of 100 μM nonlabeled GTP, GDP, GTPγS, ATP, CTP, TTP, and UTP. Samples were passed through nitrocellulose filters (BA85, 0.45 μm; Schleicher & Schuell), washed three times with 3 ml of ice-cold buffer (20 mM Tris-HC1, pH 8, 5 mM MgC1$_2$, 10 mM NH$_4$C1, 0.5 mM DTT, and 0.1 M KCl), and dried. The radioactivity bound to the filter was determined by liquid scintillation counting. Both techniques showed that Rab proteins specifically bind GTP and GDP.

GTPase Assay

Rab proteins, like all the Ras and Ras-related proteins, hydrolyze GTP very slowly.[6] We used the purified Rab proteins to test their capacities to hydrolyze GTP. Equal amounts of active Rab proteins, as determined by their GTP binding, were incubated with radioactive GTP. The GTP hydrolysis can be measured.

1. Rab proteins (1 μM) are incubated for 15 min at 37° in a total volume of 100 μl in a solution containing 50 mM Tris-HC1, pH 7.8, 2 mM EDTA, 1 mM DTT, 10 μM [α-^{32}P]GTP (2000 cpm/pmol). Under these conditions (EDTA > Mg^{2+}) GDP/GTP exchange is promoted.

2. GTPase activity is initiated by adding MgC1$_2$ to a final concentration of 10 mM. Samples of 5 μl are withdrawn from the incubation mixture at different times, mixed with 5 μl of a solution consisting of 0.2% (v/v) SDS, 5 mM EDTA, 50 mM GDP, and 50 mM GTP at 4°. Samples are then heated at 70° for 2 min to dissociate protein-bound nucleotides and 1-μl aliquots are spotted on polyethyleneimine-cellulose thin-layer chromatography plates. They are developed in 0.6 M sodium phosphate buffer, pH 3.4, for 25 min, dried, and autoradiographed. GTP-bound hydrolysis is followed here by the generation of GDP. The GTP and GDP

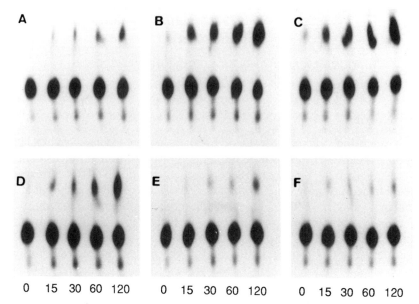

0   15   30   60   120   0   15   30   60   120   0   15   30   60   120

FIG. 3. GTP hydrolysis of human Rab proteins. Bacterially produced, purified proteins (1 μM final concentration) were incubated at 37° for 10 min with 10 μM [α^{32}P]GTP in the presence of 2 mM EDTA. The GTPase activity was started by addition of MgCl$_2$ (10 mM final concentration). Aliquots were withdrawn from the incubation mixture at the indicated times (in minutes) and analyzed by thin-layer chromatography. A, Rab1; B, Rab2; C, Rab3A; D, Rab4; E, Rab5; F, Rab6.

spots are excised and radioactivity is measured by liquid scintillation counting (Fig. 3).

The GTPase activity can also be measured by release of [^{32}P]p$_i$. In this case GTP hydrolytic activity is detected by preequilibrating Rab proteins with [γ-^{32}P]GTP, incubating the mixture at 37°, and then quantitating the decrease in radiolabeled Rab–GTP complex by nitrocellulose filtration.[17]

[17] C. Leupold, R. S. Goody, and A. Wittinghofer, *Eur. J. Biochem.* **124,** 237 (1983).

[37] Localization of Rab Family Members in Animal Cells

By Marino Zerial, Robert Parton, Philippe Chavrier, and Rainer Frank

Introduction

Among the different classes of Ras-like low molecular weight GTP-binding proteins expressed in mammalian cells, Rab proteins are the most closely related to Ypt1p and Sec4p, which are involved in control of secretion in the yeast *Saccharomyces cerevisiae*.[1,2] Due to this sequence similarity they have been functionally implicated in the regulation of membrane traffic in mammalian cells. In fact, Rab1 has been shown to functionally replace Ypt1p in *S. cerevisiae*[3] and studies on Rab5 have indicated that this protein is involved in the process of early endosome fusion *in vitro*[4] and in control of endocytosis *in vivo*.[5] The localization data so far obtained have shown that Rab proteins are associated with specific subcompartments along the exocytic and endocytic pathway. Using affinity-purified antibodies in immunofluorescence and electron microscopy studies, Rab2 was found associated with an intermediate compartment between the endoplasmic reticulum (ER) and the Golgi apparatus.[6] Using similar techniques, Rab6 was found localized to the medial- and trans-Golgi cisternae.[7] Rab3a was demonstrated by subcellular fractionation to be present in synaptic vesicles in neurons and in chromaffin granules in adrenal medulla.[8-10] Three other Rab proteins were found associated with compartments along the endocytic pathway: Rab5 was detected at the cytoplasmic surface of both the plasma membrane and early endo-

---

[1] A. Salminen and P. J. Novick, *Cell* **49**, 527 (1987).

[2] N. Segev, J. Mulholland, and D. Bostein, *Cell* **52**, 915 (1988).

[3] H. Haubruck, R. Prange, C. Vorgias, and D. Gallwitz, *EMBO J.* **8**, 1427 (1989).

[4] J.-P. Gorvel, P. Chavrier, M. Zerial, and J. Gruenberg, *Cell* **64**, 915 (1991).

[5] C. Bucci, R. G. Parton, I. Mather, H. Stunnenberg, K. Simons, B. Hoflack, and M. Zerial, *Cell* **70**, in press (1992).

[6] P. Chavrier, R. G. Parton, H. P. Hauri, K. Simons, and M. Zerial, *Cell* **62**, 317 (1990).

[7] B. Goud, A. Zahraoui, A. Tavitian, and J. Saraste, *Nature (London)* **345**, 553 (1990).

[8] G. Fischer von Mollard, G. A. Mignery, M. Baumert, M. S. Perin, T. J. Hanson, P. M. Burger, R. Jahn, and T. C. Sudhof, *Proc. Natl. Acad. Sci. USA* **87**, 1988 (1990).

[9] F. Darchen, A. Zahraoui, F. Hammel, M.-P. Monteils, A. Tavitian, and D. Scherman, *Proc. Natl. Acad. Sci. USA* **87**, 5692 (1990).

[10] A. Mizoguchi, S. Kim, T. Ueda, A. Kikuchi, H. Yorifuji, N. Hirokawa, and Y. Takai, *J. Biol. Chem.* **265**, 11872 (1990).

somes,[6] Rab4 on early endosomes,[11] whereas Rab7 was associated with late endosomes.[6]

The methods described here outline the different techniques we have used to identify and localize members of the Rab protein family.

Molecular Cloning of Rab cDNAs

Ras and Ras-related proteins are structurally recognizable by a high sequence conservation in the four regions forming the GTP-binding site[12] and by the presence of one or two cysteines at their C termini (Fig. 1). Isoprenylation of these cysteines is a requirement for membrane association. The high amino acid conservation in these regions has enormously facilitated the identification of several members of the Rab protein subfamily.[13] Rab proteins could be identified by screening cDNA libraries either with oligonucleotides corresponding to highly conserved sequences[14,15] or using Rab cDNA probes at low-stringency hybridization conditions.[16]

cDNA Library Screening

To identify cDNAs encoding Rab proteins we use a modification of the protocol described by Touchot *et al.*[14] We screen an oriented λ ZapII (Stratagene, La Jolla, CA) MDCK cDNA library with two degenerate oligonucleotides. One corresponds to the conserved sequence WDTAGQE (single-letter amino acid code) in region 2 (oligo2, Fig. 1) shared by Sec4p, Ypt1p, Rab and Rho proteins: 5'-TGGGA(C_{50}/T_{50})AC($A_{70}/C_{10}/T_{10}/G_{10}$) GC($T_{30}/A_{70}$)GG($A_{25}/G_{25}/C_{25}/T_{25}$)CA($G_{20}/A_{80}$)GAA-3' (numbers in subscript refer to the relative frequency of each base at a given position). Because Ras and Rap proteins contain instead an LDTAGQE sequence,[13] this oligonucleotide preferentially hybridizes to Rab and Rho cDNAs. The second oligonucleotide corresponds to the fourth conserved domain in Ras-like proteins (oligo4, Fig. 1) and has the following sequence: 5'-TT($T_{50}/$ C_{50})(T_{50}/A_{50})T(G_{50}/T_{50})GA(A_{25}/G_{75})(A_{75}/G_{25})C($A_{50}/C_{30}/G_{10}/T_{10}$)A$_{75}/$ T_{25}(G_{75}/C_{25})(T_{75}/C_{25})GC-3'. After a 1-hr prehybridization at 42° in 6 ×

[11] P. van der Sluijs, M. Hull, A. Zahraoui, A. Tavitian, B. Goud, and I. Mellman, *Proc. Natl. Acad. Sci. USA* **88**, 6313 (1991).

[12] E. F. Pai, W. Kabsch, U. Krengel, K. C. Holmes, J. John, and A. Wittinghofer, *Nature (London)* **341**, 209 (1989).

[13] A. Valencia, P. Chardin, A. Wittinghofer, and C. Sander, *Biochemistry* **30**, 4637 (1991).

[14] N. Touchot, P. Chardin, and A. Tavitian, *Proc. Natl. Acad. Sci. USA* **84**, 8210 (1987).

[15] P. Chavrier, M. Vingron, C. Sander, K. Simons, and M. Zerial, *Mol. Cell. Biol.* **10**, 6578 (1990).

[16] A. Zahraoui, N. Touchot, P. Chardin, and A. Tavitian, *J. Biol. Chem.* **264**, 12394 (1989).

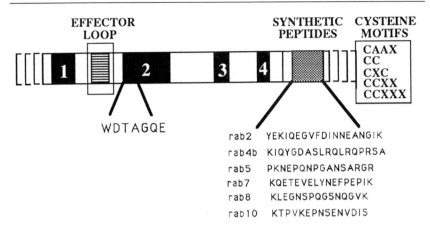

FIG. 1. Scheme of a typical Rab protein showing the variable lengths of the N and C termini, the four conserved regions participating in the formation of the GTP-binding site (black boxes), the "effector region" (striped box), and the C-terminal hypervariable region containing the diverse cysteine motifs. The synthetic peptides used for immunization were derived from this region.

SSC (1 × SSC is 0.15 M NaCl plus 0.015 M sodium citrate), 5× Denhardt's solution, 0.05% (w/v) sodium pyrophosphate, 0.5% (w/v) sodium dodecyl sulfate (SDS), 100 μg/ml boiled herring sperm DNA, duplicate filters are hybridized for 18 hr at 42° in 6× SSC, 1× Denhardt's solution, 0.05% (w/v) sodium pyrophosphate, 100 μg/ml yeast tRNA, with 25 pmol/ml of oligonucleotide 2 (^{32}P) end-labeled using T4 polynucleotide kinase. Filters are then washed for 3 hr at 44° in a 6× SSC/0.05% sodium pyrophosphate solution. Positive phage DNAs are then retested by hybridization with foligo2 at 42 or 37° with oligo4. Washing conditions are as above except that, in the case of the latter oligonucleotide, washing is performed at 25°. Phages hybridizing to both oligonucleotides are selected and characterized by DNA sequencing. For this purpose, oligo2 and its antisense counterpart can be efficiently used in dideoxy sequencing reactions.

Using this technique the frequency of isolation of Rab-encoding cDNAs is quite low. We find an average of 1 positive clone out of 50,000 screened phages for most of the Rab cDNA clones we identify. This is due both to the relatively low abundance of Rab mRNAs and also to the use of oligonucleotides that restrict the screening to cDNA clones containing most of the coding region. Therefore, to maximize the screening efficiency it is crucial to screen a library having long cDNA inserts (~2-kb average insert length).

To overcome this limitation, we have employed a more efficient method with the use of oligonucleotide 2 in a rapid amplification of cDNA ends (RACE) polymerase chain reaction (PCR).[17] Using this method, we could identify 11 novel Rab proteins and one new Rho protein.[18]

Production of Specific Rab Proteins Antisera

To determine the localization of Rab proteins we raise polyclonal antisera in rabbit either against recombinant Rab proteins produced in *Escherichia coli* (we use MSII polymerase – Rab fusion proteins) or against peptides derived from the C-terminal hypervariable region of Rab proteins.[6] This region is not conserved among the different Rab proteins identified so far[13]; therefore, the latter method circumvents the potential problem of cross-reactivity. Furthermore, peptides can be used in competition experiments to control the specificity of the antisera (see below). Using peptides, we have successfully raised antibodies against Rab2, Rab4b, Rab5, Rab7, Rab8, and Rab10 proteins. Peptides are chosen from the C-terminal sequence[15] (Fig. 1), beginning from the last amino acid residues of α helix 5, as deduced from the structure of p21^{ras}.[12] Before injection, synthetic peptides are covalently coupled to keyhole limpet hemocyanin (KLH; Calbiochem, San Diego, CA) as described by Kreis.[19] These peptides are 16 to 20 amino acid residues in length and contained 1 or 2 lysines required for the coupling reaction.

Injection and Bleeding of Rabbits

We use the immunization procedure described by Louvard[20] based on the lymph node antigen injection protocol. Typically, two rabbits are injected with each antigen, a total of 750 μg of coupled peptide or recombinant Rab protein per animal. The first injection contains 100–200 μg of antigen in 0.5 ml of Freund's complete adjuvant (Sigma, St. Louis, MO). The maximum possible is injected into the thigh lymph nodes and the remainder is injected subcutaneously in the back of the neck. After 3 weeks 50–100 μg of antigen in 0.5 ml of Freund's incomplete adjuvant is injected, one-half subscapular and one-half in the neck region. After 6 weeks 100 μg of antigen in phosphate-buffered saline (PBS) is injected intramuscularly in the leg. At 7 weeks 50 μg is injected intramuscularly and the injection is repeated intravenously 1 and 2 days later. Animals are bled and subsequently boosted at weeks 9–12. Blood is collected into a 50-ml

[17] M. A. Frohman, M. K. Dush, and G. R. Martin, *Proc Natl. Acad. Sci. USA* **85**, 8998 (1988).
[18] P. Chavrier, K. Simons, and M. Zerial, *Gene* **112**, 261 (1992).
[19] T. E. Kreis, *EMBO J.* **5**, 931 (1986).
[20] D. Louvard, *J. Cell Biol.* **92**, 92 (1982).

Falcon (Becton Dickinson, Oxnard, CA) tube containing a long wooden stick. After a 1-hr incubation at 37° the blood is left at 4° overnight. The day after, the coagulated clot around the stick is discarded and the serum spun at 4000 g for 20 min at 4° to remove remaining debris. The serum is stored at 4° after addition of 0.02% (w/v) sodium azide or at −70°, frozen in aliquots.

Affinity Purification of Antisera

Both for biochemical and morphological studies anti-Rab protein antisera are first affinity purified. In the case of antiserum raised against MSII polymerase–Rab fusion proteins, affinity purification is performed by adsorbing the antiserum onto nitrocellulose filters containing the corresponding proteins. Anti-peptide antibodies are affinity purified on the same filters or on Sepharose 4B-peptide columns. For this purpose, 1 g of CNBr-activated Sepharose 4B (Pharmacia, Piscataway, NJ) is swollen in 50 ml of 1 mM HCl for 20 min at room temperature and washed on a sintered glass filter (No. 3, Schott Duran, Mainz, Germany) with 500 ml of the same buffer. Beads are then transferred to a 15-ml Falcon tube and washed twice with coupling buffer (0.2 M NaHCO$_3$, pH 8.5, 0.5 M NaCl). After centrifugation at 1000 g for 1 min the liquid is removed and the Sepharose 4B incubated with 10 mg of coupled peptide dissolved in 7.5 ml of coupling buffer, first for 2 hr at room temperature and then overnight at 4° on a rotating wheel. Beads are washed with 30 ml of coupling buffer and the reacting groups blocked by incubation in 40 ml of 0.2 M glycine, pH 8.5, for 4 hr at 4°. After three cycles of washes consisting of 40 ml sodium acetate (0.1 M)/NaCl (0.5 M), pH 4.0, and then coupling buffer, beads are washed twice in PBS and stored in PBS containing 0.02% sodium azide at 4°.

For affinity purification, 1–2 ml of antiserum is diluted to 10 ml with PBS and bound at 4° overnight on their corresponding MSII polymerase–Rab fusion proteins immobilized on nitrocellulose filter or peptide columns. Antibodies are eluted in 100 mM glycine hydrochloride, pH 2.8, and neutralized by addition of 3 M Tris-HCl, pH 8.8. Affinity-purified antibodies are supplemented with 0.5% (w/v) IgG-free bovine serum albumin (BSA) (Sigma) and 0.02% sodium azide and stored at 4°.

Despite the affinity purification, antisera occasionally gave some background in immunofluorescence or electron microscopy studies. To avoid this, antisera are first cleared by adsorption at 4° on a Sepharose 4B matrix coupled to a bacterial lysate prior to affinity purification (we used K537 cells transformed with plasmid pEX34b and temperature induced to express the MSII polymerase polypeptide[6]; other bacterial extracts can be used instead). This step removed antibodies unspecifically bound to the peptide column. To prepare the lysate, bacteria from a 40-ml overnight culture are spun and lysed in 5 ml of 50 mM Tris-HCl, pH 8.0, 5.0 mM ethylenediaminetetraacetic acid (EDTA), 0.2 mg/ml lysozyme for 30 min at 37°. After addition of Triton X-100 to a concentration of 1% (v/v) the extract is ultrasonified. Insoluble aggregates are pelleted in a microfuge. One milliliter of the supernatant (~30 mg/ml) is used in the coupling reaction performed as described above. Coupling efficiency is monitored by analyzing samples of the lysate before and after the coupling reaction by polyacrylamide gel electrophoresis. The affinity-purified antibodies are characterized by immunoblot analysis and by immunofluorescence staining. Different amounts (1 ng–1 μg) of recombinant Rab proteins produced in *E. coli* are used to titer the affinity-purified antisera by Western blot analysis.

Immunofluorescence Localization Analysis

For this type of analysis we permeabilize cells using saponin as detergent prior to fixation. This procedure removes most of the cytosolic proteins and allows us to better visualize the membrane staining pattern. This method is particularly useful in the case of transfected cells expressing high levels of Rab proteins (see below). In these cells the excess of cytosolic Rab protein can be washed off during the permeabilization step.

Cells are grown on 10-mm round cover slips for 24 hr prior to treatment. Cells are washed once with PBS and permeabilized with 0.05% (w/v) saponin in 80 mM K$^+$-piperazine-N,N'-bis(2-ethanesulfonic acid) (PIPES), pH 6.8, 5 mM ethylene glycol-bis(β-aminoethyl ether)-N,N,N',N'-tetraacetic acid (EGTA), 1 mM MgCl$_2$ for 5 min. After a 15-min fixation in 3% (w/v) formaldehyde in PBS, pH 7.4, cells are washed with 0.05% saponin in PBS for 5 min and free aldehyde groups quenched with 50 mM NH$_4$Cl in PBS for 10 min. Cells are washed with 0.05% saponin in PBS for 5 min and then incubated with the first antibody in PBS–0.05% saponin for 20 min. Depending on the antiserum, antibodies are diluted at volume ratios of 1:10–1:300. After rinsing the cells three times (15 min in total) primary antibody binding is visualized with goat anti-rabbit fluorescein isothiocyanate (FITC) or goat anti-rabbit RITC diluted in 0.05% saponin–PBS for 20 min. After one wash in PBS–0.05% saponin and three washes in PBS (20 min in total), the cover slips can be mounted on glass sides in Mowiol (Hoechst, Frankfurt, Germany) and viewed.

Overexpression of Rab Proteins Using T7 RNA Polymerase Recombinant Vaccinia Virus System

To facilitate the localization studies of Rab proteins we make use of a transient expression system based on the T7 RNA polymerase recombi-

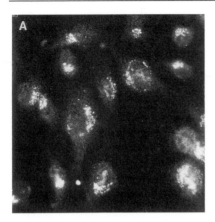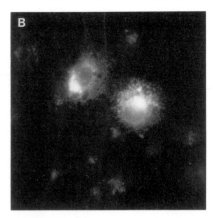

FIG. 2. Immunofluorescence localization of Rab2 in untransfected and transfected cells. (A) BHK cells labeled with affinity-purified anti-Rab2 antibodies. The staining pattern of vesicular structures restricted to the perinuclear area corresponds to an intermediate compartment between the ER and the Golgi apparatus.[6] (B) BHK cells infected with the T7 RNA polymerase–recombinant vaccinia virus and transfected with Rab2 construct. The exogenous protein gives an intense signal close to the nucleus, a staining pattern similar to that of the endogenous protein.

nant vaccinia virus.[21] Cells are infected with this vaccinia virus and transfected with a plasmid containing the Rab cDNA under the control of the T7 promoter. We currently use plasmid pGEM1 (Promega, Madison, WI) for these studies. This system has been useful in proving the specificity of the affinity-purified antibodies. We often obtain affinity-purified antibodies reacting well in Western blot experiments but giving artifactual staining by immunofluorescence analysis. When antisera are specific, in cells expressing the exogenous Rab protein, the staining pattern is similar and of higher signal intensity compared to that of nontransfected cells. Figure 2 shows an immunofluorescence analysis using anti-Rab2 antibodies of untransfected cells and cells overexpressing Rab2.

For these experiments cells are split 18–24 hr before transfection, so that on the day of transfection they are about 80% confluent. The cells are washed twice with serum-free medium and infected with T7 RNA polymerase–recombinant vaccinia virus. Infections are carried out with 3–5 pfu/cell at room temperature for 30 min with intermittent agitation. The cells are then washed twice with serum-free medium and transfected by lipofection using the DOTAP reagent according to the instructions of

[21] T. T. Fuerst, E. G. Niles, F. W. Studier, and B. Moss, *Proc. Natl. Acad. Sci. USA* **83**, 8122 (1986).

the manufacturer (Boehringer Mannheim, Indianapolis, IN). To obtain high transfection efficiencies (up to 95%)[5] we use plasmids purified twice on CsCl density gradients and dialyzed against 0.1 × TE (1.0 mM Tris, pH 7.5, 0.1 mM EDTA). Cells are incubated for 3–4 hr at 37° in a 5% CO_2 incubator and processed for immunofluorescence microscopy (Fig. 2) or immunoelectron microscopy. Longer incubation times lead to accumulation of Rab protein, posttranslationally unmodified in the cytosol. This does not affect the immunofluorescence analysis owing to the permeabilization step. However, accumulation of cytosolic Rab protein should be avoided in the case of immunoelectron microscopic analysis (see below).

Immunoelectron Microscopic Localization of Rab Proteins

Practical Details

We examined the intracellular location of Rab2, 5, and 7 in cultured cells. For this, affinity-purified anti-peptide antibodies were applied to thawed cryosections according to published procedures.[22-25]

Pellets of BHK, NRK, or HeLa cells were prepared for cryosectioning as follows.[26] Cultured cells on 6-cm culture dishes were washed with ice-cold PBS and then were removed from the culture dish by incubation on ice with 1 ml of 20- to 50-μg/ml proteinase K in PBS. This is a relatively gentle procedure for removal of the cells from the dish. The cells start to detach from the dish in sheets after about 2 min; the cells do not noticeably round up and most surface antigens are unaffected by this treatment (G. Griffiths, personal communication, 1991). The detached cells were layered onto 200 μl of 8% (w/v) paraformaldehyde in 250 mM N-2-hydroxyethylpiperazine-N'-2-ethanesulfonic acid (HEPES), pH 7.4, and were pelleted in a microfuge at 1000 rpm for 1 min at room temperature. The supernatant is then removed and replaced with fresh fixative as above without disturbing the pellet. After 30 min at room temperature the cells were spun at maximum speed in the microfuge before replacing the supernatant once more with fresh fixative.

To allow freezing of the cells without destructive ice crystal formation the cells must be cryoprotected. This is achieved by incubation of small pieces of the pellet with sucrose (2.1 M in PBS) for 15–30 min. The pieces

[22] K. T. Tokuyasu, J. Cell Biol. 57, 551 (1973).
[23] K. T. Tokuyasu, J. Ultrastruct. Res. 63, 287 (1978).
[24] G. Griffiths, K. Simons, G. Warren, and K. T. Tokuyasu, this series, Vol. 96, p. 435.
[25] G. Griffiths, A. McDowell, R. Back, and J. Dubochet, J. Ultrastruct. Res. 89, 65 (1984).
[26] J. Green, G. Griffiths, D. Louvard, P. Quinn, and G. Warren, J. Mol. Biol. 152, 663 (1981).

of pellet were then mounted on copper stubs and plunged into liquid nitrogen. the specimens were then ready for cryosectioning. This is performed at $-100°$ using glass or diamond knives. Sections were transferred from the knife to copper grids using a loop with a drop of 2.3 M sucrose in PBS. All labeling is carried out at room temperature with antibodies followed by protein A–gold (6- or 9-nm diameter) diluted in PBS containing 5% (w/v) fetal calf serum and 0.12% (w/v) glycine. After washing the grids with distilled water they are embedded in a methylcellulose/uranyl acetate (MC/UA) mixture. Excess MC/UA is then removed, leaving a thin film over the sections. After drying (a few minutes at room temperature) the grids are viewed at an accelerating voltage of 80 kV.

Specificity of Labeling

The antibodies used for all electron microscopy (EM) analyses were first checked by Western blotting and by immunofluorescence. However, there is still the possibility that artifactual "sticking" of the antibody to intracellular structures occurs. Normally the specificity is monitored by checking the amount of labeling on structures known to be negative for the antigen, for example, the nucleus. For anti-peptide antibodies, however, the specificity can often be unequivocably determined by incubating the antibody prior to applying it to sections with the specific peptide to which it was raised. In principle, this should abolish the labeling whereas incubation with the same concentration of a nonspecific peptide should have no effect. In the case of the anti-Rab antibodies, relatively high concentrations (0.1 mg/ml) of the specific peptide were required to inhibit binding. This presumably reflects the fact that the affinity-purified antibody recognizes a limited number of conformational forms that the peptide can attain in solution.[27,28] This control alone cannot rule out that the antibody recognizes different proteins sharing the same epitopes, but this can be checked by experiments on overexpressing cells and Western blotting analyses as described above.

Colocalization Studies

The unequivocal localization of any antigen to a particular compartment by morphology alone is often extremely difficult. Well-defined markers are then required for colocalization studies to identify the com-

[27] H. J. Dyson, M. Rance, R. A. Houghten, R. A. Lerner, and P. E. Wright, J. Mol. Biol. 201, 161 (1988).
[28] H. J. Dyson, M. Rance, R. A. Houghten, P. E. Wright, and R. A. Lerner, J. Mol. Biol. 201, 201 (1988).

partment of interest. Both antibodies and endocytic tracers were used as markers for endocytic and exocytic compartments. Rab2 was found to colocalize with labeling for a 53-kDa antigen previously shown to be present in the putative intermediate compartment located between the rough endoplasmic reticulum (RER) and Golgi.[29] Double labeling was achieved by one of two methods. In the protein A method the sections were treated with 1 mg/ml protein A in PBS after the first antibody and protein A–gold incubation steps. This quenches any free antibody before the incubations with the second antibody followed by a different size of protein A–gold. In the second method,[30] the sections were instead incubated with 1% glutaraldehyde in PBS at the same stage in the labeling scheme to destroy the ability of the first antibody to bind protein A–gold. For the endocytic pathway well-characterized particulate markers were used to label previously defined compartments of the cells of interest.[31,32] Bovine serum albumin–gold was internalized for either a short time (5 min) to label early compartments or for longer times to label late endosomes and lysosomes.

Experiments on Vaccinia Virus-Infected/Transfected Cells

Rab proteins are produced as soluble proteins that then appear to associate with the membrane of specific compartments. On overexpression of these proteins an increase of labeling associated with a specific compartment occurs before the cytoplasmic pool of soluble protein starts to increase. For immunoelectron microscopy a large fraction of the cell population should express the protein at a high level, as the sample size for electron microscopy is relatively small when compared to, for example, immunofluorescence. However, the time after transfection should not be too long, as a high level of cytosolic protein will obscure the signal associated with a specific membrane. One way to overcome this is to extract the cells with saponin before fixation as for immunofluorescence (see above). However, in our experience a simpler solution, which was more compatible with fine structure preservation, was to choose a transfection time when the membrane-associated signal of protein is increased but the cytosolic pool of the protein is still below detectable levels.

[29] A. Schweizer, J. A. M. Fransen, T. Bachi, L. Ginsel, and H. P. Hauri, *J. Cell Biol.* **107,** 1643 (1988).
[30] J. W. Slot, H. J. Geuze, S. Gigenack, G. E. Lienhard, and D. E. James, *J. Cell Biol.* **113,** 123 (1991).
[31] G. Griffiths, B. Hoflack, K. Simons, I. Mellman, and S. Kornfeld, *Cell* **52,** 329 (1988).
[32] G. Griffiths, R. Back, and M. Marsh, *J. Cell Biol.* **109,** 2703 (1989).

Author Index

Numbers in parentheses are footnote reference numbers and indicate that an author's work is referred to, although the name is not cited in the text.

Subject Index

involved in endosomal fusion events, 36–37

Guanosine triphosphate blot, 379

Guanosine triphosphate dot-blot, 377

Guanosine 5'-triphosphate hydrolysis, 92, 158

H

HABA. *See* 2-(-4'-Hydroxyazobenzene)benzoic acid

Halotransferrin, 251

Hamster. *See* Baby hamster kidney cells; Chinese hamster ovary cells

Hansenula holstii phosphomannan, 154

HEDTA. *See* Hydroxyethylethylenediaminetetraacetic acid

HeLa cells
  homogenization of, 100
  permeabilization of, 105

Heparan sulfate proteoglycan, 82

Hepatoma cell line, 100

25/125 HEPES/KOAc, 117

Hexosaminidase, secreted, measurement of, 183–184

High-performance liquid chromatography, preparation of nucleotide-free Ypt1 protein with, 379–380

High-salt extract
  preparation of, 40–42
    from bovine brain, 43–44
  restorative activity of
    proteins distinct from N-ethylmaleimide-sensitive factor which account for, 42–43
    with trypsinized endosomes, 40–42

High-salt stripping buffer, 295

High salt wash, 147

High-speed pellet fraction, 137
  in endoplasmic reticulum to Golgi transport assay, 133–135
  preparation of, 127–128, 141–142

High-speed supernatant, 137
  preparation of, 141–142
  recovery of, 314

HPLC. *See* High-performance liquid chromatography

HSP fraction. *See* High-speed pellet fraction

hsPG. *See* Heparan sulfate proteoglycan

HSS. *See* High-speed supernatant

HSSB. *See* High-salt stripping buffer

HTC. *See* Hepatoma cell line

Human apotransferrin
  biotinylation of, 14–15
  preparation of, 202
  2-(-4'-Hydroxyazobenzene)benzoic acid, colorimetric assay with, 15

Hydroxyethylethylenediaminetetraacetic acid, as permeabilizing agent, 180, 187

Hydroxylapatite, concentration on, in coat protomer purification, 335

Hydroxylapatite chromatography, 334–335

N-Hydroxysuccinimide-activated biotin, 15

I

ICT. *See* Intracellular transport buffer

Immature secretory granules
  separation of
    from constitutive secretory vesicles, 90–91
    from trans-Golgi network, 88–90
    trans-Golgi network, cell-free formation of, 81–93
      controls for, 91–92
      procedures for, 87–88

Immunization procedure
  in generation of anti-Ypt protein antibodies, 384
  in Rab protein antisera production, 401–402

Immunoelectron microscopic localization, of Rab proteins, 405–407

Immunofluorescence localization analysis, of Rab proteins, 403–405

Immunoglobulin A. *See* Polymeric immunoglobulin A

Immunoglobulin G. *See also* Anti-2,4-dinitrophenol IgG-rabbit anti-mouse IgG-*Staphylococcus aureus*
  biotinylation of, 15
  isolation of, 385

Immunoprecipitation
  materials for, 270
  in transcytotic vesicle fusion assay, 50

Immunostaining, 78–79

Insulin receptors, 223

10° Intermediate, chase of, 273–275

Intracellular transport buffer, 208

N

O